Sequence-Specific DNA Binding Agents

RSC Biomolecular Sciences

This Series is devoted to coverage of the interface between the chemical and biological sciences, especially structural biology, chemical biology, bio- and chemo-informatics, drug discovery and development, chemical enzymology and biophysical chemistry.

Ideal as reference and state-of-the-art guides at the graduate and post-graduate level.

Titles in the series:

Visit our website on www.rsc.org/biomolecularsciences

For further information please contact:
Sales and Customer Services, Royal Society of Chemistry, Thomas Graham House, Science Park, Milton Road, Cambridge CB4 0WF, UK
Telephone +44 (0)1223 432360, Fax +44 (0)1223 426017, Email sales@rsc.org

Sequence-Specific DNA Binding Agents

Edited by

Michael Waring
Department of Pharmacology, University of Cambridge, Cambridge, UK

RSCPublishing

ISBN-10: 0-85404-370-5
ISBN-13: 978-0-85404-370-5

A catalogue record for this book is available from the British Library

Published by The Royal Society of Chemistry,
Thomas Graham House, Science Park, Milton Road,
Cambridge CB4 0WF, UK

Registered Charity Number 207890

For further information see our web site at www.rsc.org

Typeset by Macmillan India Ltd, Bangalore, India
Printed by Henry Lings Ltd, Dorchester, Dorset, UK

Preface

Twenty or thirty years ago the concept of gene targeting as a means of alleviating disease began to gain currency, whereas it had previously been little more than a pipe dream. Thanks to huge strides in our understanding of basic biology and medicine the concept is no longer a dream but an attainable goal. The talents of traditional biologists and geneticists, but perhaps above all chemists and biochemists, have conspired to create a climate in which the fruits of current research on DNA and gene function are clearly pointing the way to new therapies that will continue the revolutionary progress of chemotherapy as a prime modality for treating cancer. For much of cancer treatment to date has had to rely on drug interactions with DNA, and that is the principal area of medicine that has stimulated research on DNA binders (though it has to be acknowledged that such drugs can find useful employ in other areas of biotechnology and medical science as well). This book represents an attempt to summarise and illustrate some key aspects of the remarkable progress that has been made towards understanding how drugs can bind specifically to nucleic acids, and thus underpin the endeavour to make gene targeting a reality.

The brief was to assemble a set of chapters written by senior scientists, who are acknowledged experts in the field, dealing with diverse aspects of the binding of antibiotics and drugs to DNA. Because of the importance of such substances in medicine, perhaps particularly the treatment of cancer, there are chapters that deal with established agents like actinomycin D, the first antibiotic found to be useful for cancer treatment, which is still in use today and forms the subject of chapter 6. After 60 years we are still learning new and surprising things about this remarkable antibiotic. The necessarily historical emphasis of this contribution is complemented by a rare and reflective chapter that describes the coming of age of theoretical and computational studies devoted to understanding how drugs interact specifically with DNA. Then there are contributions focussing on novel agents that show fairly immediate promise for the future of chemotherapy, notably topoisomerase inhibitors, telomerase inhibitors, peptide nucleic acids and triple helix-forming oligonucleotides. Research success is critically dependent upon advances in experimental methodology, so there is an important place for descriptions of new approaches that originate from the study of slow kinetics, melting curve analysis and improvements in classical medicinal chemistry brought about by discoveries originating from sophisticated forays into structural chemistry.

The book concludes with a thoughtful chapter on nucleic acid structures, mostly RNA, that might in due course become important targets for drug action.

The topics chosen by several authors are unique, or almost so, in cutting across the standard divisions of the discipline to provide a novel perspective. As a result some areas, such as topoisomerase inhibitors and telomerase inhibitors, establish a framework that allows treatment by several different authors in a complementary manner. This may lead to occasional 'overlap' in information content, but has the inestimable virtue of furnishing a varied overview of progress in research from diverse points of view. The reader who progresses systematically through the text will be rewarded with some prime examples of how science works, through glimpses of the story of topoisomerase or telomerase inhibitors all the way from theory to drug development.

This wide compass of subject matter is a feature that may commend the book to students, established research workers, teachers and even historians of science. It is hoped that everyone will find something new and stimulating to read, set within the context of a coherent and multi-faceted attack upon some of the most pressing medical problems of the day. Unifying it all is a clear message of the role played by good chemistry in solving those problems.

Contents

CHAPTER 1

DNA Recognition by Triple Helix Formation

DAVID A. RUSLING,[a] TOM BROWN[b] AND KEITH R. FOX[a]

[a] School of Biological Sciences, University of Southampton, Bassett Crescent East, SO16 7PX, Southampton, UK
[b] School of Chemistry, University of Southampton, Highfield, SO17 1BJ, Southampton, UK

1.1 Introduction

Oligonucleotides can bind in the major groove of double-stranded DNA by forming hydrogen bonds with exposed groups on the base pairs, generating a triple-helical structure (Figure 1A). This was first demonstrated nearly 50 years ago by Rich and co-workers[1] by mixing the synthetic polyribonucleotides polyU and polyA in a 2:1 ratio. Further studies showed that polyC and polyG can generate a similar structure under conditions of low pH[2] and a variety of DNA and RNA triple-stranded structures have since been identified.[3–6]

Since these complexes form in a sequence-specific fashion they can be used to target unique sequences. By knowing the rules that govern triplex formation, it should be possible to design oligonucleotides to interact with any desired DNA sequence. The formation of intermolecular triple-helical DNA therefore has a number of applications including inhibition of gene transcription, site-directed mutagenesis, and various biotechnological applications.[7–12]

1.1.1 Triplets and Triplex Motifs

Triplex-forming oligonucleotides bind in the DNA major groove and make specific contacts with the purine strand of the duplex. The binding can be either parallel or antiparallel to the target strand, depending on the base composition of the oligonucleotide.

Pyrimidine-rich oligonucleotides bind under low pH conditions in a parallel orientation to the purine strand of the target duplex, with T and protonated C forming Hoogsteen hydrogen bonds with AT and GC base pairs,

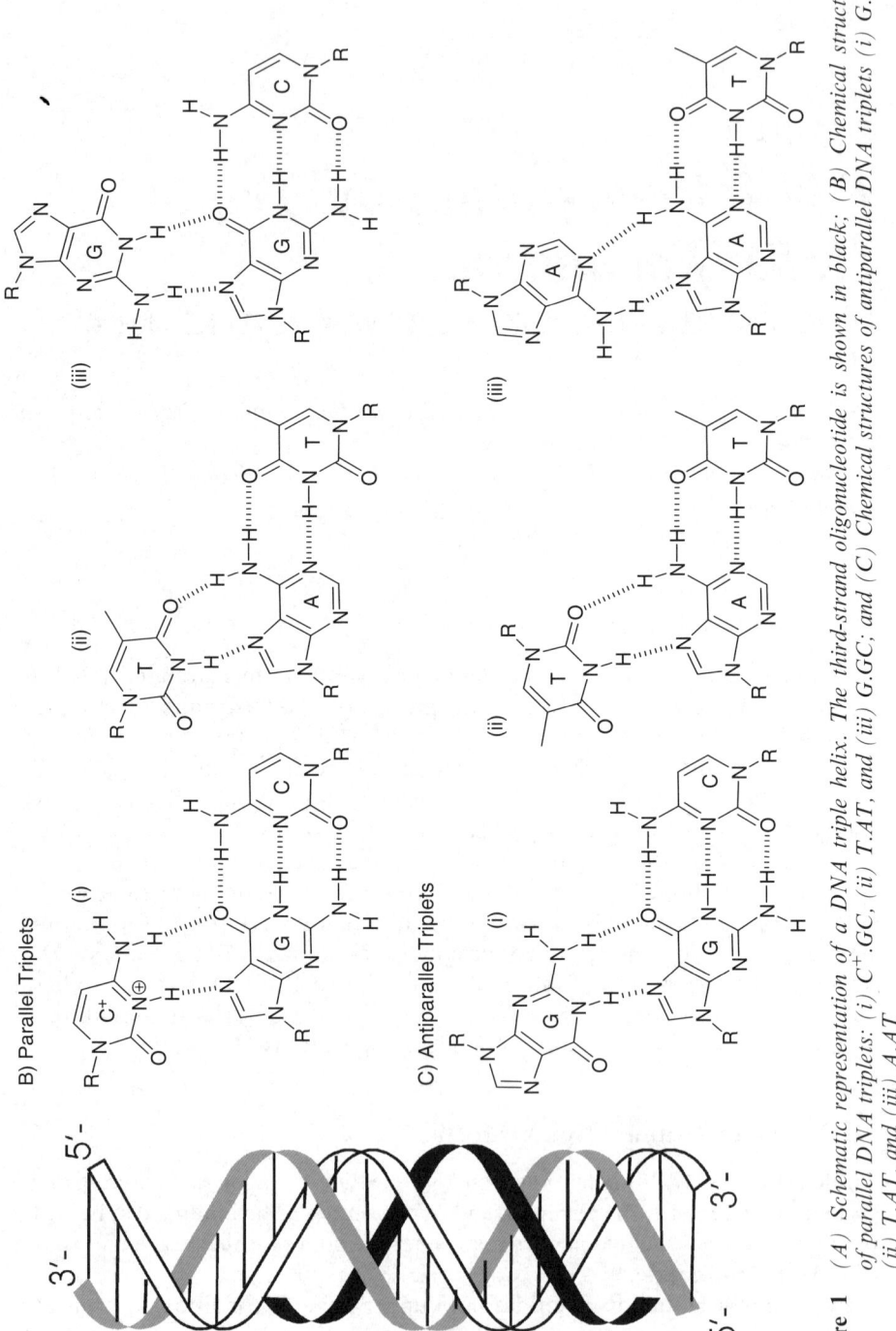

Figure 1 *(A) Schematic representation of a DNA triple helix. The third-strand oligonucleotide is shown in black; (B) Chemical structures of parallel DNA triplets: (i) C+.GC, (ii) T.AT, and (iii) G.GC; and (C) Chemical structures of antiparallel DNA triplets (i) G.GC, (ii) T.AT, and (iii) A.AT*

respectively.[13,14] This generates the base triplets T.AT and C⁺.GC as shown in Figure 1B. (The notation X.ZY refers to a triplet in which the third-strand base X interacts with the duplex base pair ZY, forming hydrogen bonds to base Z). These triplets are isomorphic, that is if the C-1′ atoms of their Watson-Crick base pairs are superimposed, the positions of the C-1′ atoms of the third strand are almost identical.[15] This minimizes backbone distortion of both the third strand and duplex between adjacent triplets. It is also possible to form a G.GC triplet within this motif, though this is not isomorphic with T.AT and C⁺.GC (Figure 1B (iii)). Most of this chapter will concentrate on the parallel motif.

Purine-rich oligonucleotides bind in an antiparallel orientation to the purine strand of the target duplex, with A and G forming reverse-Hoogsteen hydrogen bonds with AT and GC base pairs respectively, generating A.AT and G.GC triplets (see Figure 1C).[16,17] In contrast to the parallel triplets, A.AT and G.GC triplets are not isomorphic, leading to structural distortions at junctions between each triplet. As a consequence of this and other factors, the antiparallel motif is often less stable than the parallel motif. T.AT triplets can also adopt an antiparallel orientation.

Since G.GC and T.AT triplets can be formed in both the parallel and antiparallel motifs GT-containing oligonucleotides can be designed to bind in either orientation, parallel or antiparallel. The backbone distortion imposed by the non-isomorphic nature of these two triplets means that the most stable orientation is dependent on the number of GpT and TpG steps.[15,18,19]

Triplex formation in the parallel motif suffers from the requirement for low pH, which is necessary for protonation of cytosine at N3. Without protonation the C.GC triplet only contains one hydrogen bond between the exocyclic N4 of cytosine and 6-keto group of guanine. The pK_a of cytosine is around 4.5, though this is increased on triplex formation, and is higher in the centre than the termini of a triplet.[20,21] Runs of contiguous cytosine residues are destabilizing as they decrease the pK_a.[20–25] A large number of cytosine analogues have been prepared to overcome this problem and are described in a later section.

Several reports have suggested that C⁺.GC is more stable than T.AT.[26–28,20,24] This is attributed to electrostatic interactions between the positive charge of cytosine and the negatively charged phosphodiester backbone and/or favourable stacking interactions between the charged base and the π-stack.

1.2 Strategies to Increase Triplex Stability

1.2.1 Sugar Modifications

The affinity of a third strand for its target is affected by its ability to adopt N- or S-type conformations, since the former require less distortion of the duplex purine strand upon triplex formation.[29] This explains why RNA TFOs have a higher affinity for duplex DNA than those composed of DNA.[30–34] Oligonucleotide modifications that favour N-type sugars are therefore expected to produce more stable triplexes and some examples are shown in Figure 2. The addition of an electronegative group at the 2′-position of the sugar, as in RNA,

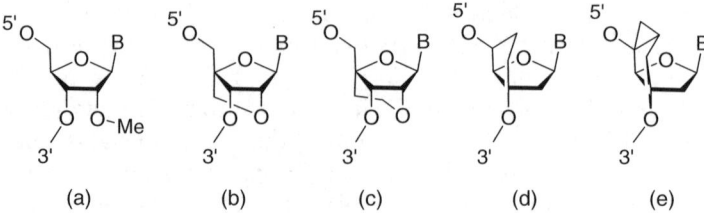

Figure 2 *Chemical structures of modified sugars with restricted conformations. (a) 2'-*
OMe; (b) LNA (BNA); (c) ENA; (d) bicyclo-DNA; and (e) tricyclo-DNA.
For each structure B signifies the DNA base

strongly favours the N-type sugar pucker due to the gauche effect[35] and the 2'-
O-methyl modification (Figure 2a) enhances triplex stability.[31,33] NMR studies
have confirmed that the 2'-methoxy group increases triplex stability by reducing the distortion of the duplex purine strand and enhancing the rigidity of the
triplex.[29]

Other modifications can also be used to restrict the sugar pucker and reduce
the rotational freedom of the sugar phosphate backbone. The best characterized of these modifications is locked nucleic acid (LNA), which is also known as
bridged nucleic acid (BNA), in which a 2'-O,4'-C methylene bridge is used to
constrain the sugar to N-type (Figure 2b). This modification was developed
independently by the Wengel and Imanishi groups for use in antisense or
antigene applications, respectively.[36,37] TFOs that contain LNA residues every
2–3 nucleotides are markedly more stable than their unmodified counterparts.[37,38] Two further derivatives have also been developed from this. ENA
contains an additional carbon in the bridge and unlike LNA can be fully
substituted into TFOs (Figure 2c).[39] In addition, 3'-amino-2',4'-LNA combines
the LNA sugar with the N3'-P5' modification (considered below), though
triplexes with this analogue are no more stable than those formed with LNA.[40]

The bicyclo and tricyclo furanose modifications developed by Leumann and
co-workers[41,42] represent further attempts to restrict the sugar conformation
(Figures 2d and 2e). Bicyclo-DNA contains a 3'-O,5'-C ethylene bridge that
locks the sugar in an S-type conformation. The tricyclo derivative contains an
additional cyclopropane unit locking the sugar in a N-type pucker and studies
with TFOs composed of tricyclothymidine showed a 2°C increase in T_m per
modification at pH 7.[43]

1.2.2   Addition of Positive Charges

Triplex stability is limited by charge repulsion between the three negatively
charged backbones. The simplest strategy to alleviate this problem is to
incorporate positively charged moieties into the TFO by either modifying the
backbone, the sugar or the base, and examples of each of these modifications
are shown in Figure 3.

A) Charged backbones

i)   ii)   iii)

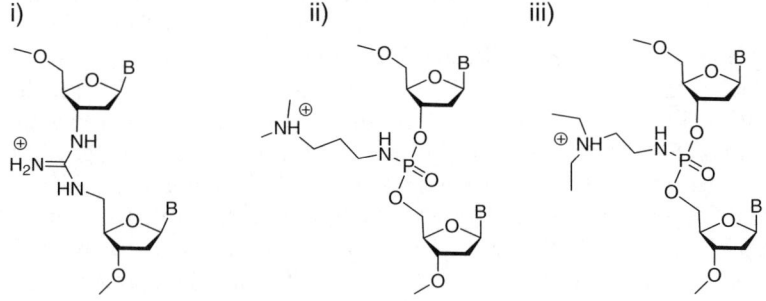

B) Charged sugars

i)   ii)   iii)

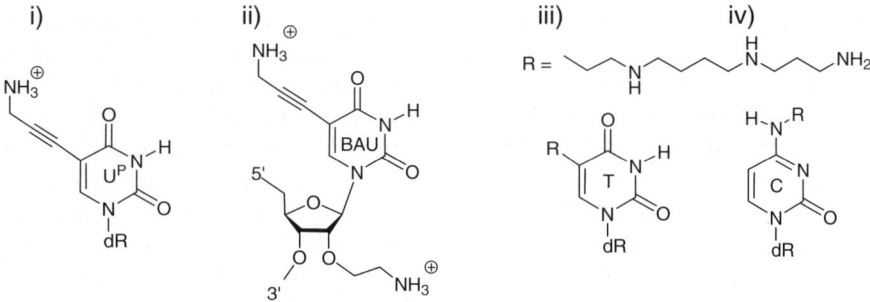

C) Charged bases

Figure 3 *Chemical structures of nucleotide analogues that contain positive charges. (A) Positively charged backbones: (i) DNG, (ii) PNHDMAP, and (iii) DEED; (B) Positively charged sugars: (i) 2′-aminoethoxy, (ii) 4′-aminoethyl, and (iii) pyrrolidine; and (C) Positively charged bases: (i) U^P, (ii) BAU, and spermine derivatives of T (iii) and C (iv)*

1.2.2.1 Addition of Positive Charge to the Backbone

Bruice and co-workers have replaced the phosphodiester backbone with positively charged guanidinium linkages. A pentameric thymidyl oligomer of deoxyriboguanidine (DNG, Figure 3A (i)) with four positive charges exhibited an unprecedented high affinity for poly(dA), forming a 2:1 thymine adenine complex.[44,45] The synthesis of the ribose derivative has recently been reported but has yet to be studied for its triplex-forming properties.[46] This derivative may combine the stabilizing effect of the additional positive charge with a favourable sugar conformation. Two further modifications that replace the phosphate residues with either cationic dimethylaminopropyl phosphorami-date linkages (PNHDMAP) or *N,N*-diethyl-ethylenediamine linkages (DEED) (Figure 3A (ii and iii)) have also been reported.[47–49] TFOs with the PNHD-MAP modification generated a triplex which was more stable than the under-lying duplex at pH 7.[49] Surprisingly the α-anomer produced a more stable triplex than the β-anomer though, as for unmodified TFOs, α-anomers bind in the opposite orientation to the normal β-anomers.[15,49]

1.2.2.2   Addition of Positive Charge to the Sugar

An important strategy has centred on the addition of positively charged groups to various positions of the sugar unit. Modification has been attempted at either the 2′ (Figure 3B (i))[50] or 4′-positions (Figure 3B (ii)).[51] In both cases, the most stable triplexes were formed by the addition of an aminoethoxy or aminoethyl side chain with a T_m increase of 3.5°C and 1°C per modification at pH 7 for the 2′ and 4′ derivatives, respectively.[50,51] The greater stabilization by the 2′-derivative has been attributed to the formation of a salt bridge between the positive charge and a pro-R oxygen of a negatively charged phosphate of the purine strand and a favourable N-type sugar pucker.[52] In experiments with psoralen-linked oligonucleotides it has been suggested that the 2′-aminoethoxy modification is more effective when the positively charged derivatives are clustered together.[53] Another modification involves substitution of the furanose oxygen with nitrogen, generating pyrrolidine oligonucleotides (Figure 3B (iii)). This positions a positive charge next to the pro-R non-bridging phosphate oxygen in the purine strand. In this instance the effect of the modification depends on the base that is attached to the modified sugar; pseudoisocytosine is stabilizing giving a T_m increase of 2°C per modification[54] while uracil is destabilizing.[55,56] The addition of guanidines instead of amino functions is another strategy,[57] which offers two further advantages. Firstly, it can be applied post-oligonucleotide synthesis and secondly, the guanidine group is protonated over a greater pH-range than the amine. This modification gives typically the same increase in stability as an amine at neutral pH, but in principle should afford greater triplex stabilization at higher pH.

1.2.2.3   Addition of Positive Charge to the Bases

As mentioned above, the C[+].GC triplet is more stable than T.AT due, in part, to the effect of the positive charge. With this in mind 5-propargylamino dU (U[P], Figure 3C (i)) was prepared in order to enhance the affinity of T for AT base pairs.[58] This analogue bears a positive charge attached to the 5-position of

U rather than in the stacked ring system (as in protonated C). TFOs containing several substitutions of this analogue are markedly more stable than unmodified TFOs, though the complexes are pH-dependent as a result of the requirement for protonation of this amino group.[58] Unlike protonated C, adjacent U^P substitutions are not destabilizing. This demonstrates that removing the charge from the π-stacked bases and placing it in the major groove is a useful approach for stabilizing triplexes. The alkylyl moiety of U^P also contributes to triplex stability by enhancing stacking interactions.

The covalent attachment of polyamine groups to different bases has also been attempted. The attachment of spermine at the 5-position of uracil[59] and to the N4 position of methylcytosine[60] (Figure 3C (iii and iv)) both increased triplex stability at physiological pH, though the complexes exhibited decreased selectivity.

Combining favourable base modifications with suitable sugar modifications has also been extremely successful. *Bis*-amino dU (BAU) combines the propargylamino modification at the 5-position of U with addition of an aminoethoxy group at the 2′-position of the sugar[61,62] (Figure 3C (ii)). This analogue contains two positive charges at physiological pH and dramatically increases triplex stability with an increase in T_m of 8°C per modification at pH 6, and the stabilization is much greater than when either modification is used alone. This is shown in the melting curves presented in Figure 4, which compare triplex stability at pH 6.0 using oligonucleotides in which the central triplet is either T.AT, U^P.AT, or BAU.AT. These two positive charges act in different ways to enhance triplex stability: the 2′-aminoethoxy group interacts with a phosphate

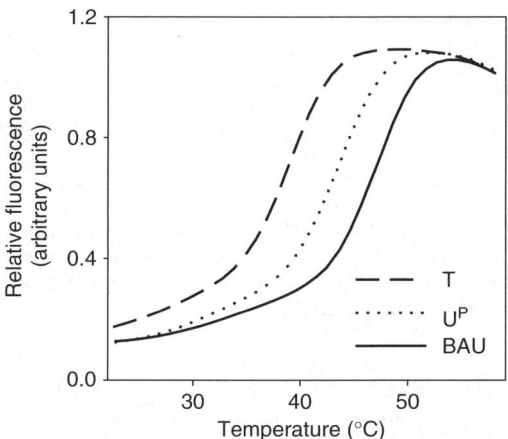

Figure 4 *Fluorescence melting curves comparing the stability of the triplex formed between 5′-Q-TCTCTCTTNTCCTCCTCC and the duplex 5′-F-AGAGAGA-AAAGGAGGAGG/5′-CCTCCTCCTTTTCTCTCT for N = T, U^P, or BAU. The experiments were performed in 50 mM sodium acetate pH 6.0 containing 200 mM NaCl. (F = fluorescein, Q = methyl red)*

on the duplex purine strand, while the 5-propargylamino group interacts with a third-strand phosphate.[61] Interestingly this analogue has greater sequence selectivity than thymidine, with enhanced discrimination against pyrimidine inversions.[62] The high stability of the BAU.AT triplet permits triplex formation at physiological pH, even for sequences that contain several GC base pairs. An example of this effect is shown in Figure 5a for which an 11-mer oligonucleotide containing 5 BAU residues, together with three cytosines and three

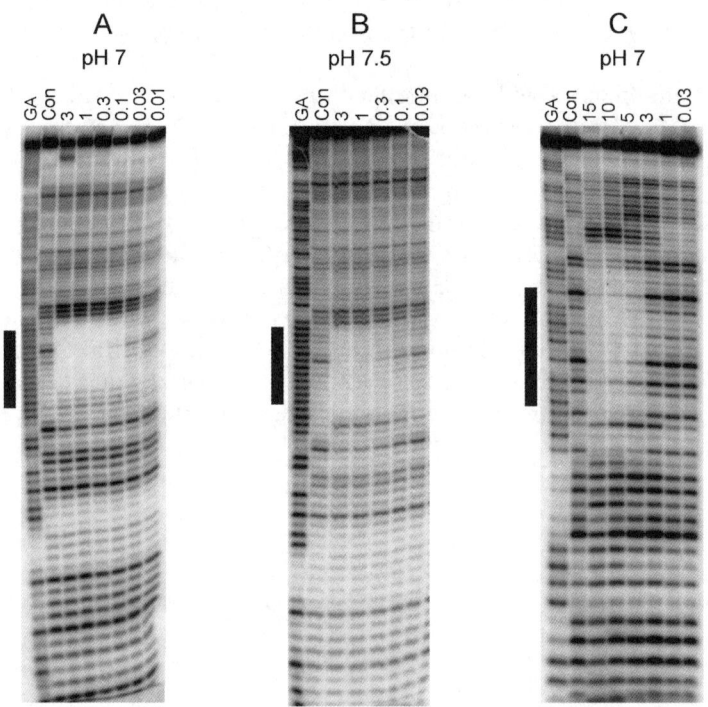

Figure 5   *(A and B) DNase I footprinting patterns for the interaction of different modified oligonucleotides with a 12-mer oligopurine tract in tyrT(43-59) (AGAGAAAAAAGA). The reactions were performed in 10 mM Tris-HCl (pH 7.0 or 7.5) containing 50 mM NaCl, by incubating the oligonucleotide with the target DNA sequence at 20°C overnight in siliconized Eppendorf tubes. (A) Binding of BCBCBTBTBTCT (B = BAU); quantitative analysis of the footprint produces a C_{50} value of 5.5 ± 1.3 nM. (B) Binding of TMTMTBTBTBMT (M = 3-methyl,2-aminopyridine; B = BAU); quantitative analysis of the footprint produces a C_{50} value of 0.2 ± 0.2 μM. (C) Binding of MBPBSBTMTMPTSMTMBT to a 19-mer target sequence GACATAA-GAGCATGAGAA; (B = BAU, M = 3-methyl,2-aminopyridine, P = ^{A}PP; S = TA recognition nucleotide S) quantitative analysis of the footprint produces a C_{50} value of 0.3 ± 0.1 μM. The boxes indicate the location of the triplex footprint. Oligonucleotide concentrations (μM) are shown at the top of each gel lane*

thymine bases is able to produce a triplex with an apparent dissociation constant of 5.5 nM at pH 7.0.

1.2.3 Backbone Modifications

An alternative method for decreasing the charge repulsion between the three-polyanionic DNA strands is to use TFOs that contain neutral backbones. Examples of these are shown in Figure 6. Several such modifications have been developed. Replacement of the phosphate linkage with a methylphosphonate group (Figure 6 (iii)) was successfully used for triplex formation using short oligonucleotides containing alternating methylphosphonate and phosphodiester linkers.[63] However, subsequent studies with longer fully substituted TFOs showed that this modification was destabilizing.[64,65] This may have been caused by the diastereoisomeric mixture of the methylphosphonate residues. Other studies showed that α-methylphosphonate TFOs produce stable triplexes but these are much more difficult to synthesize.[65]

The N3′-P5′ amidate modification, where O3′ of the internucleoside phosphate is replaced by NH[66] (Figure 6 (ii)) increases the binding constant at neutral pH by nearly two orders of magnitude.[37] Triplex binding is probably improved as this modification favours the N-type sugar conformation as discussed above. This modification has also been combined with the addition of a cationic copolymer, which cooperatively stabilizes triplex formation and increases association rates by four orders of magnitude.[67]

Morpholino oligonucleotides are an interesting class of analogues, in which the ribose sugar is replaced with a six-membered morpholino ring and the phosphodiester linkage is replaced by a phosphorodiamidate (Figure 6 (iv)). TFOs containing this modification are less stable than those containing the N3′-P5′ amidate modification at high concentration of cations but are more stabilizing at low ionic strength.[68,69] This modification has also been used with α-oligonucleotides.[70]

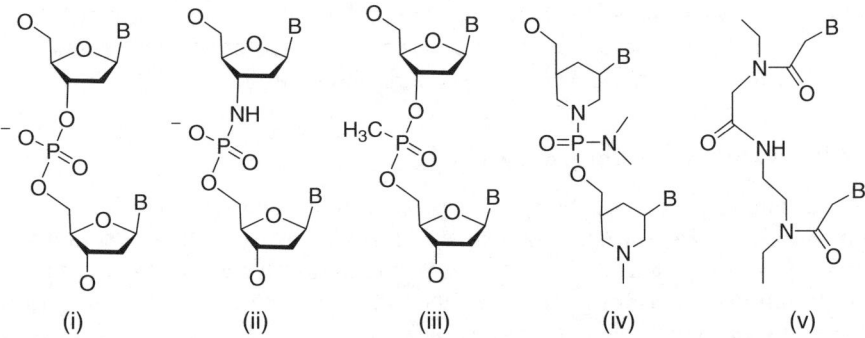

Figure 6 *Chemical structures of modified DNA backbones. (i) phosphodiester, (ii) N3′-P5′, (iii) methylphosphonate, (iv) morpholino, and (v) PNA*

The most extensively employed uncharged backbone modification is peptide nucleic acid (PNA). PNA is composed of repeating (2-aminoethyl)glycine units to which nucleobases are linked by methylene bridges[71,72] (Figure 6 (v)). PNA usually interacts with duplex DNA *via* a mechanism of strand displacement and P-loop formation, requiring two molecules of PNA,[73] generating a 2:1 PNA: DNA triplex. Two pyrimidine-containing PNA molecules form a local triplex with the purine-containing DNA strand. This leaves the pyrimidine DNA strand looped out as a single strand. The resulting triplex is more stable than the equivalent DNA triplex since there is much lower charge repulsion between the three strands. In a few instances, PNA can form a 1:2 PNA:DNA triplex by simple binding of a PNA third strand to a DNA duplex, though this is usually restricted to cytosine-containing PNAs.[74]

1.2.4 Base Stacking

Base stacking is an important factor that affects the stability and structure of both duplex and triplex DNA and increasing the aromatic surface area of a base might be expected to enhance triplex formation. Most such modifications have been based on thymine, by adding further aromatic rings across the 4–5 or 5–6 positions,[75–77] which should not affect the hydrogen-bonding groups. Surprisingly triplexes containing these modifications do not show enhanced stability, though the non-natural pyrido[2,3-*d*] pyrimidine nucleoside (F) recognizes AT base pairs with a similar affinity to T.[75] The addition of a hydrophobic substituent at the 5-position of the pyrimidine base increases hydrophobic stacking interactions within the major groove. The simplest of these is the addition of a methyl group and probably explains why T.AT is more stable than U.AT and $^{Me}C^+$.GC is more stable than C^+.GC.[21,78] The addition of a propyne group to the 5-position of pyrimidines further extends the hydrophobic surface and each propynyl-dU substitution increases the T_m by about 2.5°C relative to thymine and this modification requires lower concentrations of magnesium to achieve a stabilizing effect.[79–81] Propynyl-dC reduces triplex stability relative to C as it decreases the pK_a of the N(3) atom of the heterocycle.[80,82] A recent study on the properties of four different C5-amino modified deoxyuridines showed that the order of stability produced by 5-substitutions is alkyne > *E*-alkene > alkane > *Z*-alkene.[83] This order must result from steric factors as well as stacking interactions.

1.2.5 Triplex-Binding Ligands

Several small molecules have been developed that preferentially bind to triplex over duplex DNA and stabilize triple-stranded structures. These compounds are usually composed of aromatic rings containing heteroatoms for stacking (intercalation) between the base triplets and commonly incorporate a positive charge to partially alleviate the triplex charge repulsion. The first to be described was a benzopyridoindole (BePI),[84] though a wide range of such ligands has now been described. These have recently been reviewed.[85]

Triplex stability has also been enhanced by covalently attaching other DNA-binding ligands to the TFO. In this way sequence selectivity is achieved through the oligonucleotide, while the DNA-binding ligand is used as a non-specific anchor to enhance the affinity. Examples include acridine,[86,87] triplex-binding ligands,[88,89] minor groove-binding ligands,[90] and daunomycin.[91] These compounds have also been used to enhance the affinity of weaker triplexes that are formed at sites containing pyrimidine interruptions.[92] Triplex-forming ligands can be attached to psoralen, generating cross-links on UV irradiation.[93]

1.3 Overcoming the pH Dependency

A number of base analogues have been synthesized in order to overcome the requirement for low pH that is necessary for generating the $C^+.GC$ triplet in the parallel-binding motif. Most of these analogues have been based on either pyrimidine or purine rings (Figure 7), so as to generate the same arrangement of hydrogen bond donors as occurs in protonated cytosine. In some instances these have been used in combination with sugar or backbone modifications.

A) Pyrimidine analogues

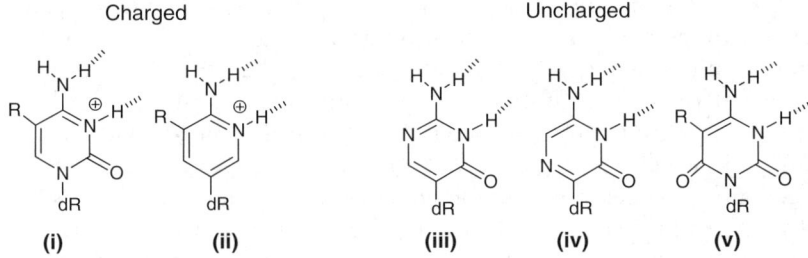

B) Purine analogues

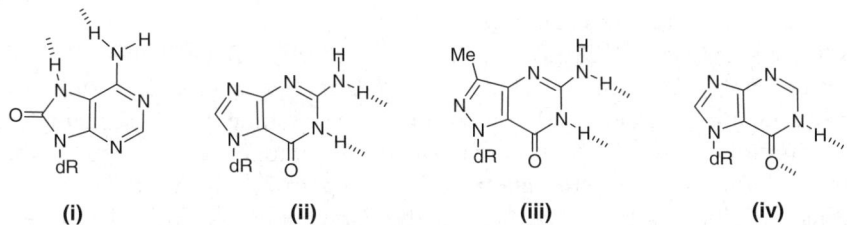

Figure 7 *Base analogues for recognition of GC base pairs. (A) Pyrimidine analogues: (i) C (R = H) and ^{Me}C (R = CH₃), (ii) 2-aminopyridine (R = H) and 3-methyl,2-aminopyridine (R = CH₃), (iii) Pseudoisocytosine (^ψC), (iv) PyD-DA, and (v) 6-oxoC (R = H) and 5-methyl-6-oxo-C (R = CH₃). (B) Purine analogues: (i) 8-oxoA, (ii) N7G, (iii) P1, and (iv) N7I. In each case dR indicates the deoxyribose sugar and the hydrogen bonds to the GC pairs are shown*

1.3.1 Pyrimidine Base Analogues

The simplest cytosine modification is the addition of a methyl group at the 5-position (^{Me}C)[22,94] (Figure 7A (i)). The lower pH dependence and higher affinity relative to cytosine was first attributed to an increase in pK_a but an increase of only 0.1–0.2 pH units seems too little to account for the improved binding. It has been suggested that the increase in stability might be entropic in origin, resulting from disruption of the surrounding water structure, greater base stacking, or hydrophobic interactions within the major groove.[21,78] Alternatively the improved stacking may increase the residence time of the non-protonated base in the uncharged C.GC triplet, thereby increasing its stability.[95]

The C-nucleoside 2-amino-pyridine-2′-deoxyriboside (AP), was synthesized independently by the Neidle, Reese, and Leumann groups[96,97] and differs from cytosine by substitution of a carbon at N1 and removal of the 2-carbonyl (Figure 7A (ii)). The β and α-anomers were both evaluated as the α-anomer is slightly more basic than the β-anomer.[96,98] Triplexes containing the β-anomer exhibited a lower pH dependency on stability than cytosine. This was attributed to its pK_a of 6.5, and β-AP generated triplexes that were stable at pH 6.5 even at target sites that contained multiple adjacent GC base pairs.[98] The 3-methyl (^{Me}P) and 2′-O-methyl (^{OMe}P) derivatives have also been prepared, but did not produce a further improvement in stability.[99] Recent studies in our laboratory have shown that 3-methyl-2-aminopyridine acts cooperatively with the doubly charged thymine analogue BAU to produce triplexes which have nanomolar-binding affinities at pH 7.5 in the absence of divalent metal cations.[100] The arrangement of the substitutions is important and oligonucleotides in which these analogues are evenly distributed throughout the third-strand bind tighter than those in which they are clustered together.[100] An example of this is shown in Figure 5b for which a 12-mer oligonucleotide containing 3 ^{Me}P, 3 BAU and six thymidines produces a footprint at pH 7.5 with an apparent dissociation constant of 0.2 μM. In contrast to the other cytosine analogues which are described below, this base retains the positive charge on the ring and so might be expected to generate more stable complexes due to stabilizing interactions with the phosphodiester backbones of the triplex.

Pseudoisocytosine (^{Ψ}C) (Figure 7A (iii)) was one of the first non-charged cytosine analogues to be developed. Its 2′-O-methyl derivative forms stable triplexes at pH 7 under conditions where deoxycytidine and 2′-O-methylcyti-dine do not. As with 2-aminopyridine this analogue can be successfully employed for targeting contiguous GC base pairs.[101,102] Several derivatives of this base have been developed and the complicated synthesis has recently been streamlined.[103] The deoxyribose derivative exhibits a lower affinity for GC than the 2′-O-methyl analogue presumably because the former adopts the less-favourable S-type sugar conformation.[104] The pyrrolidino derivative of pseu-doisocytosine produced a 2.5–3°C increase in T_m per modification, dependent on the sequence context. This modification can also be used to target contig-uous guanines.[54,103] Pseudoisocytosine has been frequently employed for the pH independent recognition of DNA by PNA.[105] A similar analogue is a pyrazine (Figure 7A (iv)), which possesses a nitrogen at the 6-position (instead of the usual 1-position). This can also be used to produce stable triplexes at pH 7.[106]

6-*oxo* cytosine (^{oxo}C) and its 5-methyl derivative (Figure 7a (v)) have been studied as potential cytosine mimics.[107,108] At low pH these analogues produce triplexes with lower stability than protonated cytosine, though binding is much less pH dependent. Indeed, at physiological pH ^{oxo}C is superior to cytosine.[108] Surprisingly contiguous substitutions of 6-*oxo* cytosine are destabilizing.[109] The lower stability of the ^{oxo}C.GC triplet relative to C$^+$.GC is attributed to unfavourable stacking interactions and/or steric hindrance owing to the 6-carbonyl group, which lies close to the furanose oxygen in the *anti*-conformation that is required in triplexes. This has been partially overcome by attaching the base to the backbone *via* an acyclic linker which gives greater flexibility.[110] 2'-*O*-Methyl and ribo derivatives of this base have also been synthesized though these produce less stable complexes.[107,111] We have recently synthesized the 2'-aminoethoxy derivative of this nucleoside, but this does not produce a major increase in affinity.

1.3.2 Purine Base Analogues

The first purine analogues to be tested for pH-independent recognition of guanine were 8-*oxo*-adenine[112] and its N6-methyl derivative.[113] The presence of the 8-*oxo*-group forces these bases to adopt a *syn*-conformation which presents the 6-amino and N7 protons in a suitable orientation for recognition of guanine (Figure 7B (i)). The 7,8-dihydro derivative has similar properties.[114] These analogues recognize GC in a pH-independent fashion and generate triplexes that have the same stability as those containing ^{Me}C at low pH. They can also be used for targeting contiguous guanines, but have reduced affinity at isolated Gs, as the triplets are not isomorphic with T.AT.

A different strategy has been developed using N7-purine derivatives. The first to be synthesized was N7G.[115] This base analogue alters the G.GC triplet so that it can be incorporated within the parallel-binding motif (Figure 7B (ii)). This base displays good selectivity and pH insensitivity but suffers from sequence constraints since it is not isomorphic with the other triplets in the parallel motif.[116] A single substitution of this analogue is as stable as 5-methyldeoxycytidine at pH 7 but is less stable at pH 5. When alternate third-strand bases are substituted with this analogue, the triplex is less stable by three orders of magnitude compared to contiguous substitutions. Similar characteristics were also exhibited by other N7 analogues, P1,[117,118] and N7-inosine[119–120] (Figure 7B (iii and iv)). N7-Inosine lacks the amino function of N7G but surprisingly shows stable recognition of guanine. This was attributed to the formation of an unconventional CH···O bond between the carbonyl group of inosine and the CH of guanine. It was postulated that this interaction gives a small, positive, direct electrostatic contribution to stability.[121] Electrostatic interactions are very important contributors to the stabilization of nucleic acid structures by influencing base stacking, but are difficult to predict.

Several strategies have been employed to overcome the sequence constraints imposed by the lack of isomorphism of the triplets formed by these bases. An acyclic glycerol derivative was employed to attach N7G to the oligonucleotide backbone, though the increase in flexibility did not alleviate this constraint.[122] An alternative method to compensate for this loss in binding energy is to add positive charges, as described for other analogues above. We have shown that a single substitution with the 2′aminoethoxy derivative of N7G is as stable as cytosine at pH 5.0, but when this nucleotide is employed at alternate positions it produces less stable complexes than [Me]C at pH 7.0. Optimum triplex formation may therefore require the use of a combination of such bases; N7-purines for binding to contiguous GC base pairs and pyrimidine analogues for binding to isolated guanines.[118] An alternative strategy would be to develop an isomorphic N7 derivative (such as N7-adenine) for targeting A, though the propensity for purines to bind in an antiparallel orientation may create problems.

1.4   Recognition of Pyrimidine Interruptions

Triplex formation is usually restricted to homopurine.homopyrimidine target sites since only the purine strand of the duplex is recognized. Although, there is an abundance of oligopurine target sites within the human genome[123-125] this limits the use of triplexes and a method for targeting any DNA sequence is highly desirable. Recognition of the pyrimidine bases, however, is hampered by several factors. First, the Hoogsteen faces of the pyrimidine bases only present the possibility of a single-hydrogen bond contact within the major groove. This therefore leads to triplets with low affinity and reduced selectivity. Bases that recognize T may also recognize G as these contain an H-bond acceptor at similar positions; the same occurs for recognition of C and A. Recognition of thymine presents a further problem due to the presence of the 5-methyl group in the major groove, imposing a steric barrier to third-strand binding. There have therefore been several strategies proposed to overcome each of these restrictions.[126]

1.4.1   Null Bases and Abasic Linkers

The binding of a third strand of DNA within the major groove is highly asymmetric. Consequently it is not possible to switch across the groove to recognize the partner base on the opposite strand without loss of co-operativity due to reduced base-stacking and conformational strain on the backbone of the TFO. One means of generating oligonucleotides that bind to targets that contain pyrimidine interruptions is simply to bypass the "offending" base by placing a null or universal base analogue opposite to the inversion site. Universal bases are non-hydrogen bonding, aromatic analogues that stabilize the helical structure by stacking interactions alone.[127] Abasic linkers, such 1,2-dideoxy-D-ribose (φ)[128] have been tested for skipping the interruption but the binding affinity is low due to the loss of stacking interactions. Neither of these approaches has yielded stable triple-helical structures and both cause a loss of specificity at the skipped base, as any base pair can be tolerated at this position.

1.4.2 Natural Bases

Several studies have been carried out on TFOs to investigate the stability of all possible combinations of natural bases opposite to each base pair.[129–131] Triplex formation is sensitive to single mismatches between the bases in the third strand and the duplex. A single-base mismatch results in a typical free-energy penalty of ~ 3 kcal mol^{-1}.[132–133] The destabilization is dependent on the nature and position of a mismatch. Central mismatches are more destabilizing than terminal ones since they disrupt the co-operative interactions between neighbouring triplets.[133,134]

These studies have demonstrated that the least destabilizing combinations for recognizing TA and CG base pairs are G.TA and T.CG (or C.CG) (Figure 8A). For each of these triplets the third strand only forms a single hydrogen

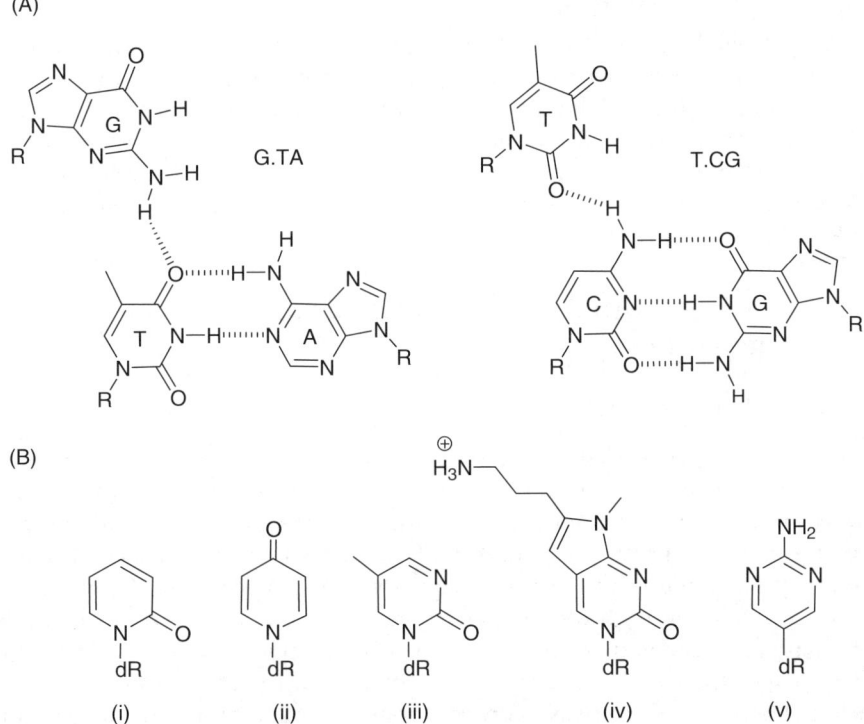

Figure 8 *(A) Structure of the G.TA and T.CG base pairs. (B) Base analogues for recognition of CG base pairs: (i) pyridine-2-one, (ii) pyridine-4-one, (iii)* ^{4H}T*, (iv)* ^{A}PP*, and (v) d2APm*

bond to the target, resulting in complexes which are less stable and less selective than the canonical triplets.

The G.TA triplet contains a single-hydrogen bond between the exocyclic amino group of guanine and the 4-carbonyl of thymine.[135] This hydrogen-bonding arrangement has been confirmed by replacement studies using several guanine analogues. Removal of the 2-amino group or the 6-oxo group, generating inosine or 2-aminopurine, respectively, produces triplets which are less stable than guanine.[136] The latter is more surprising as the 6-oxo group is not thought to be involved in binding, although 2-aminopurine also differs from guanine in lacking a hydrogen atom on N1. The stability of the G.TA triplet is affected by the sequence context and flanking T.AT triplets (especially on the 3'-side) produce more stable complexes than flanking C^+.GC triplets. This is thought to be due to the formation of a second (weaker) hydrogen bond with the T of an adjacent T.AT triplet.[135] Stable complexes can be formed when this triplet is present at every fourth position, so long as the triplex contains some C^+.GC triplets and T.AT is located on the 3'-side of each G.TA.[137] The interaction is further stabilized by the appropriate use of charged base analogues such as 5-propargylamino-dU (U^P). Duplex regions of $(AT)_n$ can be targeted with GT-containing oligonucleotides, forming alternating G.TA and T.AT triplets, though this interaction is only observed if this is anchored by a more stable triplex.[138,139]

The parallel T.CG triplet was first proposed by Yoon *et al.*[130] and has been shown to involve a single-hydrogen bond between O2 of the third-strand thymine and the free C4-amino proton on the duplex cytosine.[140] This hydrogen-bonding pattern can also be generated with a third-strand cytosine forming the C.CG triplet. Up to three consecutive T.CG or G.TA triplets can be tolerated in the centre of a triplex, if the interaction is stabilized by a triplex-binding ligand.[141]

1.4.3   Nucleotide Analogues for Recognizing Pyrimidine Interruptions

1.4.3.1   Targeting CG

A number of thymine analogues have been used to recognize CG interruptions within oligopurine tracts (see Figure 8B). Their efficacy was initially demonstrated in the antiparallel motif using pyridine-2-one and pyridine-4-one (Figure 8B (i and ii)). These compounds utilize carbonyl groups for recognition of the exocyclic amino group of cytosine.[142] By omitting either the 4-carbonyl or 3-NH groups of T, both of which are used in the recognition of adenine, the selectivity for CG could be increased. 5-methyl-pyrimidine-2-one (^{4H}T) (Figure 8B (iii)) was the first such derivative to be synthesized;[143] deletion of the 4-carbonyl group of T removes hydrogen from N3 and abolishes the recognition of adenine. The 2-carbonyl of this derivative may also form an unconventional C–H$\cdots$O bond at the 5-position of cytosine.[121] This interaction had previously been reported for the N7I.GC triplet. The ^{4H}T.CG triplet is as stable as T.CG but loses the recognition of adenine.[143] A further increase in affinity is seen if the third strand is fully modified with the 2'-aminoethoxy substitution. This increases the T_m by about 1.5°C per substitution and it can be used to recognize up to five CG interruptions at pH 6.5 (33% pyrimidine content in the target strand).[144,145]

We have recently extended this strategy by preparing of a variety of substituted pyrrolopyrimidine-2-ones.[146,147] These retain the H-bonding pattern of ^{4H}T but introduce increased aromacity and allow the introduction of substituents at the 6-position. One such variant, ^{A}PP (Figure 8B (iv)), contains a protonated primary amino group. This analogue is still selective for CG and is 2°C more stable than T at pH 6.0.

Another analogue for cytosine recognition is 2-aminopyrimidine (d2APm) (Figure 8B (v)) which has a pK_a of 3.3 and is therefore unprotonated at all practical pH values. It has a nitrogen atom (H-bond acceptor) which acts as a partner with the exocyclic amino group of cytosine. This base produces a 4°C increase in T_m at pH 7 relative to cytosine.[148]

It is possible to compensate for the loss in binding energy at single-pyrimidine inversions by combining these analogues with other sugar and/or backbone modifications that are known to increase stability. The LNA sugar modification is one such example and recognition of CG has been achieved using LNA bearing a 2-pyridone base.[149] A TFO containing a single such substitution exhibited a T_m that was 9°C higher than when this base was attached to deoxyribose. However, this modification also enhanced the recognition of AT. To overcome this an LNA bearing 1-isoquinolone (Q^B) was prepared, reasoning that binding to AT base pairs would be sterically hindered by the proximity of this bicyclic analogue to the 5-methyl group of thymine on the opposite side of the major groove.[150] In this way the binding to AT was reduced, but so too was the desired interaction with CG. We have prepared the 2'-aminoethoxy derivative of Q, but find that in this context it is not effective for recognition of CG.

A variety of imidazoles, attached to LNA sugars have been assessed for their ability to target pyrimidine interruptions, both CG and TA. Oxazole recognized CG slightly better than TA, imidazole was not selective and 2-aminoimidazole recognized GC.[151] However, all these complexes were much less stable than triplexes that contain the canonical triplets T.AT and C^+.GC.

1.4.3.2 Targetting TA

The presence of a methyl group at the 5-position of thymine presents a problem for developing analogues to interact with this base. One strategy for overcoming this problem is to use a linker that projects the analogue past the methyl group to bind to the thymine 4-oxo group. To date this has only been attempted using 3-oxo-2,3-dihydropyridazine (E), which has been attached to a PNA backbone *via* a β-alanine linker. This analogue shows a T_m increase of 5°C relative to guanine at an isolated TA interruption.[152] A derivative of this analogue, which has reduced flexibility in the linker owing to the presence of a double bond, has recently been described and is awaiting assessment.[153]

We have attempted to increase the stability of the G.TA triplet by attaching an aminoethoxy group at the 2'-position of G or by adding a propargylamino group at the 7-position of 7-deaza-G. However these modifications do not increase the affinity of G for TA, but instead they increase the affinity for GC base pairs (Yang, Booth, Rusling, Fox, and Brown, unpublished observations).

1.4.4 Nucleotide Analogues for Recognizing both Partners of the Base Pair

An ingenious strategy for recognizing pyrimidine inversions uses analogues that are able to make contact with both bases of the target, recognizing the entire base pair, rather than one or other of the bases. Example of such triplets are shown in Figure 9.

1.4.4.1 CG Recognition

4-(3-Benzamidophenyl)imidazole (D$_3$) (Figure 9A) was designed to sterically match the edges of a CG base pair.[154] It was anticipated that the imidazole moiety would form a single-hydrogen bond to cytosine while additional stacking interactions would be possible due to the presence of two aromatic rings positioned in the major groove. The rotational freedom between these two rings could maximize these non-bonding interactions. Affinity cleavage experiments showed that this base bound to CG and TA base pairs with greater affinity than to GC or AT. However, later studies showed that this base formed triplets that are less stable when C$^+$.GC is present on its 3'-side.[23] A subsequent NMR study showed that it binds by intercalation into the YpR step, thereby skipping the inversion site.[155] Two similar carbocyclic ribofuranose analogues, L1 and L2 were also developed (Figure 9A), which exhibited a preference for binding at pyrimidine inversions, and are also thought to bind by an intercalative mechanism.[156]

N4 cytosine derivatives have been the most successful for recognition of both partners of the CG base pair (Figure 9B). N^4-(3-acetamidopropyl)cytosine (Figure 9B (ii)) positions a side chain across the major groove allowing the 3-amino group to form a hydrogen bond to the O6 carbonyl group of guanine. UV melting showed this base to be more stable than C.CG but less stable than the canonical triplets.[157] A further derivative, N^4-(6-aminopyridinyl)cytosine (Figure 9B (ii)), contains a ring to constrain the rotational freedom of the side chain. This base recognizes both CG and AT and produced a broad melting transition indicative of intercalation plus H-bonding.[158,159]

Another attempt to recognize both bases of the CG base pair used the ureoisoindolin-1-one homo-*N*-nucleoside (Figure 9B (iii)). This compound effectively formed a triplet with CG in chloroform, but showed no binding

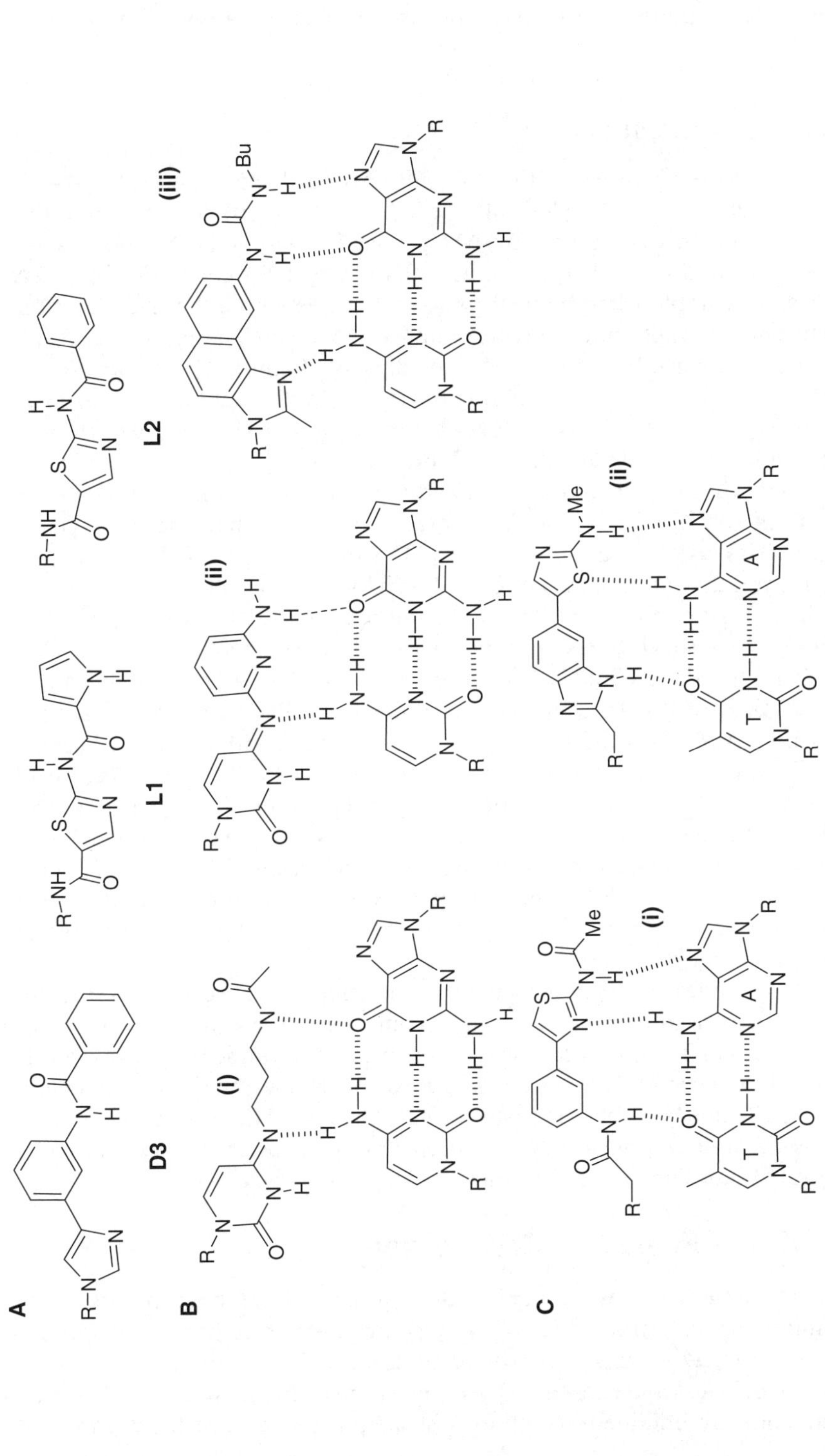

Figure 9 *Analogues for recognition of YR base pairs. (A) Nucleotides that bind by intercalation. (B) Nucleotides for recognizing CG base pairs: (i) AcPrC.CG, (ii) AmPyC.CG, and (iii) Ureido naphthimidazole.CG. (C) Nucleotides for recognizing TA base pairs: (i) S.TA and (ii) Bt.TA*

when incorporated into a TFO, probably as it is not isomorphic with the other triplets.[160]

1.4.4.2   TA recognition

The most successful base for the recognition of TA base pairs is N-(4-(3-acetamidophenyl)thiazoyl-2-yl-acetamide (S) (Figure 9C (i)). The design of this nucleoside is based on the previously reported D_3 monomer,[161] and has the potential to form three hydrogen bonds with the target TA base pair. The S.TA triplet was less stable when flanked by C^+.GC triplets on either side, and the interaction was thought not to involve intercalation. Further derivatives of this nucleoside have since been prepared, including B^t (Figure 9C (ii)), which was designed in anticipation that increasing the rigidity of S would improve triplex-binding properties.[162,163] This analogue retains a similar H-bonding pattern to S but is conformationally less flexible. Unfortunately this analogue exhibited a decrease in affinity and a loss of selectivity, suggesting that it favours an intercalating mode of binding. We have also shown that the analogue S recognizes CG as well as TA at low pH, with little or no discrimination between them, though it binds better to TA at higher pHs. The interaction and selectivity is further enhanced by addition of a 2'-aminoethoxy group.[164]

A different approach to the recognition of both partners of the base pair has arisen from the design and synthesis of a series of 2-aminoquinazolin-5-yl monomers (Figure 10). These molecules, designated TRIPsides, are designed to bind symmetrically within the major groove, unlike other triplex-forming oligonucleotides, positioning the oligonucleotide backbone in the centre of the groove.[165–168] In this strategy only the purine strand of the target is read, but because the backbone is located in the centre of the major groove, either strand can be recognized by choosing the appropriate TRIPside. To date antiCG, antiTA, and antiGC have been successfully employed at pH 7 for recognizing a 19-mer-target site in which the purines switch from one strand to the other four times.[167]

Recently a series of extended guanine analogues have been suggested for recognition of TA interruptions.[169] These compounds use aminobenzimidazole for recognition of T (mimicking the G.TA triplet). A thymine base is covalently attached to the benzimidazole ring *via* a linker to allow simultaneous recognition of the adenine by either Hoogsteen (N2) or reverse-Hoogsteen (N1) bonds. Synthesis of the free nucleosides has been reported but these have yet to be incorporated into triplex-forming oligonucleotides.

1.5   Mixed Sequence Recognition

Although a large number of nucleoside analogues have been prepared for overcoming the pH problem and for recognizing pyrimidine interruptions, there are very few examples in which these have been combined to recognize mixed-DNA sequences at physiological pH. Most studies have only targeted single-pyrimidine interruptions in the purine tract, or have employed

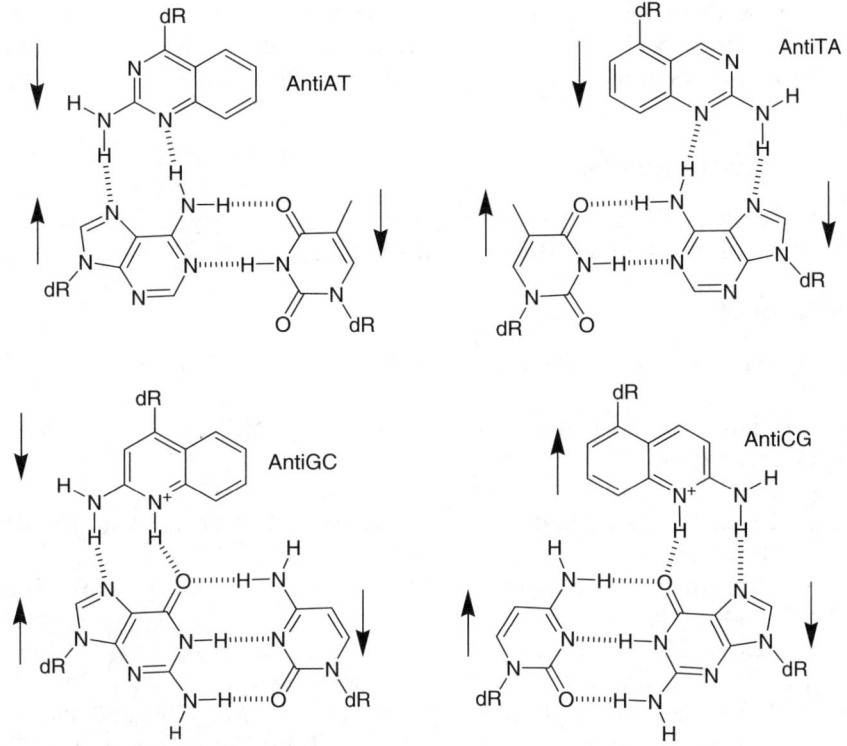

Figure 10 *Structure of the four-TRIPsides that have been proposed for recognizing all four base pairs. The arrows show the orientation of the oligonucleotide strands*

a single-cytosine analogue. One exception is the recent synthesis of the oligo-TRIPs,[167] which have been successfully employed to bind to a 19-mer target in which the purine-containing strand switches from one side to the other four times. Successful recognition of three base pairs (AT, GC, and CG, but not TA) has also been achieved using fully modified 2′-aminoethoxy RNAs containing 5-methyl-2-pyrimidone for recognition of CG.[144] This combination was able to recognize up to five CG inversions in a 15-mer-DNA target with high selectivity and good affinity.

We have also prepared oligonucleotides that contain combinations of modified nucleotides for recognizing mixed sequences. For this we have used BAU to target AT with high affinity, [Me]P for recognition of GC base pairs at elevated pHs and S and [A]PP for recognizing TA and CG base pairs, respectively.[147] With this combination we can generate a triplex footprint at a 19-mer-target site that contains four pyrimidine interruptions (see Figure 5c). This complex is sensitive to pH, since it contains the [Me]P.GC triplet, but it forms at low micromolar concentration at pH 7.0. Footprinting and melting experiments have been demonstrated that this heavily modified oligonucleotide retains its sequence specificity and that changing a single-base pair opposite to any one of the modified nucleosides leads to a large decrease in affinity. The only exception is S which appears to form stable complexes opposite to both TA and CG base pairs.[147]

Acknowledgements

Original work in the authors' laboratories is supported by Cancer Research UK, the European Union, BBSRC, and EPSRC.

References

1. G. Felsenfeld, D.R. Davies and A. Rich, *J. Am. Chem. Soc.*, 1957, **79**, 2023.
2. F.B. Howard, J. Frazier, M.N. Lipsett and T. Mills, *Biochem. Biophys. Res. Commun.*, 1964, **17**, 93.
3. A.R. Morgan and R.D. Wells, *J. Mol. Biol.*, 1968, **37**, 63.
4. M. Riley, B. Maling and M.J. Chamberlin, *J. Mol. Biol.*, 1966, **20**, 359.
5. C. Marck and D. Thiele, *Nucleic Acids Res.*, 1978, **5**, 1017.
6. S.L. Broitman, D.I. Dwight and J.R. Fresco, *Proc. Natl. Acad. Sci. USA*, 1987, **84**, 5120.
7. S. Buchini and C.J. Leumann, *Curr. Opin. Chem. Biol.*, 2003, **7**, 717.
8. M.P. Knauert and P.M. Glazer, *Hum. Mol. Genet.*, 2001, **20**, 2243.
9. K.M. Vasquez and P.M. Gazer, *Quart. Rev. Biophys.*, 2002, **35**, 89.
10. M.C. Rice, K. Czymmek and E.B. Kmiec, *Nat. Biotech.*, 2001, **19**, 321.
11. M.M. Seidman and P.M. Glazer, *J. Clin. Invest.*, 2003, **112**, 487.
12. V.N. Potaman, *Expert. Rev. Mol. Diagn.*, 2003, **3**, 481.
13. H.E. Moser and P.B. Dervan, *Science*, 1987, **238**, 645.
14. T. Le Doan, L. Perrouault, D. Praseuth, N. Habhoub, J.-L. Decout, N.T. Thuong, J. Lhomme and C. Hélène, *Nucleic Acids Res.*, 1987, **19**, 7749.
15. N.T. Thuong and C. Hélène, *Angew. Chem. Int. Ed. Engl.*, 1993, **32**, 666.
16. P.A. Beal and P.B. Dervan, *Science*, 1991, **251**, 1360.
17. R.H. Durland, D.J. Kessler, S. Gunnel, M. Duvic, B.M. Pettitt and M.E. Hogan, *Biochemistry*, 1991, **30**, 9246.
18. C. Giovannangeli, M. Rougée, T. Garestier, N.T. Thuong and C. Hélène, *Proc. Natl. Acad. Sci. USA*, 1992, **89**, 8631.
19. J.-S. Sun and C. Hélène, *Curr. Opin. Struct. Biol.*, 1993, **3**, 345.
20. J.L. Asensio, A.N. Lane, J. Dhesi, S. Bergqvist and T. Brown, *J. Mol. Biol.*, 1998, **275**, 811.
21. D. Leitner, W. Schröder and K. Weisz, *Biochemistry*, 2000, **39**, 5886.
22. J.S. Lee, M.L. Woodsworth, L.J.P. Latimer and A.R. Morgan, *Nucleic Acids Res.*, 1984, **12**, 6603.
23. L.L. Kiessling, L.C. Griffin and P.B. Dervan, *Biochemistry*, 1992, **31**, 2829.
24. J. Völker and H.K. Klump, *Biochemistry*, 1994, **33**, 13502.
25. N. Sugimoto, P. Wu, H. Hara and Y. Kawamoto, *Biochemistry*, 2001, **40**, 9396.
26. M.D. Keppler and K.R. Fox, *Nucleic Acids Res.*, 1997, **25**, 4644.

27. A.M. Soto, J. Loo and L.A. Marky, *J. Am. Chem. Soc.*, 2002, **124**, 14355.
28. R.W. Roberts and D.M. Crothers, *Proc. Natl. Acad. Sci. USA*, 1996, **93**, 4320.
29. J.L. Asensio, R. Carr, T. Brown and A.N. Lane, *J. Am. Chem. Soc.*, 1999, **121**, 11063.
30. R.W. Roberts and D.M. Crothers, *Science*, 1992, **258**, 1463.
31. M. Shimizu, A. Konishi, Y. Shimada, H. Inoue and E. Oshtsuka, *FEBS Letts.*, 1992, **302**, 155.
32. H.Y. Han and P.B. Dervan, *Nucleic Acids Res.*, 1994, **22**, 2837.
33. C. Escudé, J.-S. Sun, M. Rougée, T. Garestier and C. Hélène, *Comp. Rend. Acad. Sci. III*, 1992, **315**, 521.
34. E. Bernal-Mendez and C.J. Leumann, *J. Biol. Chem.*, 2001, **276**, 35320.
35. J. Plavec, C. Thibaudeau and J. Chattopadhyaya, *J. Am. Chem. Soc.*, 1994, **116**, 6558.
36. S.K. Singh, P. Nielsen, A.A. Koshkin and J. Wengel, *Chem. Commun.*, 1998, **455**.
37. H. Torigoe, Y. Hari, M. Sekiguchi, S. Obika and T. Imanishi, *J. Biol. Chem.*, 2001, **276**, 2354.
38. B.-W. Sun, R. Babu, M.D. Sørensen, K. Zakrzewska, J. Wengel and J.-S. Sun, *Biochemistry*, 2004, **43**, 4160.
39. M. Koizumi, K. Morita, M. Daigo, S. Tsutsumi, K. Abe, S. Obika and T. Imanishi, *Nucleic Acids Res.*, 2003, **31**, 3267.
40. S. Obika, M. Onoda, K. Morita, J. Andoh, M. Koizumi and T. Imanishi, *Chem. Commun.*, 2001, 1992.
41. M. Tarköy and C.J. Leumann, *Angew. Chem. Intl. Ed. Engl.*, 1993, **32**, 1432.
42. R. Steffens and C.J. Leumann, *J. Am. Chem. Soc.*, 1999, **121**, 3249.
43. D. Renneberg and C.J. Leumann, *ChemBiochem.*, 2004, **5**, 1114.
44. R.O. Dempcy, K.A. Browne and T.C. Bruice, *Proc. Natl. Acad. Sci. USA*, 1995, **92**, 6097.
45. A. Blaskó, O.D. Dempcy, E.E. Minyat and T. Bruice, *J. Am. Chem. Soc.*, 1996, **118**, 7892.
46. M. Park and T. Bruice, *Biorg. Med. Chem. Lett.*, 2005, **15**, 3247.
47. S. Chaturvedi, T. Horn. and R.L. Letsinger, *Nucleic Acid Res.*, 1996, **24**, 2318.
48. J.M. Dagle and D.L. Weeks, *Nucleic Acids Res.*, 1996, **24**, 2143.
49. T. Michel, F. Debart, F. Heitz and J.-J. Vasseur, *ChemBiochem.*, 2005, **6**, 1254.
50. B. Cuenoud, F. Casset, D. Hüsken, F. Natt, R.M. Wolf, K.-H. Altmann, P. Martin and H.E. Moser, *Angew. Che. Int. Ed.*, 1998, **37**, 1288.
51. N. Atsumi, Y. Ueno, M. Kanazaki, S. Shuto and A. Matsuda, *Biorg. Med. Chem.*, 2002, **10**, 2933.
52. M.J.J. Blommers, F. Natt, W. Jahnke and B. Cuenoud, *Biochemistry*, 1998, **37**, 17714.
53. N. Puri, A. Majumdar, B. Cuenod, P.S. Millar and M.M. Seidman, *Biochemistry*, 2004, **43**, 1343.

54. A. Mayer, A. Häberli and C.J. Leumann, *Org. Biomol. Chem.*, 2005, **3**, 1653.
55. A. Haberli and C.J. Leumann, *Org. Lett.*, 2002, **4**, 3275.
56. A. Haberli, A. Mayer and C.J. Leumann, *Nucleosides Nucleotides and Nucleic Acids*, 2003, **22**, 1187.
57. T.P. Prakash, A. Puschl, E. Lesnik, V. Mohan, V. Tereshko, M. Egli and M. Manoharan, *Org. Lett.*, 2004, **6**, 1971.
58. J. Bijapur, M.D. Keppler, S. Bergqvist, T. Brown and K.R. Fox, *Nucleic Acids Res.*, 1999, **27**, 1802.
59. H. Nara, A. Ono and A. Matsuda, *Bioconjug. Chem.*, 1995, **6**, 54.
60. D.A. Barawker, K.G. Rajeev, V.A. Kumar and K.N. Ganesh, *Nucleic Acids Res.*, 1996, **25**, 4187.
61. M. Sollogoub, R.A.J. Darby, B. Cuenoud, T. Brown and K.R. Fox, *Biochemistry*, 2002, **41**, 7224.
62. S. Osbourne, V.E.C. Powers, D.A. Rusling, O. Lack, K.R. Fox and T. Brown, *Nucleic Acids Res.*, 2004, **32**, 4439.
63. P.S. Miller, N. Dreon, S.M. Pulford and K.B. McParland, *J. Biol. Chem.*, 1980, **255**, 9659.
64. L. Kibler-Herzog, B. Kell, G. Zon, K. Shinozuka, S. Mizan and W.D. Wilson, *Nucleic Acids Res.*, 1990, **18**, 3545.
65. F. Debart, A. Meyer, J.-J. Vasseur and B. Rayner, *Nucleic Acids Res.*, 1998, **26**, 4551.
66. S.M. Gryaznov, D.H. Lloyd, J.K. Chen, R.G. Schultz, L.A. DeDionisio, L. Ratmeyer and W.D. Wilson, *Proc. Natl. Acad. Sci. USA*, 1995, **92**, 5798.
67. H. Torigoe and A. Maruyama, *J. Am. Chem. Soc.*, 2005, **127**, 1705.
68. L. Lacroix, P.B. Arimondo, M. Takasugi, C. Hélène and J.-L. Mergny, *Biochem. Biopys. Res. Comm.*, 2000, **270**, 363.
69. J. Basye, J.O. Trent, D. Gao and S.W. Ebbinghaus, *Nucleic Acids Res.*, 2001, **29**, 4873.
70. T. Michel, F. Debart, J.-J. Vasseur, F. Geinguenaud and E. Taillandier, *J. Biomol. Struct. Dyn.*, 2003, **21**, 435.
71. P.E. Nielsen, M. Egholm, R.H. Berg and O. Buchardt, *Science*, 1991, **254**, 1497.
72. P.E. Nielsen, *Curr. Med. Chem.*, 2001, **8**, 545.
73. P.E. Nielsen, M. Egholm and O. Buchardt, *J. Mol. Recogn.*, 1994, **7**, 165.
74. P. Wittung, P. Nielsen and B. Nordén, *Biochemistry*, 1997, **36**, 7973.
75. A.B. Staubli and P.B. Dervan, *Nucleic Acids Res.*, 1994, **22**, 2637.
76. J. Michel, J.-J. Toulmé, J. Vercauteren and S. Moreau, *Nucleic Acids Res.*, 1996, **24**, 1127.
77. F. Godde, J.-J. Toulme and S. Moreau, *Biochemistry*, 1998, **37**, 13765.
78. L.E. Xodo, G. Manzini, F. Quadrifoglio, G.A. van der Marel and J.H. Boom, *Nucleic Acids Res.*, 1991, **20**, 5625.
79. N. Colocci and P.B. Dervan, *J. Am. Chem. Soc.*, 1994, **116**, 785.
80. A.K. Phipps, M. Tarköy, P. Schultz and J. Feignon, *Biochemistry*, 1998, **37**, 5820.
81. L. Lacroix, J. Lacoste, J.F. Reddoch, J.-L. Mergny, D.D. Levy, M.M. Seidman, M.D. Matteucci and P.M. Glazer, *Biochemistry*, 1999, **38**, 1893.

82. B.C. Froehler, S. Wadwani, T.J. Terhost and S.R. Gerrard, *Tett. Lett.*, 1992, **37**, 5307.

83. J.A. Brazier, T. Shibata, J. Townsley, B.F. Taylor, E. Frary, N.H. Williams and D.M. Williams, *Nucleic Acids Res.*, 2005, **33**, 1362.

84. J.L. Mergny, G. Duval-Valentin, C.H. Nguyen, L. Perroualt, B. Faucon, M. Rougée, T. Montenay-Garestier, E. Bisagni and C. Hélène, *Science*, 1992, **256**, 1681.

85. R.A.J. Darby and K.R. Fox, in *Interaction of Small Molecules with DNA and RNA From Synthesis to Nucleic Acid Complexes*, M. Demeunynck, C. Bailly and W.D. Wilson (eds), Wiley, VCH, NJ, 2002, 360–383.

86. U. Asseline, J.F. Hau, S. Czernecki, T. Lediguarher, M.C. Perlat, J.M. Valery and N.T. Thuong, *Nucleic Acids Res.*, 1991, **19**, 4067.

87. T.J. Stonehouse and K.R. Fox, *Biochim. Biophys. Acta*, 1994, **1218**, 322.

88. G.C. Silver, J.S. Sun, C.H. Nguyen, A.S. Boutorine, E. Bisagni and C. Hélène, *J. Am. Chem. Soc.*, 1997, **119**, 263.

89. M.D. Keppler, C.M. McKeen, O. Zegrocka, L. Strekowski, T. Brown and K.R. Fox, *Biochim. Biophys. Acta*, 1999, **1447**, 137.

90. J. Robles and L.W. McLaughlin, *J. Am. Chem. Soc.*, 1997, **119**, 6014.

91. M.L. Capobianco, M. Champdore, F. Arcamone, A. Garbesi, D. Guinvarc'h and P.B. Arimondo, *Biorg. Med. Chem.*, 2005, **13**, 3209.

92. S. Kureti, J.-S. Sun, T. Garestier and C. Hélène, *Nucleic Acids Res.*, 1997, **25**, 4264.

93. M. Takasugi, A. Guendouz, M. Chassignol, J.L. Decout, J. Lhomme, N.T. Thuong and C. Hélène, *Proc. Natl. Acad Sci. USA*, 1991, **88**, 5602.

94. T.J. Povsic and P.B. Dervan, *J. Am. Chem. Soc.*, 1989, **111**, 3059.

95. S.F. Singleton and P.B. Dervan, *Biochemistry*, 1992, **32**, 4761.

96. P.J. Bates, C.A. Laughton, T.C. Jenkins, D.C. Capaldi, P.D. Roselt, C.B. Reese and S. Neidle, *Nucleic Acids Res.*, 1996, **24**, 4176.

97. S. Hildbrand and C.J. Leumann, *Angew. Chem. Int. Ed. Eng.*, 1996, **35**, 1968.

98. S.A. Cassidy, P. Slickers, J.O. Trent, D.C. Capaldi, P.D. Roselt, C.B. Reese, S. Neidle and K.R. Fox, *Nucleic Acids Res.*, 1997, **25**, 4891.

99. S. Hilbrand, A. Blaser, S.P. Parel and C.J. Leumann, *J. Am. Chem. Soc.*, 1997, **119**, 5499.

100. D.A. Rusling, L. LeStrat, V.E.C. Powers, V.J. Broughton-Head, J. Booth, O. Lack, T. Brown and K.R. Fox, *FEBS Lett.*, 2005, **579**, 6616.

101. A. Ono, P.O.P. Ts'o and L. Kan, *J. Am. Chem. Soc.*, 1991, **113**, 4032.

102. A. Ono, P.O.P. Ts'o and L. Kan, *J. Org. Chem.*, 1992, **57**, 3225.

103. A. Mayer and C.J. Leumann, *Nucleosides, Nucleotides and Nucleic Acids*, 2003, **22**, 1919.

104. T.-M. Chin, S.-B. Lin, S.-Y. Lee, M.-L. Chang, A. Y.-Y. Cheng, F.-C. Chang, L. Psaternack, D.H. Huang and L.-S. Kan, *Biochemistry*, 2000, **39**, 12457.

105. M. Egholm, L. Christensen, K.L. Dueholm, O. Buchardt, J. Coull and P.E. Nielsen, *Nucleic Acids Res.*, 1995, **23**, 217.

106. U.V. Krosigk and S.A. Benner, *J. Am. Chem. Soc.*, 1995, **117**, 5361.

107. B. Berressem and J.W. Engels, *Nucleic Acids Res.*, 1995, **23**, 3465.
108. G. Xiang, W. Soussou and L.W. McLaughlin, *J. Am. Chem. Soc.*, 1994, **116**, 11155.
109. G. Xiang, R. Bogacki and L.W. McLaughlin, *Nucleic Acids Res.*, 1996, **24**, 1963.
110. G. Xiang and L.W. McLaughlin, *Tetrahedron*, 1998, **54**, 375.
111. U. Parsh and J.W. Engels, *Chem. Eur. J.*, 2000, **6**, 2409.
112. P.S. Miller, P. Bhan, C.D. Cushman and T.L. Trapane, *Biochemistry*, 1992, **31**, 6788.
113. S.H. Krawczyk, J.F. Milligan, S. Wadwani, C. Moulds, B.C. Froehler and M.D. Matteucci, *Proc. Natl. Acad. Sci. USA*, 1992, **89**, 3761.
114. M.C. Jetter and F.W. Hobbs, *Biochemistry*, 1993, **32**, 3249.
115. J. Hunziker, S.E. Prietley, H. Brunar and P.B. Dervan, *J. Am. Chem. Soc.*, 1995, **117**, 2661.
116. K.M. Koshlap, P. Schultz, H. Brunar, P.B. Dervan and J. Feigon, *Biochemistry*, 1997, **36**, 2659.
117. J.S. Koh and P.B. Dervan, *J. Am. Chem. Soc.*, 1992, **114**, 1470.
118. S.F. Singleton and P.B. Dervan, *J. Am. Chem. Soc.*, 1995, **117**, 10376.
119. J. Marfurt, J. Hunziker and C.J. Leumann, *Biorg. Med. Chem.*, 1996, **24**, 3021.
120. J. Marfurt, S.P. Parel and C.J. Leumann, *Nucleic Acids Res.*, 1997, **25**, 1875.
121. J. Marfurt and C.J. Leuman, *Angew. Chem. Int. Ed.*, 1998, **37**, 175.
122. A. St Clair, G. Xiang and L.W. Mclaughlin, *Nucleosides and Nucleotides*, 1998, **17**, 925.
123. H. Manor, B.S. Rao and R.G. Martin, *J. Mol. Evol.*, 1988, **27**, 96.
124. M.J. Behe, *Nucleic Acids Res.*, 1995, **23**, 689.
125. J.R. Goni, X. de la Cruz and M. Orozco, *Nucleic Acids Res.*, 2004, **32**, 354.
126. D.M. Gowers and K.R. Fox, *Nucleic Acids Res.*, 1999, **27**, 1569.
127. D. Loakes, *Nucleic Acids Res.*, 2001, **29**, 2437.
128. D.A. Horne and P.B. Dervan, *Nucleic Acids Res.*, 1991, **19**, 4963.
129. L.C. Griffin and P.B. Dervan, *Science*, 1989, **245**, 967.
130. K. Yoon, C.A. Hobbs, J. Koch, M. Sardaro, R. Kutny and A.L. Weis, *Proc. Natl. Acad. Sci. USA*, 1992, **89**, 3840.
131. S.P. Chandler and K.R. Fox, *FEBS Letts.*, 1993, **332**, 189.
132. R.W. Roberts and D.M. Crothers, *Proc. Natl. Acad. Sci. USA*, 1991, **88**, 9397.
133. M. Rougée, B. Faucon, J.L. Mergny, F. Barcelo, C. Giovannangeli, T. Garestier and C. Hélène, *Biochemistry*, 1992, **31**, 9269.
134. J.L. Mergny, J.S. Sun, M. Rougée, T. Montenay-Garester, F. Barcelo, J. Chomilier and C. Hélène, *Biochemistry*, 1991, **30**, 9791.
135. I. Radhadkrishnan, X.L. Gao, C. de los Santos, D. Live and D.J. Patel, *Biochemistry*, 1991, **30**, 9022.
136. O.A. Amosova and J.R. Fresco, *Nucleic Acids Res.*, 1999, **27**, 4632.

137. D.M. Gowers, J. Bijapur, T. Brown and K.R. Fox, *Biochemistry*, 1999, **38**, 13747.
138. S.P. Chandler and K.R. Fox, *FEBS Letts.*, 1995, **360**, 21.
139. D.M. Gowers and K.R. Fox, *Nucleic Acids Res.*, 1998, **26**, 3626.
140. I. Radhadkrishnan and D.J. Patel, *J. Mol. Biol.*, 1994, **241**, 600.
141. D.M. Gowers and K.R. Fox, *Nucleic Acids Res.*, 1997, **25**, 3787.
142. R.H. Durland, T.S. Rao, G.R. Revenkar, J.H. Tinsley, M.A. Myrick, D.M. Seth, J. Rayford, P. Singh and K. Jayaraman, *Nucleic Acids Res.*, 1994, **22**, 3233.
143. I. Prévot-Halter and C.J. Leumann, *Biorg. Med. Chem. Lett.*, 1999, **9**, 2657.
144. S. Buchini and C.J. Leumann, *Tetrahedron. Letts.*, 2003, **44**, 5065.
145. S. Buchini and C.J. Leumann, *Angew. Chem.*, 2004, **116**, 4015.
146. R.T. Ranasinghe, D.A. Rusling, V.E.C. Powers, K.R. Fox and T. Brown, *Chem. Commun.*, 2005, 2555.
147. D.A. Rusling, V.E.C. Powers, R.T. Ranasinghe, Y. Wang, S.D. Osbourne, T. Brown and K.R. Fox, *Nucleic Acids Res.*, 2005, **33**, 3025.
148. D.L. Chen and L.W. McLaughlin, *J. Org. Chem.*, 2000, **65**, 7469.
149. S. Obika, Y. Hari, M. Sekiguchi and T. Imanishi, *Angew. Chem. Int. Ed. Engl.*, 2001, **40**, 2079.
150. Y. Hari, S. Obika, M. Sekiguchi and T. Imanishi, *Tetrahedron*, 2003, **59**, 5123.
151. Y. Hari, S. Obika, H. Inohara, M. Ikejiri, D. Une and T. Imanishi, *Chem. Pharm. Bull.*, 2005, **53**, 843.
152. A.B. Eldrup, O. Dahl and P.E. Nielsen, *J. Am. Chem. Soc.*, 1997, **119**, 11116.
153. A.G. Olsen, O. Dahl and P.E. Nielsen, *Nucleosides, Nucleotides and Nucleic Acids*, 2003, **22**, 1331.
154. L.C. Griffin, L.L. Kiessling, P.A. Beal, P. Gillespie and P.B. Dervan, *J. Am. Chem. Soc.*, 1992, **114**, 7976.
155. K.M. Koshlap, P. Gillespie, P.B. Dervan and J. Feigon, *J. Am. Chem. Soc.*, 1993, **115**, 7908.
156. T.E. Lehmann, W.A. Greenberg, D.A. Liberles, C.K. Wada and P.B. Dervan, *Helv. Chim. Acta*, 1997, **80**, 2002.
157. C.-Y. Huang, C.D. Cushman and P.S. Miller, *J. Org. Chem.*, 1993, **58**, 5048.
158. C.-Y. Huang and P.S. Miller, *J. Am. Chem. Soc.*, 1993, **115**, 10456.
159. C.-Y. Huang, B. Guixia and P.S. Miller, *Nucleic Acids Res.*, 1996, **24**, 2606.
160. E. Mertz, S. Mattei and S.C. Zimmerman, *Bioorg. Med. Chem.*, 2004, **12**, 1517.
161. D. Guianvarc'h, J.-L. Fourrey, R. Maurisse, J.-S. Sun and R. Benhida, *Chem. Commun.*, 2001, 1814.
162. D. Guianvarc'h, J.-L. Fourrey, R. Maurisse, J.-S. Sun and R. Benhida, *Org. Letts.*, 2002, **14**, 4209.

163. D. Guianvarc'h, J.-L. Fourrey, R. Maurisse, J.-S. Sun and R. Benhida, *Bioorg. Med. Chem.*, 2003, **11**, 2751.
164. Y. Wang, D.A. Rusling, V.E.C. Powers, O. Lack, S.D. Osbourne, K.R. Fox and T. Brown, *Biochemistry*, 2005, **44**, 5884.
165. J.-S. Li, Y.-H. Fan, L.A. Marky and B. Gold, *J. Am. Chem. Soc.*, 2003, **125**, 2084.
166. J.-S. Li, R. Shikiya, L.A. Marky and B. Gold, *Biochemistry*, 2004, **43**, 1440.
167. J.-S. Li, F.-X. Chen, R. Shikaya, L.A. Marky and B. Gold, *J. Am. Chem. Soc.*, 2005, **127**, 12657.
168. J.-S. Li and B. Gold, *J. Org. Chem.*, 2005, **70**, 8764.
169. N. Van Craynest, D. Guianvarc'h, C. Peyron and R. Benhida, *Tetrahedron Letts.*, 2004, **45**, 6243.

CHAPTER 2

Interfacial Inhibitors of Human Topoisomerase I

CHRISTOPHE MARCHAND AND YVES POMMIER

Laboratory of Molecular Pharmacology, Bldg. 37, Rm. 5068, Center for Cancer Research, National Cancer Institute, Bethesda, MD 20892-4255, USA

2.1 Introduction

Reversible competitive inhibitors were the first inhibitors to be understood at the molecular level for their "lock and key" mechanism of action consisting in blocking and competing for the binding of a natural ligand to its specific receptor site.[1] Non-competitive inhibitors do not enter into competition with natural ligands for their specific receptor but rather bind at a topologically distinct region (allosteric site) and exert their influence through a propagated effect resulting in a reduced affinity of the natural ligand for its receptor.[2] Interfacial inhibitors represent a paradigm of reversible uncompetitive inhibition.[3] These inhibitors bind at the interface of two (or more) macromolecules as the multimeric complex undergoes a conformational change resulting in the trapping of the catalytic intermediate in a specific conformation. The multimeric complexes can be constituted of proteins as in the case of tubulin inhibitors such as paclitaxel (Taxol), vinblastine,[4] and colchicine [for review see[5]] or can contain both protein and nucleic acids as in the case of the RNA polymerase inhibitor α-amanitin, the ribosome inhibitors (hygromycin B, pactamycin, cyclohexamide) and topoisomerase I (Top1) inhibitors [for review see[6]]. We recently proposed that natural compounds (alkaloids) provide a rich and diverse source of interfacial inhibitors.[5,6] In this chapter, we will focus on Top I inhibitors, which represent a paradigm for interfacial inhibitors.

Nuclear Top I (hereinafter referred to as Top1) is an essential enzyme[7,8] ubiquitous and highly conserved in eukaryotes.[9–11] A mitochondrial topoisomerase I (Top1mt) was recently discovered in vertebrates.[12,13] Top1mt is encoded in the nuclear genome and corresponds to a duplication of an ancestral Top I gene, which is still present in yeast and plants.[13] Top1 belongs to the family of tyrosine recombinases and is involved in DNA relaxation. Top1 reversibly breaks one of the DNA backbones by forming a covalent enzyme-DNA intermediate (cleavage complex) between a tyrosine residue (Y723 for human Top1) and the phosphate at the 3′-hydroxyl end of the DNA break (Figure 1A–C). DNA relaxation is driven by DNA supercoiling, which leads to the rotation of the free end of the broken DNA around the intact DNA strand (Figure 1B).[14]

The alkaloid from the Chinese tree *Camptotheca acuminata*, camptothecin[15] specifically targets Top1.[16–18] Camptothecin is a planar heterocycle consisting of 5 rings, including a critical α-hydroxy-lactone E-ring (Figure 1F). Camptothecin has several key features. First, the natural alkaloid (20-*S*-camptothecin enantiomer) is potent against Top1 and experimental cancers but the synthetic 20-R enantiomer lacks both activities.[15,18] Second, camptothecin traps the Top1-DNA cleavage complex, forming a potentially lethal DNA lesion.

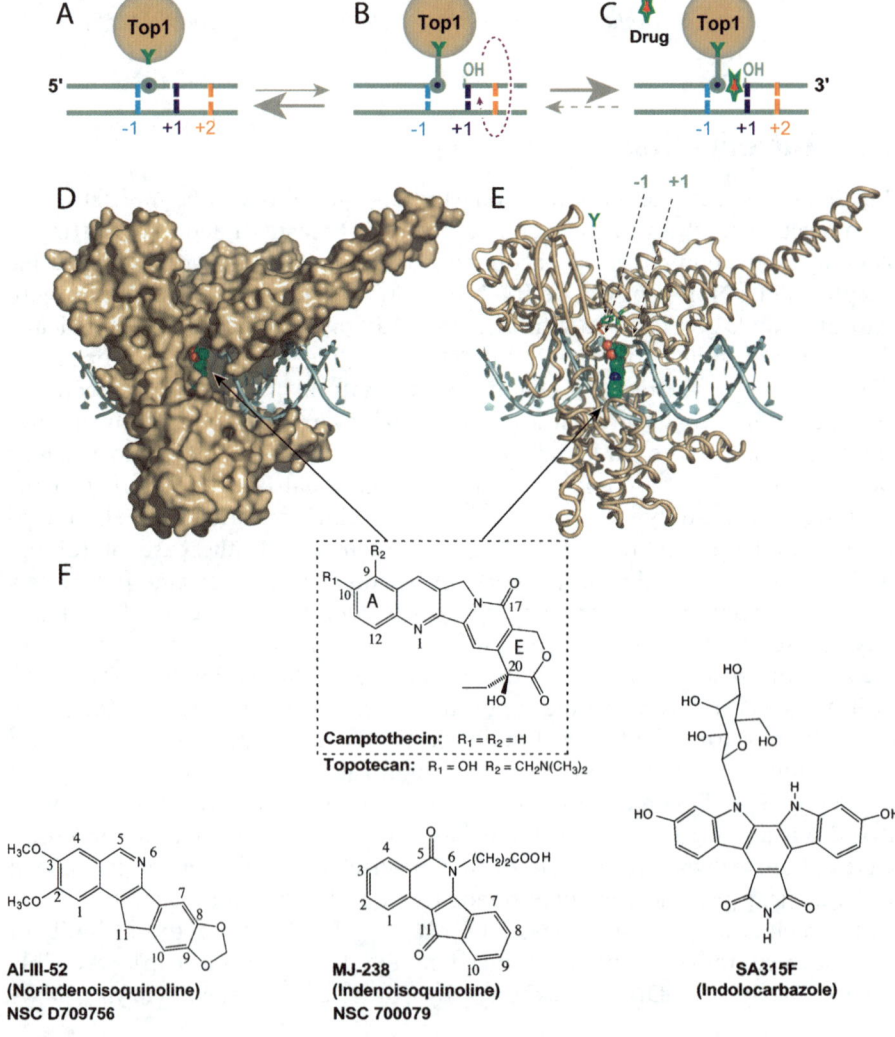

Camptothecin: R₁ = R₂ = H

Topotecan: R₁ = OH R₂ = CH₂N(CH₃)₂

AI-III-52
(Norindenoisoquinoline)
NSC D709756

MJ-238
(Indenoisoquinoline)
NSC 700079

SA315F
(Indolocarbazole)

Thus, the anticancer activity is not due to inhibition of Top1 enzyme activity *per se*. This distinction was demonstrated in Top1-deficient yeast strains, which are viable and completely lack the camptothecin-sensitivity of wild-type strains.[19–21] Third, camptothecin does not bind significantly to either DNA or purified Top1; both must be present together in the form of cleavage complexes.[15] These characteristics led to the hypothesis that camptothecin forms a ternary complex with Top1 and its DNA substrate by binding to a stereospecific site in the Top1-DNA cleavage complex[17,18] (Figure 1C). The DNA sequence at the site of cleavage exhibits a strong bias for a guanine residue at the 5′ side of the DNA break (position +1 in Figures 1A–C).[17] In view of the planar aromatic structure of camptothecin which resembles a fused-DNA base pair, it was hypothesized that camptothecin forms a ternary complex with Top1 and DNA by stacking against the +1 guanine-containing base pair on the 5′ side of the break, while at the same time binding to amino acid residues on the Top1 protein.[17] Camptothecin was, therefore, hypothesized to stabilize the DNA-Top1 interface in a structure similar to the normal intermediate of the topoisomerase reaction (Figure 1C). By convention, the base pairs flanking the break are referred to as −1 and +1 (represented in cyan and dark blue, respectively; Figure 1A). Position −1 corresponds to the nucleotide covalently linked to Top1 and +1 to the nucleotide at the free 5′-hydroxyl terminus (Figures 1A–C). The atomic structure of a ternary Top1 cleavage complex was later confirmed by X-ray crystallography with the clinical camptothecin-derivative topotecan (see structure in Figure 1F). This structure confirmed the drug-stacking

---

Figure 1 *Structure of the Top1 cleavage complex trapped by camptothecin. (A to B) Top1 nicking-closing reaction. (A) Top1 is generally bound non-covalently to DNA. The Top1 catalytic tyrosine (Y723 for human nuclear Top1) is represented in green (Y). (A to B) Top1 cleaves one strand of the duplex as it forms a covalent phosphodiester bond between the catalytic tyrosine and the 3′- DNA terminus. The other DNA terminus is a 5′-hydroxy-ribose (OH). (B) The Top1 cleavage complex allows rotation of the 5′-terminus around the intact strand, which relaxes DNA supercoiling (purple dotted circle with arrowhead). Following DNA relaxation, Top1 religates the DNA. (B to A) Under normal conditions, the religation (closing) reaction rate constant is much higher than the cleavage (nicking) rate constant. More than 90% of the Top1-DNA complexes are non-covalent[48] (as in A). (C) The Top1 inhibitors shown in panel F trap the Top1 cleavage complex by binding at the enzyme-DNA interface between the base pairs flanking the Top1-mediated DNA cleavage site (by convention positions −1 and +1). The colors for the base pairs −1 (cyan), +1 (dark blue) and +2 (orange) are the same as in Figures 2 and 3. (D and E) Lateral views of a Top1-DNA complex trapped by camptothecin (represented in green and red space filling form). (D) Top1 (light brown) is shown in a surface view to emphasize the depth of the camptothecin binding pocket. (E) Top1 is represented in a ribbon diagram to allow visualization of the catalytic tyrosine (Y; colored in green and by element) and to show the drug intercalation between the −1 and +1 base pairs. (F) Chemical structure of the five drugs co-crystallized with Top1-DNA complexes.[3,22,23,32] Molecular visualization was performed using PyMOL. The PDB code used for the camptothecin-Top1 crystal structure was 1T81[23]*

hypothesis.[22] More recently, we found the binding of the natural camptothecin alkaloid to the Top1-DNA complex to be superimposable upon the binding of topotecan.[23]

Because of the anticancer activity of camptothecin, many derivatives have been synthesized for clinical development.[24] Two of these have been approved by the U.S. Food and Drug Administration (FDA): topotecan (Hycamtin®) for ovarian and lung cancers and irinotecan (CPT11; Campto®) for colon carcinomas. Several other derivatives are in preclinical development or clinical trial (Exatecan, 9-nitrocamptothecin, BAY 38-3441[25,26]). Camptothecins however have limitations. The alpha-hydroxylactone in the E-ring (see Figure 1F) is rapidly converted in the circulation to a carboxylate (not shown) whose tight binding to serum albumin limits the available active drug.[27] Also, camptothecins are actively exported from the cell by drug efflux membrane "pumps."[28]

Non-camptothecin Top1 inhibitors are therefore of great pharmaceutical interest to overcome the limitations of camptothecins [see Chapter 3 (Dias and Bailly)]. Moreover, because drugs from different chemical families that share the same molecular target generally have different spectra of clinical activity, non-camptothecin Top1 inhibitors may well be effective against different tumors.[24] Indenoisoquinoline and indolocarbazole derivatives (Figure 1F) are currently being pursued for therapeutic development. The first indenoisoquinoline Top1 inhibitor was discovered by searching the NCI cell screen database[29] for compounds resembling camptothecin in their cytotoxicity profile against 60 human cancer cell lines.[30] In the absence of definitive structural information as to the drug target site, derivatives were initially designed with amino groups that would bind electrostatically to the DNA phosphodiester backbone and/or form hydrogen bonds with Top1 residues. Over 300 indenoisoquinoline derivatives have been synthesized to date, among which several potent and selective Top1 inhibitors have been identified as potential anticancer agents.[31] We recently reported the first crystal structure of an indenoisoquinoline (MJ-238, NSC 700079) (Figure 1F) bound to the Top1 cleavage complex.[23]

This chapter describes the crystal structures of five Top1 inhibitors complexed with DNA and Top1. The ternary complexes formed by camptothecin[23] are compared to the ones formed by topotecan,[22] the indenoisoquinoline MJ-238 (NSC 700079),[23] the norindenoisoquinoline AI-III-52 (NSC D709756)[32] and the indolocarbazole SA315F.[23] Comparison of the five crystal structures reveals a common molecular mechanism of drug action. These Top1 inhibitors all bind at the Top1-DNA interface by intercalating and stacking between the base pairs flanking the DNA cleavage site and by forming critical hydrogen bonds with Top1 amino acid residues. These same Top1 residues have been implicated previously in enzyme catalysis and/or resistance to camptothecins.[33] We discuss Top1 inhibitors as paradigms for drugs that trap macromolecular interfaces, and the generality and implications of the interfacial inhibitor concept.[5,6]

2.2 Molecular Mechanism of Action of Drugs that Trap Top1 Cleavage Complexes

2.2.1 Intercalation between the Base Pairs Flanking the Top1-Mediated DNA Break

2.2.1.1 *Camptothecin*

The overall structure of the natural camptothecin alkaloid stacked in a Top1 cleavage complex is presented in Figures 1D and E. The drug is bound deep inside Top1 (Figure 1D), intercalated between the base pairs flanking the DNA cleavage site (between positions −1 and +1) (Figure 1E). The Top1 protein encircles the DNA in a closed clamp conformation.[34–36] The polypeptide hinge is behind the DNA in Figures 1D and E and the tight narrow cleft through which the DNA has presumably entered the enzyme is represented in front of the DNA. The Top1 enzyme used in these experiments comprises the core, linker and catalytic domain. It does not include the Top1 *N*-terminal domain that encompasses residues 1-174 and which is not well conserved among eukaryotic Top1 enzymes. The *N*-terminus has been shown both *in vitro* and *in vivo* to be dispensable with respect to the biochemical and cytotoxic effects of Top1 poisons including camptothecin and topotecan.[37,38] Figure 2A shows two expanded views of camptothecin binding to the DNA in the Top1 cleavage complex. The left panel is a view in the same orientation as in Figures 1D and E with the cleavage site seen from the DNA minor groove. The right panel is a 90° rotation showing camptothecin lying under the +1 nucleotide pair. The +1 base pair (shown as capped sticks) covers and stacks against the entire drug (shown in space filling representation). Extensive stacking also occurs upon the −1 base pair, which is completely covered by the drug and therefore not shown in Figure 2A(right).

2.2.1.2 *Topotecan*

Figure 2B shows that topotecan (whose complete structure is specified in Figure 1F) like camptothecin stacks and intercalates between the base pairs −1 and +1 on each side of the cleavage site generated by Top1. Both camptothecin and topotecan stack in the same orientation. Their five-membered planar hetero-cyclic structure coincides with the long axis of the base pairs, and their shapes closely match that of a base pair. The hydroxyl substituents at the 20-position lie in close proximity to the sugar of the −1 nucleotide as well as the broken phosphodiester backbone of the DNA. This close positioning contributes to the optimum stacking of the natural 20S-camptothecin and explains why the 20R-camptothecin enantiomer fails to trap the Top1 catalytic complex.[18] At the other end of camptothecin, the A ring (Figure 1F) is not immediately in contact with the intact DNA phosphodiester backbone opposite from the cleavage site. This space can therefore accommodate the 10-hydroxy substituent of topotecan

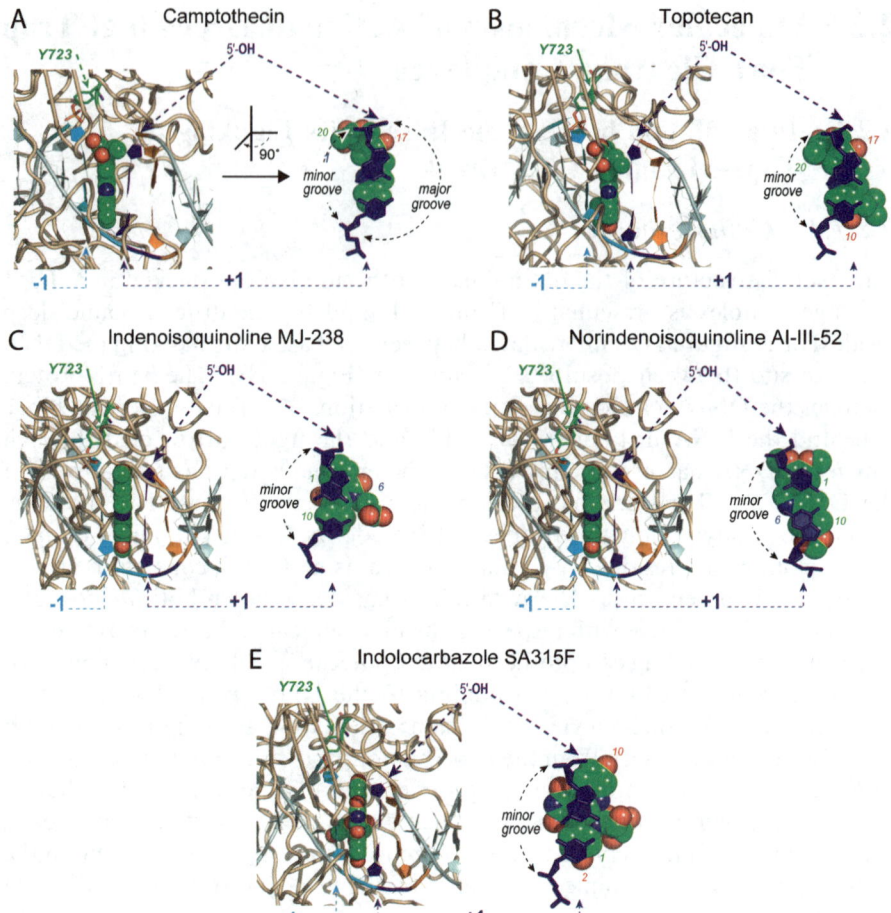

Figure 2 *Stacking of a drug molecule between two base pairs flanking the Top1 cleavage complex is a common mechanism for the camptothecins (A, B), the indenoisoquinolines (C, D) and an indolocarbazole derivative (E). Each panel shows 2 views for each drug within the Top1 cleavage complex. The left views are oriented as in Figures 1D and E with the DNA viewed from the minor groove. The right views are rotated 90° and show the +1 nucleotides covering the drug molecules. In each panel, the catalytic Top1 tyrosine is shown in green and indicated by Y723 at the top left. The −1 and +1 base pairs are marked by dashed arrows. In the right views, the colored numbers correspond to the drug atoms numbered in Figure 3. Molecular visualization was performed using PyMOL. The PDB codes for the different Top1 crystal structures were 1LT8 for norindenoisoquinoline;[32] 1SC7 for MJ-238;[23] 1SEU for indolocarbazole;[23] 1K4T for topotecan;[22] 1T81 for camptothecin[23] and 1A31 for Top1 in the absence of inhibitor[34]*

(Figure 2B right). Filling this space probably accounts for the greater affinity/potency of 10-hydroxy-substituted camptothecins and for the even greater activity of 10-11-dimethoxy-camptothecin derivatives.[18,39]

2.2.1.3 Indenoisoquinolines and Norindenoisoquinoline

Like camptothecin and topotecan, the indenoisoquinolines intercalate between the −1 and +1 base pairs at the cleavage site of the DNA-Top1 complex[3,22,23,32] (Figures 2C–D). For both drugs, the long axis of their aromatic portion lies parallel to the −1 and +1 base pair long axes. However, the two indenoisoquinolines are flipped 180° relative to each other (Figure 2, compare panels C and D). Numbering of the corresponding atoms makes this difference apparent. For AI-III-52, the sequence of atoms in a clockwise orientation is *1, 10* and *6*, whereas for MJ-238 it is *1, 6* and *10*. In other words, AI-III-52 is intercalated with the nitrogen *N6* in the minor groove. This nitrogen *N6* can hydrogen bond with the Top1 R364 residue. By contrast, MJ-238 has the nitrogen *N6* in the major groove of the DNA and the *C11* carbonyl hydrogen-bonded to R364 (see Figure 4). The side chain attached to the nitrogen *N6* of MJ-238 protrudes in the DNA major groove (Figure 2D).

2.2.1.4 Indolocarbazole

The indolocarbazole SA315F is also bound with maximum stacking interactions. The indolocarbazole aromatic portion is aligned with the axis of the −1 and +1 base pairs[23] (Figure 2E). The sugar moiety of SA315F is positioned in the major groove. For the indenoisoquinoline MJ-238, topotecan, and the indolocarbazole SA315F, the aromatic polycyclic ring of the drug covers the +1 base pair and the bulky substituents encroach upon the major groove. This common orientation would suggest that drug substitutions that impinge on the minor groove might be disfavored. Consistent with this possibility, substitutions on the camptothecins at positions 1 and 12 (see Figure 1F) preclude Top1 inhibition.[18] Also, DNA substitutions where a benzo[*a*]pyrene diol epoxide forms an adduct with deoxyguanine *N2* in the minor groove prevent Top1 cleavage.[40] However, substitutions directed toward the minor groove on indenoisoquinolines at position 11 (see Figure 1F) result in more potent compounds as regards cytotoxicity and Top1 inhibition.[41–43] This suggests that these favorable substitutions interact directly with Top1 residues and strengthen the hydrogen-bond network formed by these particular compounds.

2.2.2 DNA Untwisting by Drugs at the Top1-Mediated DNA Cleavage Site

Figure 3 examines the relative twist angles of the base pairs flanking the Top1 cleavage site (Figure 3A). In the absence of drug (Figures 3B and C), the twist angles generated by the flanking −1 and +1 base pairs (cyan and dark blue, respectively) and by the +1 and +2 base pairs (dark blue and orange) are 38° and 40°, respectively [computation derived from the published structure[34]].

A

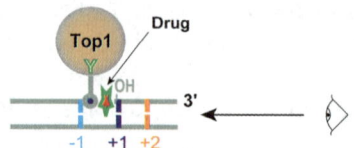

B   C

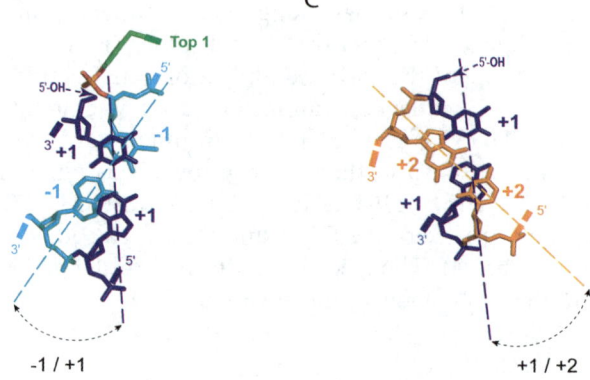

-1 / +1   +1 / +2

D

	Twist angle	
	-1 / +1	+1 / +2
No drug	38°	40°
Camptothecin	23°	27°
Topotecan	20°	32°
Indenoisoquinoline (MJ-238)	29°	29°
Norindenoisoquinoline (AI-III-52)	10°	58°
Indolocarbazole (SA315F)	21°	33°

Figure 3   *DNA unwinding by Top1 inhibitors. (A) Numbering and coloring convention for base pairs flanking the Top1 cleavage complex. (B) Illustration showing the twist angle between the base pairs flanking the Top1 cleavage site in the absence of inhibitor (from[34]). The −1 base pair is colored cyan and the +1 base pair is dark blue; the broken lines correspond to the long axes of the base pairs. (C) The +2 base pair is represented above the cleavage site and is colored orange. (D) Twist angles for the −1/+1 and +1/+2 base pairs*

In the presence of drug, the twist angle between the −1 and +1 base pairs is reduced indicating DNA unwinding at the Top1 cleavage site (Figure 3D). DNA unwinding is a characteristic of DNA intercalators, caused by the increased distance between flanking base pairs required to accommodate the intercalated molecule. For instance, intercalation of ethidium bromide reduces the twist angle between the flanking base pairs to 7°, compared to the normal 33°. In the case of the norindenoisoquinoline AI-III-52, the twist angle between the base pairs flanking the intercalated drug is reduced to 10° indicating a strong unwinding effect similar to the unwinding resulting from ethidium intercalation (Figure 3D). The twist angle reduction observed with AI-III-52 is not due to Top1 cleavage, because the twist angle between the base pairs flanking the Top1 cleavage site in the absence of drug remains near normal (Figure 3B).[34] The twist angle of 58° between the +1 and +2 base pairs in the norindenoisoquinoline structure indicates some overwinding of the DNA immediately adjacent to the drug binding (Figure 3D).[3] This overwinding probably partially compensates for the unwinding produced by the norindenoisoquinoline. In contrast, for the four other drugs the unwinding effect seems to be propagated to the +1 and +2 base pairs as their twist angle remains below the value of 40° reported for the twist angle in the absence of drug (Figure 3D).

2.2.3 Common Hydrogen-Bond Network for Top1 Inhibitors Bound in the Ternary Complex

In addition to intercalation within the Top1 cleavage site, Top1 inhibitors form a network of hydrogen bonds with conserved Top1 amino acid residues. A molecular visualization of the hydrogen-bond network for camptothecin, topotecan, the two indenoisoquinolines and the indolocarbazole is presented in Figure 4. The molecules in the ternary complexes are visualized through the DNA double helix in a similar orientation to that used for Figure 3 (+3 base pair above the plane of the cleavage site). Two of the Top1 catalytic residues are shown. Y723 forms the covalent bond with DNA. K532 serves a dual function, forming a hydrogen bond (not shown) on the minor groove side with the −1 base on the scissile strand[14,22] and also functioning as a general acid during cleavage to protonate the leaving 5′ oxygen.[44] Four other Top1 residues form hydrogen bonds with the intercalated drugs: N722, R364, D533, and N352.[3] None of these residues are essential for Top1 catalytic activity. However, N722, R364, and D533 have been implicated in resistance to camptothecins,[45] indolocarbazoles,[46] and indenoisoquinolines.[31] In all structures, D533 and R364 are stabilized by a shared hydrogen bond (Figure 4).

For camptothecin, three hydrogen bonds have been identified[23] (Figure 4A). Two are direct. One is between the camptothecin 20-hydroxy and D533, and the other between the camptothecin N1 and R364. The third H-bond has been proposed to be water-mediated between the 17-carbonyl on the camptothecin D-ring and N722 of Top1. For topotecan, only two of the three H-bonds (direct with D533 and water-mediated with N722) were initially reported[22] and are shown in Figure 4B. Reexamination of the topotecan ternary structure reveals that the R364 side chain is also in close proximity with the N1 of topotecan and well positioned for hydrogen bonding.[22]

The indenoisoquinoline MJ-238 forms only one hydrogen bond with R364 (Figure 3C).[23] For the norindenoisoquinoline AI-III-52, two hydrogen bonds

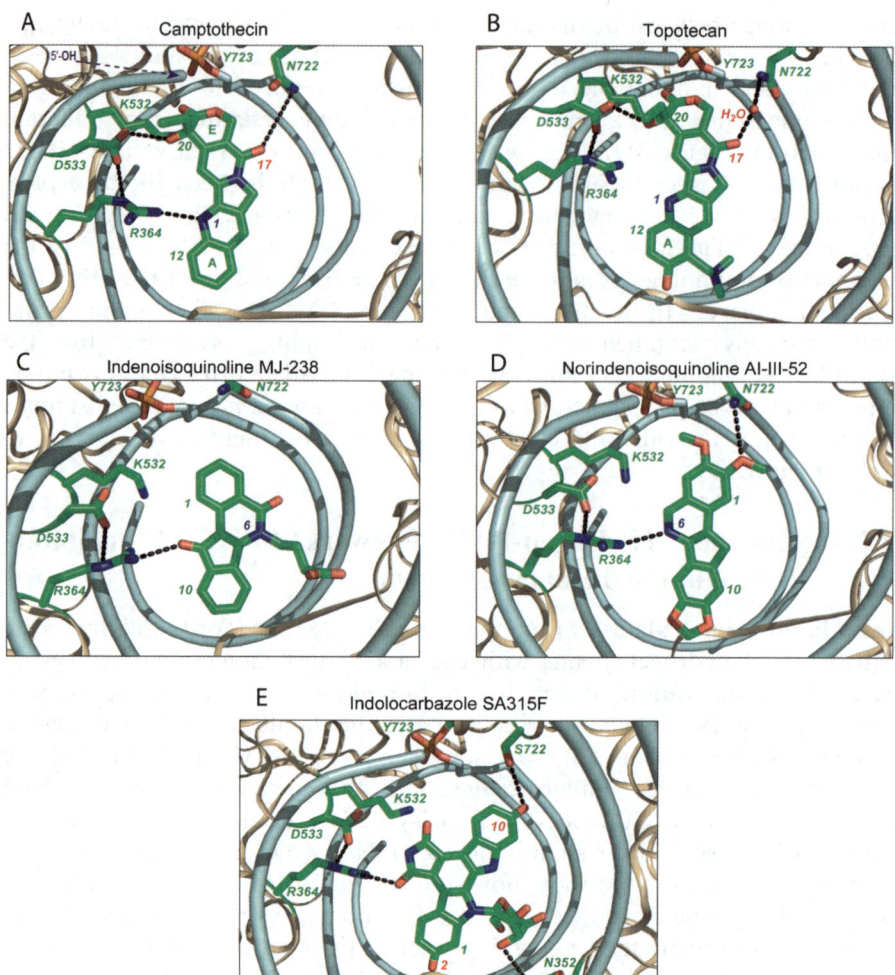

Figure 4 *Hydrogen bond networks between drugs and Top1 amino acid residues in the drug-Top1-DNA ternary complexes. (A) The natural camptothecin forms 3 hydrogen bonds with D533, N722 and R364.[23] (B) Two hydrogen bonds have been reported for topotecan with N722 and D533.[22] The third hydrogen bond with R364 has been proposed (see Discussion). Note that the water-mediated N722 hydrogen bond is represented with the amino group of N722. It is also plausible to be with the carbonyl of N722 (not shown). (C) MJ-238 forms only one hydrogen bond with Arg364.[23] (D) AI-III-52 forms two direct hydrogen bonds with N722 and R364. (E) The indolocarbazole SA315F forms three hydrogen bonds with S722, R364, and N352.[23] In all structures, there is a hydrogen bond between D533 and R364*

can be observed: one with R364 and one with N722 (Figure 4D). The weaker Top1 inhibitory activity of MJ-238 compared to AI-III-52[23,32] can therefore be attributed to its reduced hydrogen bonding with Top1 and to reduced stacking because of the smaller surface of its aromatic portion (compare panels C and D in Figures 2 and 4).

In addition to the hydrogen bonds with N722 and R364, the indolocarbazole SA315F is stabilized by an extra hydrogen bond with the top1 residue N352 (Figure 4E). This residue has been predicted to lie in close proximity to substituents present on the N6 nitrogen of indenoisoquinoline derivatives.[23]

2.3   Generalization of the Interfacial Inhibitor Concept

To our knowledge camptothecins were the first inhibitors hypothesized[17] and demonstrated[22] to bind at the "interface of two macromolecules" (Top1 and its DNA substrate) as these macromolecules undergo a catalytic conformational change, which is then stabilized by the drug (the "cleavage complex" – see Figure 1A–C). This mechanism can be generalized as it also applies to three non-camptothecin inhibitors: two indenoisoquinolines and an indolocarbazole (see Figure 2). The implications of these structures are several fold. First, they can materially aid the rational synthesis of novel Top1 inhibitors. However, the example of the two indenoisoquinolines that bind in flipped orientations (see Figures 2C and D) demonstrate the potential limitations of simple extrapolations from crystal structures, and the need to correlate predictions arising from structural biology with biological testing and ultimately additional data at atomic resolution; hence the need of an iterative process from chemistry to biology to structural biology, and back again.

A second implication of the interfacial binding of Top1 inhibitors is its apparent generality. The "interfacial inhibition" paradigm can be extended to a wide range of drugs and macromolecular targets.[3,5,6] We therefore define "interfacial inhibitors" as drugs that bind at the interface of two (or more) macromolecules as the multimeric complex undergoes a structural transition. These multimeric complexes can present a drug binding site at their protein–protein interface as in the case of brefeldin A and tubulin inhibitors (colchicine, vinblastine, paclitaxel [Taxol®])[5] or else drug-binding sites at nucleic acid–protein interfaces as in the case of Top1 inhibitors (present chapter), the RNA polymerase II inhibitor α-amanitin and the antibiotic ribosome inhibitors.[3,6]

Interfacial inhibition is reversible and uncompetitive.[5,6] Interfacial inhibitors bind to and trap a catalytic intermediate of the macromolecular complex in a specific configuration (Figure 5A). Interfacial inhibition represents a paradigm for "uncompetitive inhibition" (Figures 5B and C: Lineweaver–Burk plot equations). When data for uncompetitive inhibition are graphed (Figures 5D and E), it appears that both V_{max} and K_M decrease in the presence of the inhibitor. A series of parallel lines is obtained in the double-reciprocal Lineweaver–Burk plot (Figure 5E). The apparent K_M decreases due to reduction of available enzyme-substrate complex. V_{max} decreases because the inhibited enzyme is less catalytically active. The pharmacological activity of interfacial

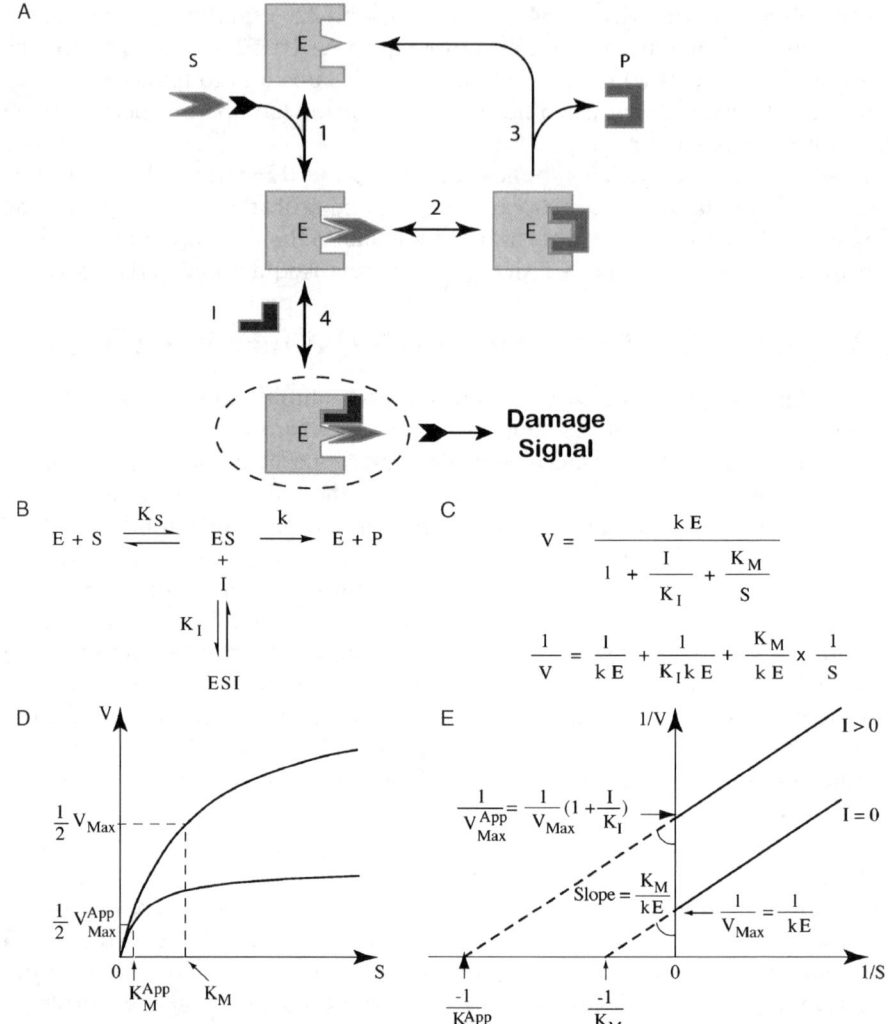

Figure 5 *Generalization of the interfacial inhibitor paradigm. (A) Schematic cartoon representating the interfacial inhibition paradigm applied to an enzymatic system: (1) A substrate (S) binds to an enzyme (E); (2) The enzyme converts the substrate to product (S → P); (3) P dissociates and the enzyme is regenerated for a new catalytic cycle; (4) An interfacial inhibitor (I) binds at the enzyme-substrate "interface" and traps the enzyme-substrate complex, thereby reversibly preventing conversion of S to P. More importantly, the trapped ternary complex is sensed as cellular damage. (B) Equilibrium representation of interfacial inhibition. K_S: substrate equilibrium dissociation constant; K_I: inhibitor equilibrium dissociation constant; k: catalytic constant. (C) Equations corresponding to uncompetitive scheme shown in panel B. V: reaction velocity; K_M: Michaelis-Menten equilibrium dissociation constant ($K_M = K_S$ when k is rate limiting). (D) Velocity curves for interfacial uncompetitive inhibition. By contrast to competitive and non-competitive inhibition, K_M decreases in the presence of an interfacial inhibitor. (E) Lineweaver-Burk plot. The lines with inhibitor (I > 0) are characteristically parallel to the line without inhibitor (I = 0)*

inhibitors is due, at least in part, to the conversion of the trapped macromolecular intermediate into lethal lesions by other macromolecules that interact with the normally transient ternary drug-macromolecular complex. For instance, in the case of Top1 inhibitors, the cleavage complexes are only lethal when they are converted into replication double-strand breaks by the collision of a DNA polymerase with a Top1 cleavage complex reversibly trapped by the drug (http://discover.nci.nih.gov/pommier/pommier.htm).[47] This explains why the antiproliferative activity of Top1 inhibitors increases with the levels of Top1. Conversely, a common resistance mechanism to Top1 inhibitors is downregulation of Top1.[45] Hence, interfacial inhibitors differ fundamentally from competitive inhibitors. Increasing the amount or availability of drug target increases resistance to competitive inhibitors, whereas it sensitizes to interfacial inhibitors.

The interfacial inhibitor paradigm has strong implications for drug discovery because interfacial inhibitors stabilize rather than inhibit the formation of macromolecular complexes. Assays should therefore be designed to look for drugs that stabilize rather than inhibit the formation of macromolecular complexes. Interfacial inhibitor screening using various methods to measure the binding of macromolecules has the potential to identify highly selective inhibitors for a particular receptor site and/or enzyme substrate.

Acknowledgments

This research was supported by the Intramural Research Program of the NIH, National Cancer Institute, Center for Cancer Research. The authors wish to thank Dr. Kurt W. Kohn for his insight and mentorship. The authors also wish to acknowledge their collaborators Drs. M. C. Wani, M. E. Wall, L. Stewart, M. R. Redimbo and M. Cushman.

References

1. J. Sterzl and J. Krecek, *Nature*, 1949, **164**, 700.
2. J. Monod, J.P. Changeux and F. Jacob, *J. Mol. Biol.*, 1963, **6**, 306.
3. C. Marchand, S. Antony, K.W. Kohn, M. Cushman, A. Ioanoviciu, B.L. Staker, A.B. Burgin, L. Stewart and Y. Pommier, *Molecular Cancer Therapeutics*, 2006, **5**, 1–9.
4. B. Gigant, C. Wang, R.B. Ravell, F. Roussi, M.O. Steinmetz, P.A. Curmi, A. Sobel and M. Knossow, *Nature*, 2005, **435**, 519.
5. Y. Pommier and J. Cherfils, *Trends Pharmacol. Sci.*, 2005, **28**, 136.
6. Y. Pommier and C. Marchand, *Curr. Med. Chem. Anti-Canc. Agents*, 2005, **5**, 421.
7. M.P. Lee, S.D. Brown and T.-S. Hsieh, *Proc. Natl. Acad. Sci. USA*, 1993, **90**, 6656.
8. S. Morham, K.D. Kluckman, N. Voulomanos and O. Smithies, *Mol. Cell. Biol.*, 1996, **16**, 6804.
9. Y. Pommier, P. Pourquier, Y. Fan and D. Strumberg, *Biochim. Biophys. Acta*, 1998, **1400**, 83.

10. J.J. Champoux, *Annu. Rev. Biochem.*, 2001, **70**, 369.
11. J.C. Wang, *Nat. Rev. Mol. Cell Biol.*, 2002, **3**, 430.
12. H. Zhang, J.M. Barcelo, B. Lee, G. Kohlhagen, D.B. Zimonjic, N.C. Popescu and Y. Pommier, *Proc. Natl. Acad. Sci. USA*, 2001, **98**, 10608.
13. H. Zhang, L.H. Meng, D.B. Zimonjic, N.C. Popescu and Y. Pommier, *Nucleic Acids Res.*, 2004, **32**, 2087.
14. L. Stewart, M.R. Redinbo, X. Qiu, W.G.J. Hol and J.J. Champoux, *Science*, 1998, **279**, 1534.
15. M.E. Wall and M.C. Wani, *Cancer Res.*, 1995, **55**, 753.
16. Y.H. Hsiang, R. Hertzberg, S. Hecht and L.F. Liu, *J. Biol. Chem.*, 1985, **260**, 14873.
17. C. Jaxel, G. Capranico, D. Kerrigan, K.W. Kohn and Y. Pommier, *J. Biol. Chem.*, 1991, **266**, 20418.
18. C. Jaxel, K.W. Kohn, M.C. Wani, M.E. Wall and Y. Pommier, *Cancer Res.*, 1989, **49**, 1465.
19. W.K. Eng, L. Faucette, R.K. Johnson and R. Sternglanz, *Mol. Pharmacol.*, 1988, **34**, 755.
20. J. Nitiss and J.C. Wang, *Proc. Natl. Acad. Sci. USA*, 1988, **85**, 7501.
21. M.-A. Bjornsti, P. Benedetti, G.A. Viglianti and J.C. Wang, *Cancer Res.*, 1989, **49**, 6318.
22. B.L. Staker, K. Hjerrild, M.D. Feese, C.A. Behnke, A.B. Burgin Jr. and L. Stewart, *Proc. Natl. Acad. Sci. USA*, 2002, **99**, 15387.
23. B.L. Staker, M.D. Feese, M. Cushman, Y. Pommier, D. Zembower, L. Stewart and A.B. Burgin, *J. Med. Chem.*, 2005, **48**, 2336.
24. L.-H. Meng, Z.-H. Liao and Y. Pommier, *Curr. Top. Med. Chem.*, 2003, **3**, 305.
25. J.F. Pizzolato and L.B. Saltz, *Lancet*, 2003, **361**, 2235.
26. C.J. Thomas, N.J. Rahier and S.M. Hecht, *Bioorg. Med. Chem.*, 2004, **12**, 1585.
27. T.G. Burke and Z. Mi, *J. Med. Chem.*, 1994, 37, 40.
28. M. Brangi, T. Litman, M. Ciotti, K. Nishiyama, G. Kohlhagen, C. Takimoto, R. Robey, Y. Pommier, T. Fojo and S.E. Bates, *Cancer Res.*, 1999, **59**, 5938.
29. K.D. Paull, R.H. Shoemaker, L. Hodes, A. Monks, D.A. Scudiero, L. Rubinstein, J. Plowman and M.R. Boyd, *J. Natl. Cancer Inst.*, 1989, **81**, 1088.
30. G. Kohlhagen, K. Paull, M. Cushman, P. Nagafufuji and Y. Pommier, *Mol. Pharmacol.*, 1998, **54**, 50.
31. S. Antony, M. Jayaraman, G. Laco, G. Kohlhagen, K.W. Kohn, M. Cushman and Y. Pommier, *Cancer Res.*, 2003, **63**, 7428.
32. A. Ioanoviciu, S. Antony, Y. Pommier, B.L. Staker, L. Stewart and M. Cushman, *J. Med. Chem.*, 2005, **48**, 4803.
33. J.E. Chrencik, B.L. Staker, A.B. Burgin, P. Pourquier, Y. Pommier, L. Stewart and M.R. Redinbo, *J. Mol. Biol.*, 2004, **339**, 773.
34. M.R. Redinbo, L. Stewart, P. Kuhn, J.J. Champoux and W.G.J. Hol, *Science*, 1998, **279**, 1504.

35. J.F. Carey, S.J. Schultz, L. Sisson, T.G. Fazzio and J.J. Champoux, *Proc. Natl. Acad. Sci. USA*, 2003, **100**, 5640.
36. M.H. Woo, C. Losasso, H. Guo, L. Pattarello, P. Benedetti and M.A. Bjornsti, *Proc. Natl. Acad. Sci. USA*, 2003, **100**, 13767.
37. L. Stewart, G. Ireton and J.J. Champoux, *J. Biol. Chem.*, 1996, **271**, 7602.
38. L. Stewart, G. Ireton, L. Parker, K.R. Madden and J.J. Champoux, *J. Biol. Chem.*, 1996, **271**, 7593.
39. Y. Pommier, G. Kohlhagen, K.W. Kohn, F. Leteurtre, M.C. Wani and M.E. Wall, *Proc. Natl. Acad. Sci. USA*, 1995, **92**, 8861.
40. Y. Pommier, G. Kohlhagen, G.S. Laco, H. Kroth, J.M. Sayer and D.M. Jerina, *J. Biol. Chem.*, 2002, **277**, 13666.
41. B.M. Fox, X. Xiao, S. Antony, G. Kohlhagen, Y. Pommier, B.L. Staker, L. Stewart and M. Cushman, *J. Med. Chem.*, 2003, **46**, 3275.
42. X. Xiao, S. Antony, G. Kohlhagen, Y. Pommier and M. Cushman, *J. Org. Chem.*, 2004, **69**, 7495.
43. X. Xiao, S. Antony, G. Kohlhagen, Y. Pommier and M. Cushman, *Bioorg. Med. Chem.*, 2004, **12**, 5147.
44. H. Interthal, P.M. Quigley, W.G. Hol and J.J. Champoux, *J. Biol. Chem.*, 2004, **279**, 2984.
45. Y. Pommier, P. Pourquier, Y. Urasaki, J. Wu and G. Laco, *Drug Resistance Update*, 1999, **2**, 307.
46. Y. Urasaki, G. Laco, Y. Takebayashi, C. Bailly, G. Kohlhagen and Y. Pommier, *Cancer Res.*, 2001, **61**, 504.
47. D. Strumberg, A.A. Pilon, M. Smith, R. Hickey, L. Malkas and Y. Pommier, *Mol. Cell Biol.*, 2000, **20**, 3977.
48. S. Antony, P.B. Arimondo, J.S. Sun and Y. Pommier, *Nucleic Acids Res.*, 2004, **32**, 5163.

CHAPTER 3

Diversity of Topoisomerase I Inhibitors for Cancer Chemotherapy

NATHALIE DIAS AND CHRISTIAN BAILLY

INSERM U-814, Institut de Recherche sur le Cancer de Lille, 59045 Lille, France

Abstract

Topoisomerase I is an essential nuclear enzyme for the regulation of DNA topology and a well-established target for anticancer drugs. Camptothecin is the prototypic topoisomerase I inhibitor capable of stimulating DNA cleavage through the stabilization of DNA-enzyme covalent complexes. Beyond camptothecin and its numerous synthetic analogues that are currently under clinical investigation, an expanding number of topoisomerase I poisons have been identified including indenoisoquinolines and indolocarbazoles. Significant progress has also been made recently in identifying naturally occurring topoisomerase I inhibitors, isolated from plants or marine organisms. This review presents a survey of selected classes of small molecule topoisomerase I inhibitors that have been characterized as potential anticancer drug candidates and discusses the attractiveness of the target in terms of drug discovery.

3.1 Introduction

DNA topoisomerase I is a ubiquitous enzyme essential for cell proliferation as it catalyzes the relaxation of supercoiled DNA in the course of nucleic acid metabolism, during such processes as transcription, replication, or chromatin assembly.[1,2] Human topoisomerase I triggers transient single-strand cleavage of duplex DNA and transfers the intact strand through the cleavage site before religation. This occurs in the cell nucleus with an accumulation of the enzyme in the nucleolus to ensure tight regulation of the topology of ribosomal DNA.[3] The reaction mechanism involves a transesterification owing to the nucleophilic attack by the catalytic tyrosine 723 hydroxyl group of topoisomerase I on the DNA phosphodiester at the site of cleavage, leading to scission of the phosphodiester backbone and the production of a covalent DNA-enzyme 3'-phosphorotyrosyl adduct, usually referred to as the "cleavable complex" (Figure 1).[4] Under physiological conditions, this covalent complex is hardly detectable because it is unstable and rapidly evolves towards the uncleaved DNA, relaxed or unrelaxed, to maintain the double-strand integrity of the genetic material. But this covalent intermediate is in fact the essential "Target", the key molecular entity that must be blocked to obtain the desired cellular effect.

The topoisomerase I enzymes can be divided into two subclasses, topoisomerases IA and topoisomerases IB. Though essentially similar in their action, these enzymes are characterized by different specificity. Topoisomerase IA binds locally to single-stranded stretches of DNA through a 5' phosphate and is only able to relax negatively supercoiled DNA, whereas topoisomerase IB binds locally to double-stranded DNA through a 3'-phosphate and can relax both negatively and positively supercoiled DNA.[5] Once the relaxation step is achieved through the rotation of the broken DNA strand around the unbroken

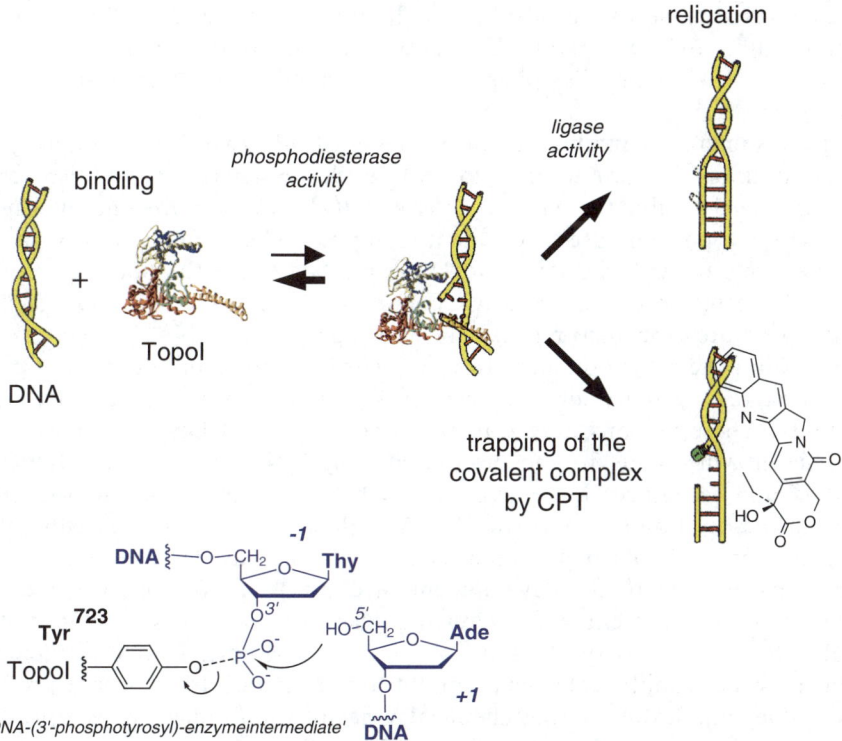

Figure 1 *Schematic illustration of DNA cleavage by topoisomerase I and the trapping of DNA-topoisomerase I covalent complexes by CPT*

one, religation occurs by reformation of the phosphodiester backbone after a second transesterification reaction in which the 5′-hydroxyl group on the nicked DNA end attacks the 3′-phosphotyrosine linkage, followed by release of the enzyme (Figure 1).

Great interest in topoisomerases was generated when the key roles they play in many cellular processes implicating DNA manipulation were recognized (including replication, transcription, recombination) along with their roles as molecular targets for naturally occuring antimicrobial and anticancer agents.[6] Here, we shall focus on human topoisomerase I, which is a well-recognized target for a number of antitumour drugs. For the past 20 years this enzyme, which belongs to the type IB group, has attracted the attention of medicinal chemists for the design and synthesis of specific inhibitors and onco-pharmacologists for the regulation of cancer cell physiology. The single-strand breaks induced by topoisomerase I are generally considered weakly toxic to the cells because they can be efficiently and rapidly repaired. But their conversion into double-strand breaks, formed when the cleaved complex encounters replication forks or transcription complexes, generates potentially lethal lesions.[7–10] A number of drugs can convert topoisomerase I into a cell poison by blocking the religation step, triggering replication fork arrest and breakage of the replication fork to generate persistent DNA double-strands breaks responsible for cell death.[6,10] The most potent and by far the most-studied inhibitor is the plant alkaloid camptothecin (CPT, Figure 1), which is at the heart of extensive pharmacological studies of all topoisomerase I inhibitors.[11,12]

Topoisomerase I inhibitors can be classified into two distinct families depending upon their mode of action. Molecules designated as topoisomerase I *suppressors* inhibit the enzyme activity but do not stabilize the intermediate DNA-topoisomerase I covalent complex. Their interaction with the free enzyme or with the unbound DNA inhibits either the binding reaction or the cleavage process. In contrast, compounds referred to as *poisons*, such as CPT, act after the binding and the cleavage of the DNA by the enzyme and inhibit the religation step. In other words, suppressors inhibit the phos-phodiesterase activity of the enzyme, whereas poisons block its ligase activity. These poisons may trap the topoisomerase I-DNA complex when (i) the enzyme binds to the preformed drug-DNA complex; (ii) the drug specifically recognizes the enzyme-DNA binary complex; or (iii) the drug-enzyme association interacts with DNA.[13] Here, we shall focus essentially on topoisomerase I poisons. Suppressors are generally less or non-specific and have not led to the development of clinically useful anticancer drugs, though they represent an interesting class of inhibitors having poten-tial antitumour activity. This is the case, for example, with thiocolchicine dimers, which, unlike colchicine, inhibit topoisomerase I but do not produce cleavable complexes.[14] Other chemical series of topoisomerase I suppressors have been reported.[15] See also Chapter 2 by Marchand and Pommier in this volume.

3.2 Camptothecins

Topoisomerase I-targeting agents that can stabilize the cleavable complex have proved to be very effective for the treatment of cancer. This is the case in particular for the quinoline natural-product CPT, which was the first compound identified, 20 years ago,[11] as a specific blocker of the DNA-topoisomerase I covalent complex. This monoterpenoid indole-alkaloid was first isolated in 1858 from extracts of the Asian tree *Camptotheca acuminata* (Xi Shu or the tree of joy), but it was later found in a fairly large variety of related or unrelated species. In 2004, over 12 plant species producing CPT had been catalogued[16] and the list continues to expand with the identification of further plants and endophytic fungi capable of production CPT.[17] Trapping of DNA-topoisomerase I covalent complexes by CPT results in extensive DNA damage during the S-phase of the cell cycle.[7,18] In the 1990s, two derivatives of CPT, topotecan (Hycamtin) and irinotecan (Camptosar), were approved for the treatment of ovarian and colorectal cancers respectively, and are now routinely used in the clinic for colorectal cancers as well as for other malignancies, such as lung cancer, pancreatic cancer, and brain metastases.[19–26] Beyond these two approved "tecan" drugs, a fairly large family of CPT derivatives have been developed, many of which are currently under clinical investigation: exatecan, lurtotecan, belatecan, silatecan, rubitecan, gimatecan,[27] diflomotecan, *etc.*[28–32] In the context of producing topoisomerase I poisons with improved antitumour activity, it is worth mentioning the extensive work on CPT-prodrug derivatives, including carbamate-based 20(S)-CPT[33] or polyethylene glycol CPT ester for example.

The X-ray crystal structure of the human topoisomerase I-DNA complex bound with CPT or its analog topotecan has been reported[34,35] [see Chapter 2]. In both cases, the drug mimics a DNA base pair and binds at the site of DNA cleavage by intercalating between the upstream (+1) and downstream (−1) base pairs. Both the closed lactone and open carboxylate forms of topotecan have been observed in the crystal structure. A subset of hydrogen bonds between Asp533 of the enzyme and the 20(S)-hydroxyl of topotecan, together with water-bridged contacts with the phosphotyrosine-723 and Asn722, stabilizes the drug-DNA-enzyme ternary complex. There are however some subtle differences between the topotecan and CPT crystal structures. Once intercalated, the downstream base pair is shifted, increasing the distance between the 5′-end of the DNA substrate and the 3′-phosphotyrosine thus preventing the religation step. Even though CPT and topotecan are structurally related, CPT is oriented and stands differently within the binding site: the drug heterocycle is closer to the minor groove side of the intercalation binding site and exhibits a slight difference in its orientation along the vertical DNA axis. These differences suggest some flexibility in the topoisomerase binding site [Chapter 2]

Although CPT displays interesting anticancer activities, its poor solubility in water and the interconversion of the lactone form into an inactive opened-carboxylate form prevented its use in the clinic (Figure 2). In human blood, the lactone/carboxylate equilibrium is shifted towards the inactive opened form owing to the preferential binding of the latter to serum albumin.[36] Another problem associated with CPT is the rapid reversion of the stabilization of the cleavage complexes after drug removal, leading to the need for long and continuous infusions of the drug to achieve maximal antitumour effect. Clinical studies with CPT have revealed reversible bone narrow depression and heamorrhagic cystitis as the major dose-limiting toxicity.[37–39] In addition, CPT is a substrate for the ABC transporter BCRP (Breast Cancer Resistance Protein) expressed in many human tumours.[40,41] These problems warranted the design of new CPT analogs to improve therapeutic properties. Several strategies have been developed.

The vast majority of the CPT analogues designed thus far preserves the conjugation and planarity of the tetracyclic A-to-D rings and maintain the essential A–B quinoline moiety.[42,43] The substitution of the quinoline with another planar heterocycle generally alters or abolishes the topoisomerase I inhibitory activity.[42] Numerous structure–activity relationship studies have shown that the introduction of different substituents, such as amino, halogen, hydroxyl, or lipophilic groups at positions 7, 9, or 10 on the quinoline ring (Figure 2), have led to highly potent substituted CPT analogues with reduced affinity for serum albumin and potentiated topoisomerase inhibitory and antitumour activities.[42–44] These modifications interfere minimally with the drug binding. The 3D model of the topotecan–topoisomerase I-DNA ternary complex indicates that the A–B portion of the molecule points towards the minor groove of DNA with minimal interference with the enzyme.[14] Several silylated derivatives such as silatecan (DB-67) with a 7-terbutyldimethylsilyl group at position 7 have demonstrated greater lipophilicity and potent anti-cancer activity against glioma in animal models.[45,46] The addition of another ring at positions 10-11, 9-10, or 7-9 can also lead to highly potent antitumour hexacyclic compounds such as exatecan (DX-8951f) (Figure 2), which has demonstrated higher topoisomerase I inhibitory activity than CPT and is currently undergoing phase II-III clinical trials.[47–50] However, not all positions on the CPT skeleton tolerate substitutions. For example, the introduction of a bulky substituent at position 11 generally leads to considerably decreased activity against topoisomerase I.[51] Similarly, in most cases CPT analogues substituted on the C or D rings were found to be inactive against topoisomerase I and thus devoid of antitumour activity.[52–54] For a long time, the lactone E ring of the CPT parent molecule was also found to be absolutely required for topoisomerase I inhibition. Several attempts to substitute the E ring[55,56] or to modify the stereochemistry of the 20-hydroxyl group[57,58] or to replace the lactone function by a stable lactam ring[58] all led to totally inactive compounds. As a consequence, the lactone moiety was considered as a key element for the antitumour activity of the CPT molecule. This initial view is no longer accepted since the discovery of tumour-active E ring-modified CPT analogues during the past five years. Bigg and co-workers were the first to successfully design an E ring-modified CPT derivative that totally preserved the capacity to inhibit topoisomerase I. The family of homocamptothecins (hCPT), built by the insertion of a methylene spacer between the carboxylic and alcoholic functions,

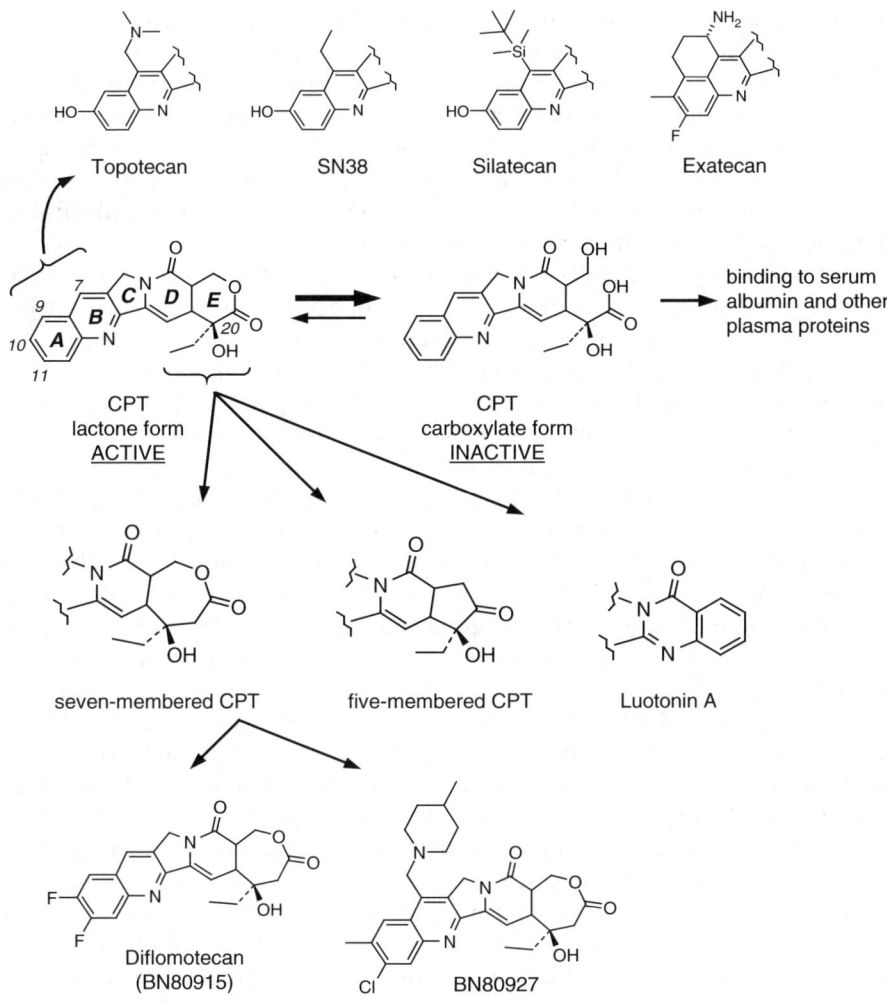

Figure 2 *Structures of CPT derivatives*

is characterized by a stabilized seven-membered lactone ring in place of the unstable six-membered lactone ring of CPT (Figure 2).[28,29]

The hCPT lead compound (BN80765) is a potent topoisomerase I poison,[60] and substitution of its quinoline moiety with two fluoro groups (BN80915, known as diflomotecan)[61] or with a methylpiperidine cationic chain (BN80927) markedly enhances the cytotoxic potential of hCPT so as to reinforce its antitumour activity.[59–68] BN80915 exerts potent cytotoxic[65] and proapoptotic properties.[68] Induction of apoptotic cell death in leukaemia cells by this compound involves activation of caspases-3 and -8, mitochondrial activation and DNA fragmentation.[68] *In vivo*, this compound has demonstrated promising properties in inhibiting the growth of colon and prostate adenoma tumours in xenograft models.[69] BN80915, as well as its newer version BN80927, which is more water soluble,[70] is currently being tested in phase II human clinical trials. These hCPT derivatives are considerably more stable than the corresponding CPT. The hydrolysis of the lactone is limited but, unlike CPT, it is irreversible. The hCPT E-ring β-hydroxylactone is the active form of each molecule and the E-ring carbonyl oxygen and adjacent unsubstituted/ unprotonated ring atom are required for full activity.[71]

If the size of the CPT lactone E ring can be expanded, it can be also reduced. Applying the same logic of preserving the topoisomerase inhibitory activity while reducing the instability of the lactone, Lavielle and co-workers have recently designed another interesting series of CPT analogues bearing a five-membered E ring. In this case, the oxygen atom of the lactone has been deleted and surprisingly these lactone-free compounds maintain a remarkable degree of activity against topoisomerase I.[72]

The structure of CPT has been gradually modified to design analogues with different properties. It is now clear that the core of CPT can be markedly changed while preserving its basic activity against topoisomerase I. The most recent example is 14-azaCPT[73] and a novel series of analogues derived from the alkaloid luotonin A. This pyrroloquinazolinoquinoline derivative originally extracted from the Chinese medicinal plant *Peganum nigellastrum* was characterized as a modest but effective topoisomerase I poison.[74,75] It shares identical A-to-C rings with CPT but differs in structural features of rings D and E (Figure 2). The 20(*S*)-α-hydroxylactone ring is replaced by a benzene ring, while the CH at position 14 is replaced by a nitrogen heteroatom. Despite these structural differences, the two drugs stabilize the topoisomerase I-DNA cleavable complex, share the same pattern of topoisomerase I-mediated DNA cleavage, and are cytotoxic towards a yeast cell line lacking the homologous yeast topoisomerase I but harbouring a plasmid containing the human topoisomerase I gene.[74] However, luotonin A is considerably less efficient than CPT at stabilizing covalent complexes and inducing DNA breaks. To understand which structural elements in the E ring are important for topoisomerase-poisoning and the ensuing biological activities, different luotonin A analogs were synthesized. The introduction of substituents at positions 16, 18, and 19 on the E ring leads to less potent or totally ineffective compounds in stabilizing the topoisomerase I-DNA complex and in inhibiting the growth of the engineered yeast strain harbouring the human topoisomerase I gene. The presence of different substituents such as fluoro, hydroxyl, or amino groups at position 17 does not modify topoisomerase I poisoning compared to luotonin A.[76] A 17-aminoluotonin A analogue maintains cytotoxic and topoisomerase I poisoning capacities comparable to that of the parental compound, and this is also the case for isoluotonin A, an isomer of luotonin A bearing the N heteroatom in ring B at position 7 instead of position 1[77–79] but not for the 14-deaza analogue of luotonin A denoted rosettacin, which has only a minor effect on topoisomerase I.[76] Thus, work in the hCPT and luotonin series has invalidated the initial dogma stating that the lactone E ring of CPT was necessary for topoisomerase I inhibition. These studies have opened the way for the design of novel chemo-types for topoisomerase I inhibition.

3.3 Indenoisoquinolines

The crystal structure of the topotecan-topoisomerase I-DNA ternary complex has revealed that the drug intercalates at the site of DNA cleavage, forming π–π base-stacking interactions with the up- and downstream base pairs[34,80] [*cf.* Chapter 2]. On this basis, it appeared logical to develop specific DNA inter-calators with a planar CPT-like chromophore capable of stabilizing the cova-lent complex further. This was achived with different classes of intercalating agents, in particular with tetracyclic indolizino-[1,2-b]quinoline[81] and in-denoisoquinoline derivatives that resemble the architecture of CPT or luoton-ins, without the lactone E ring. The compound NSC 314622 was the first indenoisoquinoline derivative identified as a topoisomerase I poison but with sequence-selectivity distinct from that of CPT.[82] Topoisomerase I-mediated DNA breaks induced by NSC 314622 are more stable than those induced by CPT. Its potent cytotoxic activities, as well as its capacity to inhibit topoisom-erase I and to bind to DNA, have encouraged the design and synthesis of several series of structural analogues by Cushman and co-workers.[83–94] Deriv-atives equipped with cationic side chains, such as MJ-III-65 (also designated NSC 706744) (Figure 3), exhibit a higher affinity for DNA and generally show enhanced topoisomerase I-inhibitory and antitumour activities.[90]

In search of more potent indenoisoquinolines, a series of derivatives featur-ing a variety of substituents on the lactam nitrogen were synthesized. The rationale of this structure–activity study was based on the X-ray model of the indenoisoquinoline-topoisomerase I-DNA complex, where the drug was shown to intercalate into DNA by mimicking a DNA base pair at the site of topoisomerase I cleavage like CPT.[34,35,83] In this model, π–π stacking interac-tions are formed between rings C and D and the non-cleaved strand bases, while rings A and B stack within the scissile strand bases. The model predicts that the C-ring carbonyl of the drug should be oriented towards the minor groove side with the formation of a hydrogen bond between the keto oxygen of the drug and Arg364, while the substituents on the lactam nitrogen are projected towards the major groove to interact with Asn352 and Ala351. Some of the most potent topoisomerase poisons with high cytotoxicity compared to the lead compound NSC 314622 possess halogen- or aminoalkyl side chains.[84–87] The aminopropyl analog has demonstrated high cytotoxic activity against several human cancer cell lines and its cytotoxicity was correlated to its great potency as a topoisomerase I poison.[85,88] According to the published crystal structure model, this aminoalkyl chain points towards the DNA major groove and stabilizes the drug interaction with topoisomerase and DNA, which might explain its higher cytotoxic and topoisomerase inhibition potencies. However, for other analogs a discrepancy between the cytotoxic effect and topoisomerase poisoning activity was observed. Substitution on the lactam nitrogen by two consecutive amine groups does not modify the topoisomerase I poisoning activity but does decrease the cytotoxicity, while the introduction of three amines results in total loss of topoisomerase inhibition without any changes in cytotoxic effect.[88] These results might suggest a mechanism of action of those compounds different from topoisomerase inhibition.

MJ-III-65 (NSC 706744) (Figure 3) containing an amino alcohol side chain at the N-6 position is as potent as CPT in poisoning topoisomerase I but traps the enzyme at sites with cytosine, not thymine as for CPT, immediately 5' from the cleavage site.[85,89] Like the lead compound NSC 314622, MJ-III-65 stabilizes topoisomerase I-cleavage complexes to a greater extent (4-fold) after its removal compared with CPT. This enhanced activity might result from

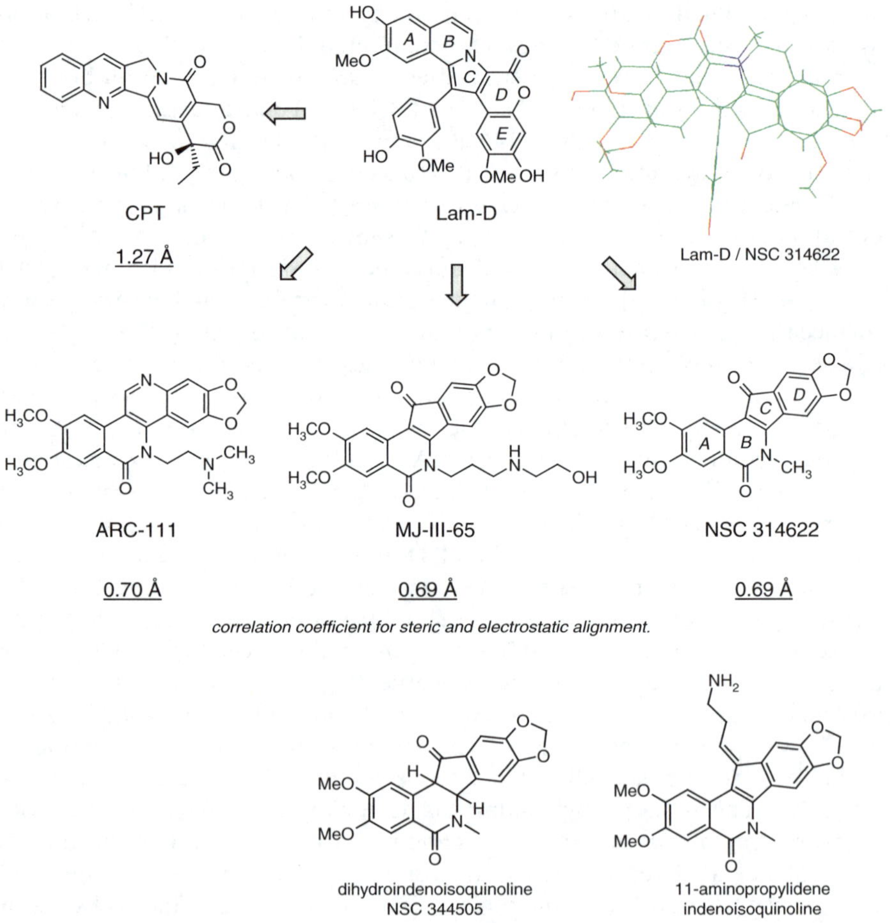

correlation coefficient for steric and electrostatic alignment.

Figure 3   *Superimposition of lamellarin D with other toposiomerase I poisons shows structural and electronic similarities, in particular with the indenoisoquinoline derivative NSC 314622. The structures of different topoisomerase I poisons are shown*

additional hydrogen-bonding interactions between the amino group, the protein and DNA. MJ-III-65 is cytotoxic towards the human T-lympoblastoid leukaemia CEM cell line as well as two resistant cell lines: the human leukaemic cells CEM/C2 having the point mutation N722S in topoisomerase I, which silences the normal allele, and the murine leukaemia cells P388/CPT45 with no detectable expression of topoisomerase I. MJ-III-65 and its parent compound NSC 314622 have also demonstrated antitumour activity and toxicity in nude mice bearing human A253 and FaDu head and neck xenografts.[90]

In the NSC 314622-DNA-topoisomerase I model, the C-ring carbonyl is oriented towards the minor groove of the DNA double helix. Therefore, an aminoalkyl substituent on the carbon-11 position should project towards the minor groove in order to reinforce DNA binding and enzyme interaction, promoting topoisomerase I and cytotoxic activities. A series of derivatives with several substituents on the C-11 position were synthesized. Although the cytotoxicity results did not correlate perfectly with topoisomerase I inhibition for all derivatives tested, it was shown that the addition of an aminopropylidene group (Figure 3) at this position did lead to enhanced topoisomerase I poisoning and high cytotoxicity.[91] The length of the alkyl side chain modulates the biological activity. Shortening the chain from 3 to 2 carbons slightly decreases the activity of the compound, whereas an increase of the length of the chain to 4, 5, or 6 carbons markedly reduces the cytotoxicity.[92]

Many active isoquinoline compounds contain methylenedioxy and methoxy groups on the aromatic D and A rings. Structure–activity studies have demonstrated that these substituents influence the biological activity of the different derivatives. For example, removal of the methylenedioxy substituent or its replacement with a naphthalene ring is detrimental for both topoisomerase inhibitory and cytotoxicity activities.[86,87] However, the methylenedioxy and methoxy substituents are not absolutely necessary for anticancer activity and topoisomerase I poisoning, even though they seem to ameliorate the global effect.[67] The substituent on the lactam nitrogen seems to play a more critical role for the biological activity of the drug.

Among the many indenoisoquinolines synthesized by Cushman and co-workers, some compounds, particularly dihydroquinoline derivatives, proved particularly cytotoxic to human cancer cells, through not necessarily more potent than NSC 314622 at inhibiting topoisomerase I.[84–87] For example, the dihyroindenoisoquinoline derivative NSC 344505 (Figure 3) selectively targets topoisomerase I *in vitro* and strongly inhibits growth of murine P388 leukaemia or human colon carcinoma HCT116 cells (IC_{50} in the 10 nM range). Considerable resistance was observed with the corresponding topoisomerase I-deficient clones P388/CPT45 and HCT116-*Top1*.[94] To understand the role of the central double bond and to design more potent drugs, further isoquinolones, indenoisoquinolium salts and indenoisoquinoline derivatives bearing halogen, amino, or alkyl side chains were synthesized. Most of these compounds exhibit potent cytotoxic activities against a large panel of cancer cells but enhanced topoisomerase activity was only characterized for the indenoisoquinolinium salt analogues. The reduction of the central double bond into a dihydro carbon–carbon liaison appears to be detrimental for topoisomerase inhibition, presumably due to the loss of planarity of the tetracyclic chromophore, which might hinder its insertion between base pairs.[86] It has been postulated that NSC 344505 might function as a prodrug of NSC 314622, requiring oxidative activation in cells to restore full topoisomerase I poisoning activity.[94] The concept of dihydroindenoisoquinoline analogues acting as prodrugs also emerged when the most potent dihydroindenoisoquinoline derivatives were evaluated for their anticancer activity in an *in vivo* model based on polyvinylidene fluoride hollow fibres containing various cancer cell lines implanted intraperitoneally or subcutaneously into mice.[86] Dihydroindenoisoquinoline derivatives, injected intraperitoneally, were more active at the subcutaneous implant than the intraperitoneal one, which is characteristic of an inactive compound that is converted to an active form after administration.

3.4   Benzimidazoles

Benzimidazole derivatives represent another structurally unique class of topoisomerase I poisons through their binding to the DNA minor groove. The well-known fluorescent benzimidazole dye Hoechst 33258 (NSC 32291, pibenzimol) binds preferentially and strongly to the DNA minor groove at AT-rich sequences and inhibits topoisomerase I by trapping the cleavable complex.[95,96] The head-to-tail orientation of the two benzimidazole units of Hoechst 33258 is important for topoisomerase I inhibition. The reverse head-to-head configuration can lead to potent cytotoxic agents but devoid of topoisomerase I poisoning activity.[97–99] Several head-to-tail terbenzimidazole derivatives have been identified as topoisomerase I poisons.[100–102] The presence of a lipophilic group on the first benzimidazole unit, such as a phenyl or bromo substituent, can enhance their topoisomerase I poisoning and cytotoxicity activities.[100–103] However, these minor groove binders generally represent weak topoisomerase I poisons. An intercalative binding mode seems to be preferable to elicit formation of durable and potently effective topoisomerase I-DNA covalent complexes.

3.5   Indolocarbazoles

The initial discovery that the indolocarbazole antibiotic BE-13793C isolated from *Streptoverticillium mobaraense* and its glycosylated analogue rebeccamycin produced by *Saccharothrix aerocolonigenes* exhibit potent antitumour activity[104,105] has stimulated the design and synthesis of several series of derivatives. The antiproliferative activity of rebeccamycin has been related to its capacity to poison topoisomerase I. This compound intercalates between two consecutive base pairs in double-stranded DNA, and is capable of stabilizing the topoisomerase I-DNA cleavable complex by preventing the religation step.[106–108] Based on the structure of rebeccamycin, a number of topoisomerase I poisons were synthesized and the glucosyl derivative NB-506 (Figure 4) has emerged as a promising anticancer drug, advanced to human clinical trials. Recently, a newer version of this compound designated J-107088 (also known

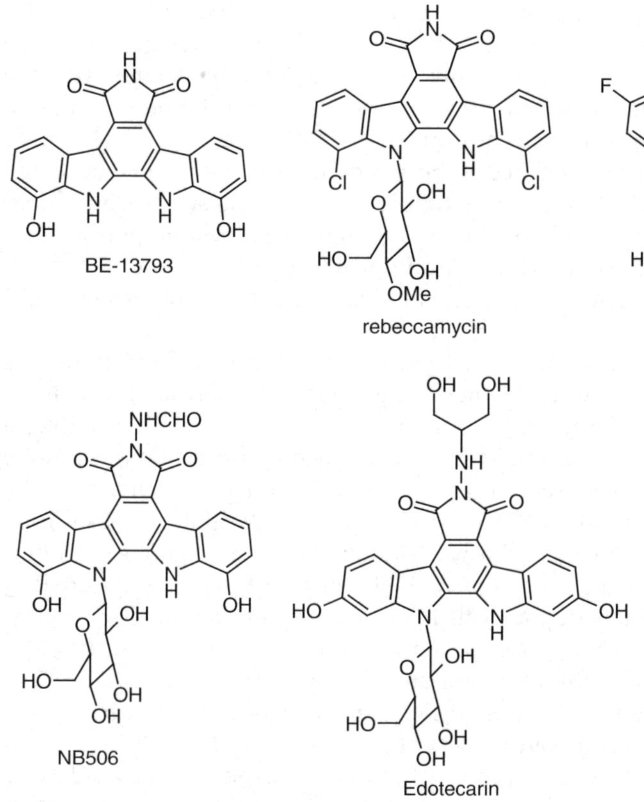

Figure 4 *Indolocarbazole inhibitors of topoisomerase I*

as ED-749, PHA-782615 or Edotecarin) has been advanced to phases I and II clinical trials for colorectal, oesophageal, gastric and breast cancer and phase III trials for glioblastoma.[109–111]

The carbohydrate moiety on rebeccamycin derivatives is a key element for both DNA binding and topoisomerase I inhibition. The removal of the sugar or interference with its linkage to the indolocarbazole *via* a α-glycosidic liaison noticeably reduces DNA binding, topoisomerase I inhibition and cytotoxic properties.[106,107,112] Altering the pattern of glycosylation also considerably changes topoisomerase I inhibition and cytotoxicity profiles. Exchange of the glucose residue for a galactose or mannose does not alter the properties of the compound providing that the sugar anomery is preserved.[113] The introduction of an amino substituent considerably enhances the affinity of the molecule for binding to DNA and increases its cytotoxicity.[114,115] Deletion of 2′-, 3′-, or/and 6′-hydroxyl groups on the carbohydrate residue abolishes the topoisomerase I activity and this is correlated with a decrease of the cytotoxic activity.[116] Rebeccamycin derivatives adopt a closed conformation, involving hydrogen bonding between the pyranose oxygen and the indole NH, to interact with DNA and to inhibit topoisomerase I.[117–119] The 2′- or 6′-OH group on the carbohydrate residue is presumably involved in a hydrogen bond with the indole NH to maintain the carbohydrate in a fixed conformation for optimal activity and interaction with the topoisomerase I-DNA complex.[119] The sugar residue can also be directly attached to both indole nitrogens, as in compound MP003 (Figure 4), which is also a topoisomerase I poison,[120] but other staurosporine-like derivatives in this series did not interfere with topoisomerase I despite enhanced affinity for DNA and improved cytotoxic potential.[121–123] The sugar residue is evidently a critical determinant of topoisomerase I inhibition.

The indolocarbazole chromophore is also amenable to substitution with methyl, amino or hydroxyl substituents capable of improving the activity towards the enzyme and the antitumour potential.[106,115,124] For example, a series of potent topoisomerase I poisons was designed by Bristol–Myers Squibb *via* the incorporation of fluorine atoms on the indolocarbazole moiety.[125,126] This is the case with compound BMS-251873, where a 3,9-difluoro substitution of the indolocarbazole moiety and a benzothiophene core (Figure 4) combine to create high topoisomerase I-mediated DNA cleavage activity, cytotoxicity against a panel of human tumour cell lines, and topoisomerase I selectivity. This fluoro derivative displays a pronounced antitumour efficacy in the PC3 xenograft model.[126] Other fluoro compounds in this series (*e.g.*, BMS-250749) were also found to exhibit highly potent topoisomerase I-inhibitory activity and in some cases a broad antitumour activity superior to that of the CPT reference compound irinotecan.[127] A vast number of indolocarbazole derivatives, having an altered geometry of the chromophore or distinct substitution patterns on the maleimide moiety have been synthesized and tested,[128,129] but to date only NB-506 and its back-up derivative Edotecarin have been advanced to clinical trials. This latter glucosyl indolocarbazole derivative, which shows synergistic activity when combined with cisplatin and oxaliplatin in colon cancer as well as additivity with 5-fluorouracil, appears a very promising topoisomerase I-targeted drug for the treatment of solid tumours. In a few years, it may become the first non-CPT topoisomerase I poison on the market.

3.6   Phenanthridines and Related Compounds

Benzophenanthridine and dibenzocinnoline derivatives have been identified as topoisomerase I inhibitors endowed with cytotoxic activities against several human cell lines.[130–133] The presence of a methylenedioxy group on the benzo ring adjacent to a nitrogen heteroatom and the addition of two methoxy groups on the aromatic ring distal from the methylenedioxy unit, as for the two compounds illustrated in Figure 5, enhance topoisomerase I inhibitory activity and cytotoxicity of the derivatives.[133,134] The related compound ARC111 (8,9-dimethoxy-5-(2-*N,N*-dimethylaminoethyl)-5,12-diazachrysen-6-one or Topovale) (Figure 3) equipped with a cationic side chain to enhance both DNA interaction and water solubility has revealed significantly higher topoisomerase

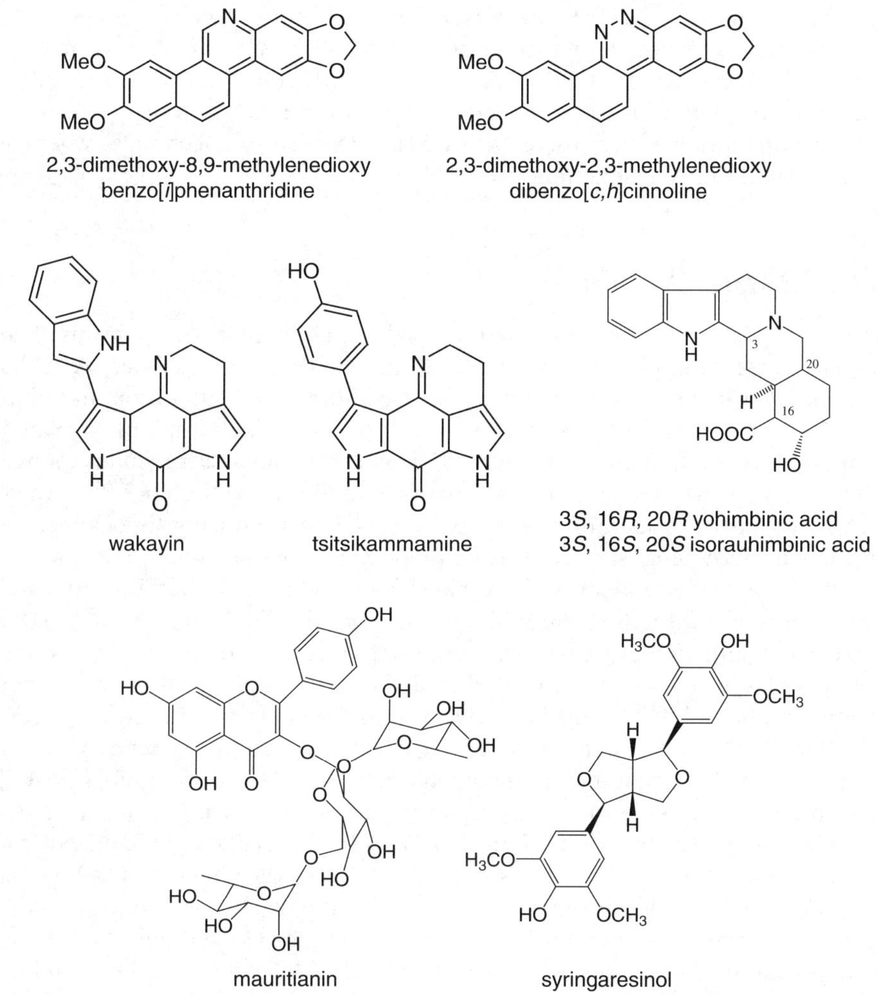

2,3-dimethoxy-8,9-methylenedioxy
benzo[*i*]phenanthridine

2,3-dimethoxy-2,3-methylenedioxy
dibenzo[*c,h*]cinnoline

wakayin

tsitsikammamine

3*S*, 16*R*, 20*R* yohimbinic acid
3*S*, 16*S*, 20*S* isorauhimbinic acid

mauritianin

syringaresinol

Figure 5 *Selected examples of natural products that act as topoisomerase I poisons*

I-inhibitory activity than CPT and the clinically useful drugs irinotecan and topotecan.[130,135,136] This compound is highly cytotoxic to human RPMI8402 lymphoblast tumour cells and murine P388 leukaemia cells, with IC_{50} in the nanomolar range. The high cytotoxic potential correlates with its capacity to inhibit topoisomerase I. The reduced activity of ARC111 towards the CPT-resistant CPT-K5 and P388/CPT45 cell lines strongly suggests that topoisomerase I is effectively the main target of this molecule.[136,137] Treatment of RPMI8402 cells with ARC111 induces cleavable complex formation, DNA strand breaks and apoptosis that is manifested by caspase activation.[137] Anti-tumour activities *in vivo* were also demonstrated using HCT-8 colon adenocarcinoma or SKNEP anaplastic Wilm's tumour xenografts in nude mice. In these models, the antitumour activity of ARC111 is comparable to that of irinotecan. In another study, ARC111 was found to be considerably more potent than irinotecan at inducing tumour regression in nude mice bearing non-estrogen responsive MDA-MD-435 breast cancer cells, whether administrated orally or parenterally.[130,135]

3.7   Marine Alkaloids

Several inhibitors of topoisomerase I are elaborated by marine organisms. The *bis*-pyrroloiminoquinone marine alkaloid wakayin (Figure 5) isolated from the ascidian *Clavelina* spp. has demonstrated topoisomerase I inhibition and high cytotoxic activity against murine cancer cell lines.[138] The structurally related compound tsitsikammamine (Figure 5) isolated from a latrunculid sponge exhibits similar cytotoxic and topoisomerase I inhibitory activities.[139,140] However, wakayin is considerably less potent than CPT at inducing topoisomerase I-mediated DNA breaks.[138] Several synthetic analogues have been made and a few structure–activity relationships have been delineated, but thus far the topoisomerase I-inhibitory and biological activities have not been greatly improved. Synthetic derivatives with the pyrrole ring of the pyrroloquinoline moiety replaced by a pyrazole or a phenyl ring are less potent than the parent natural compound[140–142] but this series has not been extensively investigated.

Within the family of marine alkaloids, the most promising series without doubt is that containing the lamellarins (6*H*-[1]-benzopyrano[4′,3′;4,5] pyrrolo[2,1-α]isoquinoline), which represent a very promising group of anti-cancer agents.[143] Lamellarin D (Lam-D in Figure 3) was first characterized as a potent poison of the human topoisomerase I.[144] Like the CPTs, this hexacyclic compound induces topoisomerase I-mediated breaks but exhibits a DNA cleavage pattern distinct from that of CPT. The plant akaloid and the marine alkaloid manifest different sequence selectivity. CPT and Lam-D stimulate DNA cleavage by topoisomerase I at certain T$^\downarrow$G sites, whereas other C$^\downarrow$G sites are uniquely cleaved by Lam-D. The structure of Lam-D differs markedly from that of CPT as well as other aforementioned topoisomerase I poisons, but when their structures are superimposed and compared a notable degree of steric and electronic analogy can be established between Lam-D and indenoisoquinolines like NSC 314622 or the benzophenanthridine ARC111 (Figure 3). Common structure–activity relationship rules can be established, in particular the absolute requirement for a planar chromophore. The 5-6 double bond in the B-ring of Lam-D has a significant impact on topoisomerase inhibitory and cytotoxic activity. A non-planar 5,6-dehydro analogue of lamellarin D is totally devoid of activity against topoisomerase I and considerably less cytotoxic than Lam-D.[144,145] As in the indenoisoquinoline series, the planarity of the chromophore is essential to permit stacking interaction with DNA base pairs and interference with topoisomerase I catalytic activity. The hydroxyl groups on the Lam-D skeleton also play a determinant role for topoisomerase I inhibition.[146]

Molecular dynamics simulations predict a model where the planar heterocyclic ring intercalates and stacks between the upstream and downstream base pairs at the site of DNA cleavage. In this model, the 20- and 8-hydroxyl groups establish key hydrogen-bonding interactions with Asn722 and Glu356 residues of the enzyme and the keto group (O17) is facing the guanidium group of Arg364. There are considerable structural similarities between CPT and Lam-D bound to the topoisomerase I-DNA interface.[146] Although lamellarins provide a novel chemotype for topoisomerase I inhibition, their mechanism of action is not as simple as that of the CPTs, which seem to target this enzyme uniquely. Lam-D and certain analogues such as Lam-H potently inhibit topoisomerase I but other lamellarins, in particular those with a 5-6 single bond (*e.g.*, Lam-K), do not interfere with topoisomerase I, while exhibiting quite potent cytotoxic potential. Topoisomerase I contributes to the cytotoxic and antitumour potential of Lam-D, but it is certainly not the unique molecular target of the alkaloid. In a very recent study, the drug was found to interfere directly with cancer cell mitochondria.[147]

3.8   Plant Natural Products

Apart from CPT and derivatives, a few topoisomerase I poisons have been isolated from plant extracts such as the flavonol glycoside mauritianin and the lignan syringaresinol (Figure 5), extracted from the plant *Rinorea anguifera* (Satan, Thailand). Albeit much less active than CPT, they both stabilize topoisomerase-DNA covalent complexes.[148] Certain flavonoids, polyphenols widely distributed in plants, interfere with topoisomerase I activity. Quercetin, myricetin, fisetin and morin have demonstrated dual inhibitory activity against topoisomerases I and II, but their cytotoxic action is rarely dependent on topoisomerase I inhibition.[149–152] The most interesting flavone in this series is luteolin, which appears equally potent to CPT at inhibiting topoisomerase I[153] and also triggers topoisomerase II-mediated DNA cleavage.[154] Unlike CPT, luteolin can bind to free (unbound) topoisomerase I and it forms intercalation complexes with DNA at very high concentrations. However, these direct interactions between luteolin and DNA or topoisomerase I do not affect the assembly of enzyme-DNA covalent complexes, which are fully stabilized by luteolin.[153] A structure–activity study indicated that among 34 polyphenolic compounds tested for topoisomerase I inhibition and DNA binding, flavones and flavonols are the most potent topoisomerase I poisons. No correlation was found between the topoisomerase I-poisoning activity and the degree of DNA intercalation.[155] Other examples of plant-derived topoisomerase I inhibitors have recently been described, such as the cytotoxic indole alkaloids yohimbic and isorauhimbic acids (Figure 5) extracted from the root of *Rauwolfia serpentina*[156] and the biflavonoid 2′′,3′′-diidroochnaflavone isolated from the leaves of *Luxemburgia nobilis*.[157] The development of high-throughput screening procedures will continue to accelerate the discovery of novel topoisomerase I poisons.

3.9   Conclusion

Small molecules that interfere with the correct functioning of topoisomerase I have historically been designed from the plant alkaloid CPT owing to the high potency, sharp selectivity and relatively easy chemical access of this natural product. CPT remains a robust lead compound for the development of drug candidates and novel tumour-active analogues have been reported almost every year for the past decade. In the CPT series, numerous structure–activity relationship studies have been published, and it is now clear that the lactone functionality of CPT is not absolutely required for stabilization of the enzyme-DNA covalent complexes. This observation, together with the elucidation of the structure of topoisomerase I-DNA-drug ternary complexes, opens new perspectives for the rational design of inhibitors. In terms of drug design, it is now obvious that mere structural resemblance to CPT is not an absolute requirement for poisoning of topoisomerase I, and a subset of potent inhibitors with prominent structural dissimilarity to the prototype has been identified. A diversity of new leads structurally distinct from the CPTs is now emerging with a few promising anticancer drug candidates, particularly in the indenoisoquinoline and indolocarbazole series. Natural products have also been identified, such as the lamellarins, which arguably represent the most innovative chemical series of topoisomerase I poisons discovered during the last five years. Natural substances represent a seemingly inexhaustible source of topoisomerase I poisons. Here, we have focused on the inhibitors of DNA relaxation and cleavage, but it is important to mention that topoisomerase I also has a kinase activity,[158] which can be targeted by small molecules. Significant progress has recently been made towards identifying specific inhibitors of this kinase activity of topoisomerase I,[159] which may also prove of considerable pharmaceutical interest in oncology and other therapeutic domains. These novel types of inhibitors herald an upsurge of interest in topoisomerase I which, more than 20 years after its discovery, continues to afford an important source of new anticancer agents. All the facets of the pleiotropic cellular and molecular effects of topoisomerase I have yet to be fully defined. There is still a lot to be learnt about this fascinating enzyme.

References

1. J.C. Wang, *Nat. Rev. Mol. Cell. Biol.*, 2002, **3**, 430–440.
2. J.J. Champoux, *Ann. Rev. Biochem.*, 2002, **70**, 369–413.
3. J.B. Leppard and J.J. Champoux, *Chromosoma*, 2005, **114**, 75–85.
4. L. Stewart, M.R. Redinbo, X. Qui, W.G. Hol and J.J. Champoux, *Science*, 1998, **279**, 1534–1541.
5. L. Stewart, G.C. Ireton and J.J. Champoux, *J. Mol. Biol.*, 1997, **269**, 355–372.
6. Y. Pommier, P. Pourquier, Y. Fan and D. Strumberg, *Biochem. Biophys. Acta*, 1998, **1400**, 83–106.

7. Y.H. Hsiang, M.G. Lihou and L.F. Liu, *Cancer Res.*, 1989, **49**, 5077–5082.
8. Y.P. Tsao, A. Russo, G. Nyauswa, R. Silber and L.F. Liu, *Cancer Res.*, 1993, **53**, 5908–5914.
9. J. Wu and L.F. Liu, *Nucleic Acid Res.*, 1997, **25**, 4181–4186.
10. T.K. Li and L.F. Liu, *Ann. Rev. Pharmacol. Toxicol.*, 2001, **41**, 53–77.
11. Y.H. Hsiang, R. Hertzberg, S. Hecht and L.F. Liu, *J. Biol. Chem.*, 1985, **260**, 14873–14878.
12. M.D. Megonigal, J. Fertala and M.A. Bjornsti, *J. Biol. Chem.*, 1997, **272**, 12801–12808.
13. D.A. Burden and N. Osheroff, *Biochim. Biophys. Acta*, 1998, **1400**, 139–154.
14. G. Raspaglio, C. Ferlini, S. Mozzetti, S. Prislei, D. Gallo, N. Das and G. Scambia, *Biochem. Pharmacol.*, 2005, **69**, 113–121.
15. C. Bailly, *Curr. Med. Chem.*, 2000, **7**, 39–58.
16. A. Lorence and C.L. Nessler, *Phytochemistry*, 2004, **65**, 2735–2749.
17. S.C. Puri, V. Verma, T. Amna, G.N. Qazi and M. Spiteller, *J. Nat. Prod.*, 2005, **68**, 1717–1719.
18. R.P. Hertzberg, M.J. Caranfa and S.M. Hecht, *Biochemistry*, 1989, **28**, 4629–4638.
19. N. Masuda, S. Kudoh and M. Fukuoka, *Crit. Rev. Oncol. Hematol.*, 1996, **24**, 3–26.
20. Y. Xu and M.A. Villalona-Calero, *Ann. Oncol.*, 2002, **13**, 1841–1851.
21. J. Carmichael and R.F. Ozols, *Opin. Invest. Drugs*, 1997, **6**, 593–608.
22. C. Kollmannsberger, K. Mross, A. Jakob, L. Kanz and C. Bokemeyer, *Oncology*, 1999, **56**, 1–12.
23. I.E. Kuppens, J. Beijnen and J.H. Schellens, *Clin. Colorectal. Cancer*, 2004, **4**, 163–180.
24. L.C. Cho and H. Choy, *Oncology*, 2004, **18**, 29–39.
25. E.T. Wong and A. Berkenblit, *Oncology*, 2004, **9**, 68–79.
26. J.F. Pizzolato and L.B. Saltz, *Expert Rev. Anticancer Ther.*, 2003, **3**, 587–593.
27. G. Pratesi, G.L. Beretta and F. Zunino, *Anticancer Drugs*, 2004, **15**, 545–552.
28. C. Bailly, *Crit. Rev. Oncol./Hematol.*, 2003, **45**, 91–108.
29. F. Zunino and G. Pratesi, *Expert Opin. Inv. Drug.*, 2004, **13**, 269–284.
30. D. Sriram, P. Yogeeswari, R. Thirumurugan and T.R. Bal, *Nat. Prod. Res.*, 2005, **19**, 393–412.
31. M. Kreditor, M. Fink and H.S. Hochster, *Chemother. Biol. Response Modif.*, 2005, **22**, 61–100.
32. J.F. Pizzolato and L.B. Saltz, *Lancet*, 2003, **361**, 2235–2242.
33. N. Pessah, M. Reznik, M. Shamis, F. Yantiri, H. Xin, K. Bowdish, N. Shomron, G. Ast and D. Shabat, *Bioorg. Med. Chem.*, 2004, **12**, 1859–1866.
34. B.L. Staker, K. Hjerrild, M.D. Feese, C.A. Behnke, A.B. Burgin and L. Stewart, *Proc. Natl. Acad. Sci. USA*, 2002, **99**, 15387–15392.
35. B.L. Staker, M.D. Feese, M. Cushman, Y. Pommier, D. Zembower, L. Stewart and A.B. Burgin, *J. Med. Chem.*, 2005, **48**, 2336–2345.

36. Z. Mi and T.G. Burke, *Biochemistry*, 1994, **33**, 10325–10336.
37. F.M. Muggia, P.J. Creaven, H.H. Hansen, M.H. Cohen and O.S. Selawry, *Cancer Chemother. Rep.*, 1972, **56**, 551–521.
38. U. Schaeppi, R.W. Fleischman and D.A. Cooney, *Cancer Chemother. Rep.*, 1974, **58**, 25–36.
39. K. Seiter, *Expert Opin. Drug Saf.*, 2005, **4**, 45–53.
40. M. Maliepaard, M.A. van Gastelen, A. Tohgo, F.H. Hausheer, R.C. van Waardenburg, L.A. de Jong, D. Pluim, J.H. Beijnen and J.H. Schellens, *Clin. Cancer Res.*, 2001, **7**, 935–941.
41. K. Lackey, J.M. Besterman, W. Fletcher, P. Leitner, B. Morton and D.D. Sternbach, *J. Med. Chem.*, 1995, **38**, 906–911.
42. R.W. Driver and L.X. Yang, *Mini Rev. Med. Chem.*, 2005, **5**, 425–439.
43. C.J. Thomas, N.J. Rahier and S.M. Hecht, *Bioorg. Med. Chem.*, 2004, **12**, 1585–1604.
44. S. Dallavalle, A. Ferrari, B. Biasotti, L. Merlini, S. Penco, G. Gallo, M. Marzi, M.O. Tinti, R. Martinelli, C. Pisano, P. Carminati, N. Carenini, G. Beretta, P. Perego, M. De Cesare, G. Pratesi and F. Zunino, *J. Med. Chem.*, 2001, **44**, 3264–3274.
45. I.F. Pollack, M. Erff, D. Bom, T.G. Burke, J.T. Strode and D.P. Curran, *Cancer Res.*, 1999, **59**, 4898–4905.
46. D. Bom, D.P. Curran, S. Kruszewski, S.G. Zimmer, J. Thompson Strode, G. Kohlhagen, W. Du, A.J. Chavan, K.A. Fraley, A.L. Bingcang, L.J. Latus, Y. Pommier and T.G. Burke, *J. Med. Chem.*, 2000, **43**, 3970–3980.
47. F.J. Giles, J.E. Cortes, D.A. Thomas, G. Garcia-Manero, S. Faderl, S. Jeha, R.L. De Jager and H.M. Kantarjian, *Clin. Cancer Res.*, 2002, **8**, 2134–2141.
48. J.P. Braybrooke, M. Ranson, C. Manegold, K. Mattson, N. Thatcher, P. Cheverton, M. Sekiguchi, M. Suzuki, R. Oyama and D.C. Talbot, *Lung Cancer*, 2003, **41**, 215–219.
49. S. Baka, M. Ranson, P. Lorigan, S. Danson, K. Linton, I. Hoogendam, K. Mettinger and N. Thatcher, *Eur. J. Cancer*, 2005, **41**, 1547–1550.
50. H. Gao, X. Zhang, Y. Chen, H. Shen, T. Pang, J. Sun, C. Xu, J. Ding, C. Li and W. Lu, *Bioorg. Med. Chem.*, 2005, **15**, 3233–3236.
51. Y. Fan, J.N. Weinstein, K.W. Kohn, L.M. Shi and Y. Pommier, *J. Med. Chem.*, 1998, **41**, 2216–2226.
52. A.W. Nicholas, M.C. Wani, G. Manikumar, M.E. Wall, K.W. Kohn and Y. Pommier, *J. Med. Chem.*, 1990, **33**, 972–978.
53. R.T. Crow and D.M. Crothers, *J. Med. Chem.*, 1992, **35**, 4160–4164.
54. H.K. Wang, S.Y. Lin, K.M. Hwang, A.T. McPhail and K.H. Lee, *Bioorg. Med. Chem. Lett.*, 1995, **5**, 77–82.
55. M.E. Wall, M.C. Wani, A.W. Nicholas, G. Manikumar, C. Tele, L. Moore, A. Truesdale, P. Leitner and J.M. Besterman, *J. Med. Chem.*, 1993, **36**, 2689–2700.
56. L. Snyder, W. Shen, W.G. Bornmann and S.J. Danishefsky, *J. Org. Chem.*, 1994, **59**, 7033–7037.

57. C. Jaxel, K.W. Kohn, M.C. Wani, M.E. Wall and Y. Pommier, *Cancer Res.*, 1989, **49**, 1465–1469.
58. R.P. Hertzberg, M.J. Caranfa, K.G. Holden, D.R. Jakas, G. Gallagher, M.R. Mattern, S.M. Mong, J. O'Leary Bartus, R.K. Johnson and W.D. Kingsbury, *J. Med. Chem.*, 1989, **32**, 715–720.
59. O. Lavergne, L. Lesueur–Ginot, F. Pla Rodas and D.C. Bigg, *Bioorg. Med. Chem. Lett.*, 1997, **7**, 2235–2238.
60. C. Bailly, A. Lansiaux, L. Dassonneville, D. Demarquay, H. Coulomb, M. Huchet, O. Lavergne and D.C.H. Bigg, *Biochemistry*, 1999, **38**, 15556–15563.
61. N. Osheroff, *IDrugs*, 2004, **7**, 257–263.
62. L. Lesueur-Ginot, D. Demarquay, R. Kiss, P.G. Kasprzyk, L. Dassonneville, C. Bailly, J. Camara, O. Lavergne and D.C. Bigg, *Cancer Res.*, 1999, **5**, 2939–2943.
63. O. Lavergne, L. Lesueur-Ginot, F. Pla Rodas, P.G. Kasprzyk, J. Pommier, D. Demarquay, G. Prevost, G. Ulibarri, A. Rolland, A.M. Schiano-Liberatore, J. Harnett, D. Pons, J. Camara and D.C.H. Bigg, *J. Med. Chem.*, 1998, **41**, 5410–5419.
64. P. Philippart, L. Harper, C. Chaboteaux, C. Decaestecker, Y. Bronckart, L. Gordover, L. Lesueur-Ginot, H. Malonne, O. Lavergne, D.C. Bigg, P.M. da Costa and R. Kiss, *Clin. Cancer Res.*, 2000, **6**, 1557–1562.
65. O. Lavergne, D. Demarquay, C. Bailly, C. Lanco, A. Rolland, M. Huchet, H. Coulomb, N. Muller, N. Baroggi, J. Camara, C. Le Breton, E. Manginot, J.B. Cazaux and D.C.H. Bigg, *J. Med. Chem.*, 2000, **43**, 2285–2289.
66. D. Chauvier, H. Morjani and M. Manfait, *Breast Cancer Res. Treat.*, 2002, **73**, 113–125.
67. A.K. Larsen, C. Gilbert, G. Chyzak, S.Y. Plisov, I. Naguibneva, O. Lavergne, L. Lesueur-Ginot and D.C. Bigg, *Cancer Res.*, 2001, **61**, 2961–2967.
68. A. Lansiaux, M. Facompre, N. Wattez, M.P. Hildebrand, C. Bal, D. Demarquay, O. Lavergne D.C. Bigg and C. Bailly, *Mol. Pharmacol.*, 2001, **60**, 450–461.
69. D. Demarquay, M. Huchet, H. Coulomb, L. Lesueur-Ginot, O. Lavergne, P.G. Kasprzyk, C. Bailly, J. Camara and D.C. Bigg, *Anticancer Drugs*, 2001, **12**, 9–19.
70. D. Demarquay, M. Huchet, H. Coulomb, L. Lesueur-Ginot, O. Lavergne, J. Camara, P.G. Kasprzyk, G. Prevost and D.C. Bigg, *Cancer Res.*, 2004, **64**, 4942–4949.
71. G.S. Laco, W. Du, G. Kohlhagen, J.M. Sayer, D.M. Jerina, T.G. Burke, D.P. Curran and Y. Pommier, *Bioorg. Med. Chem.*, 2004, **12**, 5225–5235.
72. P. Hautefaye, B. Cimetière, A. Pierré, S. Léonce, J. Hickman, W. Laine, C. Bailly and G. Lavielle, *Bioorg. Med. Chem. Lett.*, 2003, **13**, 2731–2735.
73. K. Cheng, N.J. Rahier, B.M. Eisenhauer, R. Gao, S.J. Thomas and S.M. Hecht, *J. Am. Chem. Soc.*, 2005, **127**, 838–839.

74. A. Cagir, S.H. Jones, R. Gao, B.M. Eisenhauer and S.M. Hecht, *J. Am. Chem. Soc.*, 2003, **125**, 13628–13629.

75. K. Cheng, N.J. Rahier, B.M. Eisenhauer, R. Gao, S.J. Thomas and S.M. Hecht, *J. Am. Chem. Soc.*, 2004, **127**, 838–839.

76. S.M. Hecht, *Curr. Med. Chem. Anti-Cancer Agents*, 2005, **5**, 353–362.

77. A. Cagir, B.M. Eisenhauer, R. Gao, S.J. Thomas and S. Hecht, *Bioorg. Med. Chem.*, 2004, **12**, 6287–6299.

78. A. Cagir, S.J. Thomas, B.M. Eisenhauer, R. Gao and S. Hecht, *Bioorg. Med. Chem. Lett.*, 2004, **14**, 2051–2054.

79. S. Dallavalle, S. Merlini, G.L. Beretta, S. Tinelli and F. Zunino, *Bioorg. Med. Chem. Lett.*, 2004, **14**, 5757–5761.

80. M.R. Redinbo, L. Stewart, P. Kuhn, J.J. Champoux and W.G.J. Hol, *Science*, 1998, **279**, 1504–1513.

81. A. Perzyna, C. Marty, M. Facompre, J.F. Goossens, N. Pommery, P. Colson, C. Houssier, R. Houssin, J.P. Hénichart and C. Bailly, *J. Med. Chem.*, 2002, **45**, 5809–5812.

82. G. Kohlhagen, K. Paull, M. Cushman, P. Nagafuji and Y. Pommier, *Mol. Pharmacol.*, 1998, **54**, 50–58.

83. X. Xiao, S. Antony, Y. Pommier and M. Cushman, *J. Med. Chem.*, 2005, **48**, 3231–3238.

84. D. Strumberg, Y. Pommier, K. Paull, M. Jayaraman, P. Nagafuji and M. Cushman, *J. Med. Chem.*, 1999, **42**, 446–457.

85. M. Cushman, M. Jayaraman, J.A. Vroman, A.K. Fukunaga, B.M. Fox, G. Kohlhagen, D. Strumberg and Y. Pommier, *J. Med. Chem.*, 2000, **43**, 3688–3698.

86. M. Jayaraman, B.M. Fox, M. Hollingshead, G. Kohlhagen, Y. Pommier and M. Cushman, *J. Med. Chem.*, 2002, **45**, 242–249.

87. A. Morell, S. Antony, G. Kohlhagen, Y. Pommier and M. Cushman, *Bioorg. Med. Chem. Lett.*, 2004, **14**, 3659–3663.

88. M. Nagarajan, X. Xiao, S. Antony, G. Kohlhagen, Y. Pommier and M. Cushman, *J. Med. Chem.*, 2003, **46**, 5712–5724.

89. S. Antony, M. Jayaraman, G. Laco, G. Kohlhagen, K. Kohn, M. Cushman and Y. Pommier, *Cancer Res.*, 2003, **63**, 7428–7435.

90. S. Antony, G. Kohlhagen, K. Agama, M. Jayaraman, S. Cao, F.A. Durrani, Y.M. Rustum, M. Cushman and Y. Pommier, *Mol. Pharmacol.*, 2005, **67**, 523–530.

91. B.M. Fox, X. Xiao, S. Antony, G. Kohlhagen, Y. Pommier, B.L. Staker, L. Stewart and M. Cushman, *J. Med. Chem.*, 2003, **46**, 3275–3282.

92. X. Xiao, S. Antony, G. Kohlhagen, Y. Pommier and M. Cushman, *Bioorg. Med. Chem.*, 2004, **12**, 5147–5160.

93. M. Nagarajan, A. Morrell, B.C. Fort, M.R. Meckley, S. Antony, G. Kohlhagen, Y. Pommier and M. Cuhman, *J. Med. Chem.*, 2004, **47**, 5651–5661.

94. X. Xiao, Z.H. Miao, S. Antony, Y. Pommier and M. Cushman, *Bioorg. Med. Chem. Lett.*, 2005, **15**, 2795–2798.

95. T.A. Beerman, M.M. McHugh, R.D. Sigmud, J.W. Lown, K.E. Rao and Y. Bathini, *Biochim. Biophys. Acta*, 1992, **1131**, 53–61.

96. A.Y. Chen, C. Yu and B. Gatto, Liu L.F, *PNAS*, 1993, **90**, 8131–8135.

97. S. Neidle, J. Mann, E.L. Rayner, A. Baron, Y. Opoku-Boahen, I.J. Simpson, N.J. Smitj and K.R. Fox, *J. Chem. Soc. Chem. Commun.*, 1999, 929–930.

98. J. Mann, A. Baron, Y. Opoku-Boahen, E. Johansson, G. Parkinson, L.R. Kelland and S. Neidle, *J. Med. Chem.*, 2001, **44**, 138–144.

99. A. Seaton, C. Higgins, J. Mann, A. Baron, C. Bailly, S. Neidle and H. van den Berg, *Eur. J. Cancer*, 2003, **39**, 2548–2555.

100. J.S. Kim, C. Yu, A. Liu, L.F. Liu and E.J. LaVoie, *J. Med. Chem.*, 1997, **40**, 2818–2824.

101. J.S. Kim, Q. Sun, C. Yu, A. Liu, L.F. Liu and E.J. LaVoie, *Bioorg. Med. Chem.*, 1998, **6**, 163–172.

102. M. Rangarajan, S.J. Kim, S. Jin, S.P. Sim, A. Liu, D.S. Pilch and L.F. Liu, Ej. LaVoie, *Bioorg. Med. Chem.*, 2000, **8**, 1371–1382.

103. M. Rangarajan, J.S. Kim, S.P. Sim, A. Liu, L.F. Liu and E.J. LaVoie, *Bioorg. Med. Chem.*, 2000, **8**, 2591–2600.

104. D.E. Nettleton, T.W. Doyle, B. Krishnan, G.K. Matsumoto and J. Clardy, *Tetrahedron Lett.*, 1985, **26**, 4011–4014.

105. K. Kojiri, H. Kondo, T. Yoshinari, H. Arakawa, S. Nakajima, F. Satoh, K. Kawamura, A. Okura, H. Suda and M. Okanishi, *J. Antibiot.*, 1991, **44**, 723–728.

106. E. Rodrigues Pereira, L. Belin, M. Sancelme, M. Prudhomme, M. Ollier, M. Rapp, D. Severe, J.F. Riou, D. Fabbro and T. Meyer, *J. Med. Chem.*, 1996, **39**, 4471–4477.

107. C. Bailly, J.F. Riou, P. Colson, C. Houssier, E. Rodrigues-Pereira and M. Prudhomme, *Biochemistry*, 1997, **36**, 3917–3929.

108. C. Bailly, C. Carrasco, F. Hamy, H. Vezin, M. Prudhomme, A. Saleem and E. Rubin, *Biochemistry*, 1999, **38**, 8605–8611.

109. L.H. Meng, Z.Y. Liao and Y. Pommier, *Curr. Top. Med. Chem.*, 2003, **3**, 305–320.

110. Y. Yamada, T. Tamura, N. Yamamoto, T. Shimoyama, Y. Ueda, H. Murakami, H. Kusaba, Y. Kamiya, H. Saka, Y. Tanigawara, J.P. McGovren and Y. Natsumeda, *Cancer Chemother. Pharmacol.*, 2006, **58**, 173–182 .

111. M.W. Saif and R.B. Diasio, *Clin. Colorectal. Cancer*, 2005, **5**, 27–36.

112. C. Bailly, P. Colson, C. Houssier, E. Rodrigues-Pereira, M. Prudhomme and M.J. Waring, *Mol. Pharmacol.*, 1998, **53**, 77–87.

113. M. Ohkubo, T. Nishimura, H. Kawamoto, M. Nakano, T. Honma, T. Yoshinari, H. Arakawa, H. Suda, H. Morishima and S. Nishimura, *Bioorg. Med. Chem. Lett.*, 2000, **10**, 419–422.

114. C. Bailly, X. Qu, F. Anizon, M. Prudhomme, J.F. Riou and J.B. Chaires, *Mol. Pharm.*, 1999, **55**, 377–385.

115. F. Anizon, P. Moreau, M. Sancelme, W. Laine, C. Bailly and M. Prudhomme, *Bioorg. Med. Chem.*, 2003, **11**, 3709–3722.

116. G. Zhang, J. Shen, H. Cheng, L. Zhu, L. Fang, S. Luo, M.T. Muller, G.E. Lee, L. Wei, Y. Du, D. Sun and P.G. Wang, *J. Med. Chem.*, 2005, **48**, 2600–2611.

117. P. Moreau, F. Anizon, M. Sancelme, M. Prudhomme, C. Bailly, C. Carrasco, M. Ollier, D. Severe, J.F. Riou, D. Fabbro, T. Meyer and A.M. Aubertin, *J. Med. Chem.*, 1998, **41**, 1631–1640.

118. E.J. Gilbert, J.D. Chisholm and D.L. Van Vranken, *J. Org. Chem.*, 1999, **64**, 5670–5676.

119. M. Facompré, C. Carrasco, H. Vezin, J.D. Chisholm, J.C. Yoburn, D.L. Van Vranken and C. Bailly, *ChemBioChem*, 2003, **4**, 386–395.

120. F. Anizon, P. Moreau, M. Sancelme, A. Voldoire, M. Prudhomme, M. Ollier, D. Severe, J.F. Riou, C. Bailly, D. Fabbro, T. Meyer and A.M. Aubertin, *Bioorg. Med. Chem.*, 1998, **66**, 1597–1604.

121. P. Moreau, F. Anizon, M. Sancelme, M. Prudhomme, C. Bailly, D. Severe, J.F. Riou, D. Fabbro, T. Meyer and A.M. Aubertin, *J. Med. Chem.*, 1999, **42**, 584–592.

122. C. Marminon, F. Anizon, P. Moreau, S. Leonce, A. Pierre, B. Pfeiffer, P. Renard and M. Prudhomme, *J. Med. Chem.*, 2002, **45**, 1330–1339.

123. M. Facompre, B. Baldeyrou, C. Bailly, F. Anizon, M. Marminon, M. Prudhomme, P. Colson and C. Houssier, *Eur. J. Med. Chem.*, 2002, **37**, 925–932.

124. F. Anizon, L. Belin, P. Moreau, M. Sancelme, A. Voldoire, M. Prudhomme, M. Ollier, D. Severe, J.F. Riou, C. Bailly, D. Fabbro and T. Meyer, *J. Med. Chem.*, 1997, **40**, 3456.

125. B.H. Long, W.C. Rose, D.M. Vyas, J.A. Matson and S. Forenza, *Curr. Med. Chem.*, 2002, **2**, 255–266.

126. B.N. Balasubramanian, D.R. St. Laurent, B.H. Long, C. Bachand, F. Beaulieu, W. Clarke, M. Deshpande, J. Eummer, C.R. Fairchild, D.B. Frennesson, R. Kramer, F.Y. Lee, M. Mahler, A. Martel, B.N. Naidu, W.C. Rose, J. Russell, E. Ruediger, M.G. Saulnier, C. Solomon, K.M. Stoffan, H. Wong, K. Zimmermann and D.M. Vyas, *J. Med. Chem.*, 2004, **47**, 1609–1612.

127. M.G. Saulnier, B.N. Balasubramanian, B.H. Long, D.B. Frennesson, E. Ruediger, K. Zimmermann, J.T. Eummer, D.R. St Laurent, K.M. Stoffan, B.N. Naidu, M. Mahler, F. Beaulieu, C. Bachand, F.Y. Lee, C.R. Fairchild, L.K. Stadnick, W.C. Rose, C. Solomon, H. Wong, A. Martel, J.J. Wright, R. Kramer, D.R. Langley and D.M. Vyas, *J. Med. Chem.*, 2005, **48**, 2258–2261.

128. A. Voldoire, M. Sancelme, M. Prudhomme, P. Colson, C. Houssier, C. Bailly, S. Leonce and S. Lambel, *Bioorg. Med. Chem.*, 2001, **9**, 357–365.

129. C. Bailly, in *Small Molecule DNA and RNA Binders*, M. Demeunynck and C. Bailly (eds), Vol. 2, Wilson WD, Wiley-VCH, 2003, 538–575.

130. A.L. Ruchelman, S.K. Singh, X. Wu, A. Ray, J.M. Yang, T.K. Li, A. Liu, L.F. Liu and E.J. LaVoie, *Bioorg. Med. Chem. Lett.*, 2002, **12**, 3333–3336.

131. D. Makhey, D. Li, B. Zhao, S.P. Sim, T.K. Li, A. Liu, L.F. Liu and E.J. LaVoie, *Bioorg. Med. Chem.*, 2003, **11**, 1809–1820.

132. D. Li, B. Zhao, S.P. Sim, T.K. Li, A. Liu, L.F. Liu and E.J. LaVoie, *Bioorg. Med. Chem.*, 2003, **11**, 521–528.
133. Y. Yu, S.K. Singh, A. Liu, T.K. Li, L.F. Liu and E.J. LaVoie, *Bioorg. Med. Chem.*, 2003, **11**, 1475–1491.
134. D. Li, B. Zhao, S.P. Sim, T.K. Li, A. Liu, L.F. Liu and E.J. LaVoie, *Bioorg. Med. Chem.*, 2003, **11**, 3795–3805.
135. A.L. Ruchelman, S.K. Singh, A. Ray, X.H. Wu, J.M. Yang, T.K. Li, A. Liu, L.F. Liu and E.J. LaVoie, *Bioorg. Med. Chem.*, 2003, **11**, 2061–2073.
136. A.L. Ruchelman, P.J. Houghton, N. Zhou, A. Liu, L.F. Liu and E.J. LaVoie, *J. Med. Chem.*, 2005, **48**, 792–804.
137. T.K. Li, P.J. Houghton, S.D. Desai, P. Daroui, A.A. Liu, E.S. Hars, A.L. Ruchelman, E.J. LaVoie and L.F. Liu, *Cancer Res.*, 2003, **63**, 8400–8407.
138. J.M. Kokoshka, T.L. Capson, J.A. Holden, C.M. Ireland and L.R. Barrows, *Anti-Cancer Drugs*, 1996, **7**, 758–765.
139. E.M. Antunes, D.R. Beukes, M. Kelly, T. Samaai, L.R. Barrows, K.M. Marshall, C. Sincich and M.T. Davies-Coleman, *J. Nat. Prod.*, 2004, **67**, 1268–1276.
140. R. Barret and N. Roue, *Tetrahedron Lett.*, 1999, **40**, 3889–3890.
141. V. Beneteau, A. Pierre, B. Pfeiffer, P. Renard and T. Besson, *Bioorg. Med. Chem. Lett.*, 2000, **10**, 2231–2234.
142. L. Legentil, B. Lesur and E. Delfourne, *Bioorg. Med. Chem. Lett.*, 2006, **16**, 427–429.
143. C. Bailly, *Curr. Med. Chem. – Anti-Cancer Agents*, 2004, **4**, 363–378.
144. M. Facompré, C. Tardy, C. Bal-Mahieu, P. Colson, C. Perez, I. Manzanares, C. Cuevas and C. Bailly, *Cancer Res.*, 2003, **63**, 7392–7399.
145. C. Tardy, M. Facompre, W. Laine, B. Baldeyrou, D. Garcia-Gravalos, A. Francesch, C. Mateo, A. Pastor, J.A. Jimenez, I. Manzanares, C. Cuevas and C. Bailly, *Bioorg. Med. Chem.*, 2004, **12**, 1697–1712.
146. E. Marco, W. Laine, C. Tardy, A. Lansiaux, M. Iwao, F. Ishibashi, C. Bailly and F. Gago, *J. Med. Chem.*, 2005, **48**, 3796–3807.
147. J. Kluza, M.A. Gallego, A. Loyens, J.C. Beauvillain, J.M. Fernandez Sousa-Faro, C. Cuevas, P. Marchetti and C. Bailly, *Cancer Res.*, 2006, **66**, 3177–3187.
148. J. Ma, S.H. R. Marshall, X. Wu and S.M. Hecht, *Bioorg. Med. Chem. Lett.*, 2005, **15**, 813–816.
149. A. Constantinou, R. Mehta, C. Runyan, K. Rao, A. Vaughan and R. Moon, *J. Nat. Prod.*, 1995, **58**, 217–225.
150. F. Boege, T. Straub, A. Kehr, C. Boesenberg, K. Christiansen, A. Andersen, F. Jakob and J. Kohrle, *J Biol. Chem.*, 1996, **271**, 2262–2270.
151. S. Kawaii, Y. Tomono, E. Katase, K. Ogawa and M. Yano, *Biosci. Biotechnol. Biochem.*, 1999, **63**, 896–899.
152. F. Casagrande and J.M. Darbon, *Biochem. Pharmacol.*, 2001, **61**, 1205–1215.
153. A.R. Chowdhury, S. Sharma, S. Mandal, A. Goswami, S. Mukhopadhyay and H.K. Majumder, *Biochem. J.*, 2002, **366**, 653–661.
154. N. Yamashita and S. Kawanishi, *Free Radical Res.*, 2000, **33**, 623–633.

155. M.R. Webb and S.E. Ebeler, *Biochem. J.*, 2004, **384**, 527–541.
156. A. Itoh, T. Kumashiro, M. Yamaguchi, N. Nagakura, Y. Mizushina, T. Nishi and T. Tanahashi, *J. Nat. Prod.*, 2005, **68**, 848–852.
157. M.C. Oliveira, M.G. de Carvalho, N.F. Grynberg and P.S. Brioso, *Planta Med.*, 2005, **71**, 561–563.
158. F. Rossi, E. Labourier, T. Forné, G. Divita, J. Derancourt, J.F. Riou, E. Antoine, G. Cathala, C. Brunel and J. Tazi, *Nature*, 1996, **381**, 80–82.
159. J. Tazi, N. Bakkour, J. Soret, L. Zekri, B. Hazra, W. Laine, B. Baldeyrou, A. Lansiaux and C. Bailly, *Mol. Pharmacol.*, 2005, **67**, 1186–1194.

CHAPTER 4

Slow DNA Binding

L. MARCUS WILHELMSSON, PER LINCOLN AND
BENGT NORDÉN

Physical Chemistry Section, Department of Chemistry and Bioscience, Chalmers University of Technology, Kemivägen 10, 41296 Gothenburg, Sweden

4.1 Introduction – Kinetics *vs.* Thermodynamics of DNA Binding

In vitro DNA–drug interaction studies can give information that is important both for therapeutic and diagnostic purposes, such as DNA sequence selectivity and binding geometry, *i.e.* the thermodynamic equilibrium state of the reaction. However, also the path along which the drug reaches this state, the mechanism of binding, and the binding kinetics, could be equally important as the rates of association and dissociation might be crucial for the application of the drug. Furthermore, investigating the kinetics of the binding and dissociation reactions could substantially increase the understanding about the structure and dynamics of DNA itself and can, for cases where the interaction kinetics is slow, reveal intermediate binding sites that might be significant in the action of a drug.

In the thermodynamic equilibrium state of binding (Equation (1)), the number of molecules that are bound to DNA and the number that are free in solution are on average constant (Figure 1A), which means that the average number of molecules that dissociate (Equation (2)) from DNA at any time is the same as the average number that associate (Equation (3)).

$$K = \frac{[\text{DNA} - \text{L}]}{[\text{DNA} - \text{site}][\text{L}]} \tag{1}$$

$$\frac{d[\text{DNA} - \text{L}]}{dt} = -k_{d}[\text{DNA} - \text{L}] \tag{2}$$

$$\frac{d[\text{DNA} - \text{L}]}{dt} = k_{a}[\text{L}][\text{DNA} - \text{site}] \tag{3}$$

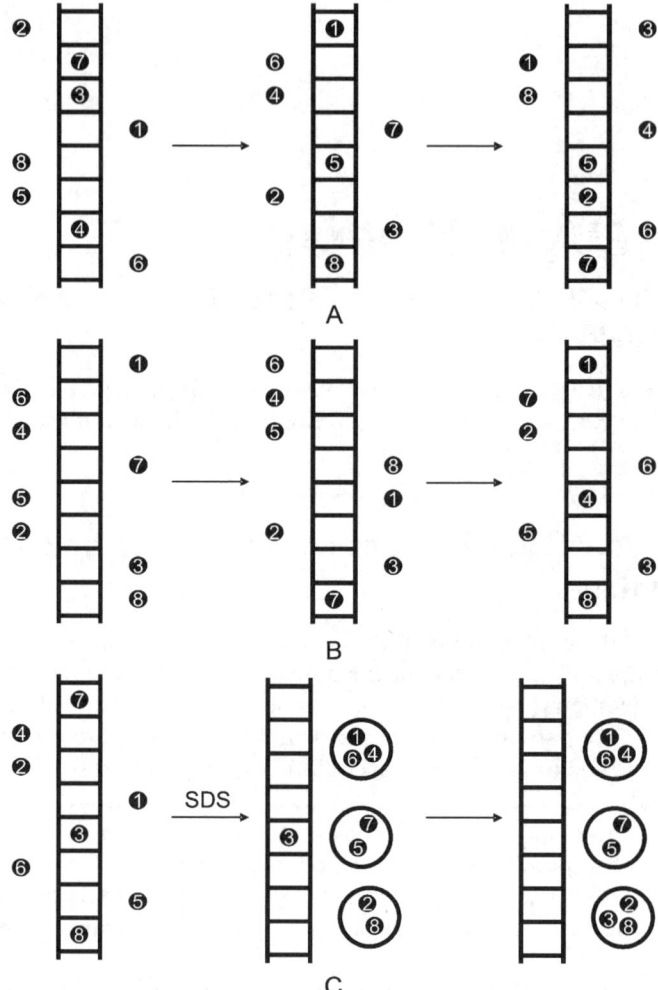

Figure 1   *Schematic representation of DNA-binding processes. (A) Thermodynamic equilibrium state of binding. Filled circles with numbers symbolize a DNA-binding molecule. Pictures from left to right are different snapshots after the DNA-binding molecule has reached its equilibrium distribution between free and bound drug. (B) Association process. Pictures from left to right are consecutive snapshots during the DNA-association process towards the thermodynamic equilibrium. (C) Dissociation process. Pictures from left to right are consecutive snapshots during the DNA-dissociation process towards the thermodynamic equilibrium. SDS (large open circles) forming micelles that sequester free drugs is added before middle snapshot*

By contrast, it should be noted that the behaviour of a single molecule may be very dynamic. The rate at which molecules associate with DNA to reach their equilibrium state of binding may vary enormously ($k_a \sim 10^3–10^4\ \mathrm{M}^{-1}\ \mathrm{s}^{-1}$ for actinomycin D[1,2] and $\sim 10^8\ \mathrm{M}^{-1}\ \mathrm{s}^{-1}$ for several groove binders[3-5] at ambient temperature and approximately 0.2 M Na^+; approximately ambient temperature and 0.1–0.2 M Na^+ are the conditions implied throughout the text unless stated otherwise) as does the rate of dissociation ($k_d \sim 10^{-3}$ s^{-1} for actinomycin $D^{6-8,1-2}$ and $\sim 10^2$ s^{-1} for the intercalator proflavine[9] under similar conditions). The equilibrium distribution of molecules between DNA and solution is governed by the relationship between the rate of association and dissociation as shown in Equation (4), when the rate of Equation (2) equals that of Equation (3)

$$K = \frac{k_a}{k_d} = \frac{[DNA - L]}{[DNA - site][L]} \qquad (4)$$

In the literature it has been suggested that not only a high binding constant but also, more importantly, a slow rate of dissociation is crucial for a DNA-binding drug to be able to block binding or movement of a protein along DNA and, as a consequence, interfere with the mechanism of, *e.g.* DNA replication or transcription.[8]

On the way towards thermodynamic equilibrium the rate at which the molecules associate with DNA differs from the rate at which they dissociate (Figures 1B and C) so that in this case the state is really dynamic and shows relaxation kinetics which may be studied, *e.g.* with the help of stopped-flow spectrophotometric techniques, depending on the rate of the reaction. Association with DNA *in vitro* is normally studied by mixing a solution of drug with a solution of DNA and observing a change in, *e.g.* drug fluorescence or absorption intensity with time. Using the same changes in intensity it may also be possible to study the dissociation from DNA by detergent-sequestering, dilution, or dialysis methods.

This chapter will focus on small molecules that bind relatively slowly to DNA. Examples of some established small DNA binders with slow kinetics, actinomycin D, and nogalamycin, will be briefly considered, but the main focus will be on dimeric ruthenium complexes that exhibit extremely slow binding kinetics as well as several other interesting properties. As we shall argue, an extremely slow dissociation rate may be useful for both therapeutic and diagnostic purposes. In addition, slow DNA binders may be used as model systems to provide a "slow motion" picture of how fast DNA binders may redistribute between solution and DNA as well as between different binding sites on DNA. Tables of characteristic rates of association and dissociation for several of the drugs mentioned in the text (Tables 1 and 2). To begin with we will briefly describe the main types of DNA binding modes, their respective time-scales of interaction and mechanisms of binding.

4.2 Different DNA-Binding Modes – Different DNA-Binding Kinetics

A molecule may interact with DNA either covalently or non-covalently. Although several drugs that bind covalently to DNA are also very interesting,

Table 1 *Rates of Association of Drugs with DNA. Unless stated otherwise, conditions are around physiological salt concentration and pH and at ambient temperature*

Drug	Rates of Association k_a		DNA	References
DAPI	$6.3 \cdot 10^7$ M^{-1} s^{-1}		AT	3
Distamycin	$7 \cdot 10^7$ M^{-1} s^{-1}	Drug:site 1:1	AT-region	4
	$7 \cdot 10^7$ M^{-1} s^{-1} and 5 s^{-1}	Drug:site 2:1		
Hairpin polyamide	$7 \cdot 10^7$ M^{-1} s^{-1}	Calculated from K	5'-AGTACT	5
Daunomycin	$\approx 10^7$ M^{-1} s^{-1}		Calf thymus	47
Daunomycin	$3 \cdot 10^6$ M^{-1} s^{-1}, 92 s^{-1}, and 4 s^{-1}		Calf thymus	48
Adriamycin	$\approx 10^7$ M^{-1} s^{-1}		Calf thymus	47
Ethidium	$\approx 10^7$ M^{-1} s^{-1}		AT and GC	38
Actinomycin D	$\approx 2 \cdot 10^4$ M^{-1} s^{-1}	Various 12-mers	5'-TGCA	1
Actinomycin D	$1 \cdot 10^3$ M^{-1} s^{-1}	10-mer GC center	5'-GGCC	2
Actinomycin D	$1 \cdot 10^4$ M^{-1} s^{-1}	10-mer GC center	5'-TGCA	2
Actinomycin D	$1 \cdot 10^4$ M^{-1} s^{-1}	10-mer GC center	5'-AGCT	2
Actinomycin D	$1 \cdot 10^4$ M^{-1} s^{-1}	10-mer GC center	5'-CGCG	2
Nogalamycin	$\approx 10^5$ M^{-1} s^{-1} (fastest)	Triphasic	Calf thymus	53
Nogalamycin	$\approx 10^5$ M^{-1} s^{-1} (fastest)	Biphasic	AT	53
Nogalamycin	$2.5 \cdot 10^3$ M^{-1} s^{-1}		GC	53
ΔΔ-[μ-c4(cpdppz)$_2$(phen)$_4$Ru$_2$]$^{4+}$	29 s^{-1} and 4.4 s^{-1} (two fastest)	Multiphasic 45 μM base	AT	102
ΔΔ-[μ-(11,11'-bidppz)(phen)$_4$Ru$_2$]$^{4+}$	$1.2 \cdot 10^{-3}$ s^{-1} and $2 \cdot 10^{-4}$ s^{-1} [a] (threading)	160 μM base	Calf thymus	104
ΔΔ-[μ-(11,11'-bidppz)(phen)$_4$Ru$_2$]$^{4+}$	10^{-4} s^{-1} [b] (shuffling)	80 μM base	AT	107
ΛΛ-[μ-(11,11'-bidppz)(phen)$_4$Ru$_2$]$^{4+}$	$3.4 \cdot 10^{-3}$ s^{-1} (threading)	120 μM base	AT	106
ΛΛ-[μ-(11,11'-bidppz)(bipy)$_4$Ru$_2$]$^{4+}$	$1.2 \cdot 10^{-2}$ s^{-1} (1st order; threading)	120 μM base	AT	106
	$5.0 \cdot 10^{-2}$ s^{-1} (2nd order; threading)			

[a] Performed at 50°C.
[b] Performed at 50°C, 2 mM Mg^{2+} and 5 mM K$^+$.

Table 2 *Rates of Dissociation of Drugs from DNA. Unless stated otherwise, conditions are around physiological salt concentration and pH and at ambient temperature*

Drug	Rates of Dissociation k_d s^{-1}		DNA	Method – References
DAPI	8.5		AT	SDS[3]
Distamycin	3	Drug:site 1:1	AT-region	SDS[4]
	0.1	Drug:site 2:1		
Hairpin polyamide	$2 \cdot 10^{-3}$		5'-AGTACT	SDS[5]
Daunomycin	10		Calf thymus	T-jump relaxation[47]
Daunomycin	170, 10, and 1		Calf thymus	SDS[48]
Adriamycin	3.5		Calf thymus	T-jump relaxation[47]
Ethidium	$\approx 10^{1}$		AT and GC	P-jump relaxation[38]
Actinomycin D	$8 \cdot 10^{-2}$, $2 \cdot 10^{-2}$, and $2 \cdot 10^{-3}$		Calf thymus	SDS[8]
Actinomycin D	Between $2 \cdot 10^{-2}$ and $5 \cdot 10^{-4}$ a		Different	Footprinting[7]
Actinomycin D	$\approx 5 \cdot 10^{-4}$	Various 5'-TGCA	5'-TGCA	SDS[1]
Actinomycin D	$6.7 \cdot 10^{-4b}$		lac DNA	Transcription assay[6]
Actinomycin D	$3.3 \cdot 10^{-3b}$		Calf thymus	SDS[6]
Actinomycin D	$1.6 \cdot 10^{-3}$	10-mer GC center	5'-TGCA	SDS[2]
Actinomycin D	$3.4 \cdot 10^{-3}$	10-mer GC center	5'-AGCT	SDS[2]
Actinomycin D	$2.3 \cdot 10^{-3}$	10-mer GC center	5'-CGCG	SDS[2]
Nogalamycin	$3.3 \cdot 10^{-3}$, $9.8 \cdot 10^{-4}$, and $2.3 \cdot 10^{-4c}$		Calf thymus	SDS[54]
Nogalamycin	$4.0 \cdot 10^{-2c}$		AT	SDS[54]
Nogalamycin	$2.7 \cdot 10^{-4c}$		GC	SDS[54]
ΔΔ-[μ-c4(cpdppz)$_2$(phen)$_4$Ru$_2$]$^{4+}$	$1 \cdot 10^{-3d}$		AT	Calf thymus DNA[67]
ΔΔ-[μ-c4(cpdppz)$_2$(phen)$_4$Ru$_2$]$^{4+}$	$7.3 \cdot 10^{-3}$		AT	SDS[101]
ΛΛ-[μ-c4(cpdppz)$_2$(phen)$_4$Ru$_2$]$^{4+}$	$4.2 \cdot 10^{-2}$		AT	SDS[101]
Meso-[μ-c4(cpdppz)$_2$(phen)$_4$Ru$_2$]$^{4+}$	$1 \cdot 10^{-2}$		AT	SDS[101]
ΔΔ-[μ-c4(cpdppz)$_2$(phen)$_4$Ru$_2$]$^{4+}$	$4.5 \cdot 10^{-3}$	95% of process	GC	SDS[102]
Meso-[μ-c4(cpdppz)$_2$(phen)$_4$Ru$_2$]$^{4+}$	$6.4 \cdot 10^{-3}$ and $1.5 \cdot 10^{-3}$		Calf thymus	SDS[101]
ΔΔ-[μ-(11,11'-bidppz)(phen)$_4$Ru$_2$]$^{4+}$	$1.9 \cdot 10^{-3}$ s^{-1} and $8 \cdot 10^{-4}$ s^{-1d}		Calf thymus	SDS[104]
ΔΔ-[μ-(11,11'-bidppz)(phen)$_4$Ru$_2$]$^{4+}$	$2.4 \cdot 10^{-5}$ s^{-1d}		Calf thymus	AT; (Nordell, P., unpublished)

a Salt concentration 10–20 mM Na^{+}.
b Performed at 37°C.
c Performed at 40°C.
d Performed at 50°C.

we shall confine ourselves here to non-covalent, reversible modes of binding. Such types of binding in the vast majority of strong-binding cases involve positively charged molecules and may for small molecules be divided into three different classes of binding: external electrostatic binding (Figure 2A), groove binding (Figure 2B), and intercalation (Figure 2C). A special case of intercalation is threading intercalation (Figure 2D), which will be treated separately since it is especially interesting in the context of slow DNA binding kinetics.

4.2.1 External Electrostatic Binding

Small molecules or ions that bind externally are usually non-specific both with respect to nucleotide sequence and location, their distribution being mainly governed by electrostatic interaction with the negatively charged phosphates of the DNA backbone. Ions like Na^+ (Figure 2A), K^+, and Li^+ all bind in this non-specific way, but with soft transition metal ions there are also examples that can bind specifically to the actual nucleobase (*e.g.* Ag^+ and Hg^{2+}).[9] Examples of small molecules that in certain studies have been suggested to bind externally to DNA are some of cationic polyamines.[10–12] External binding, being electrostatic in nature and generally requiring insignificant changes of

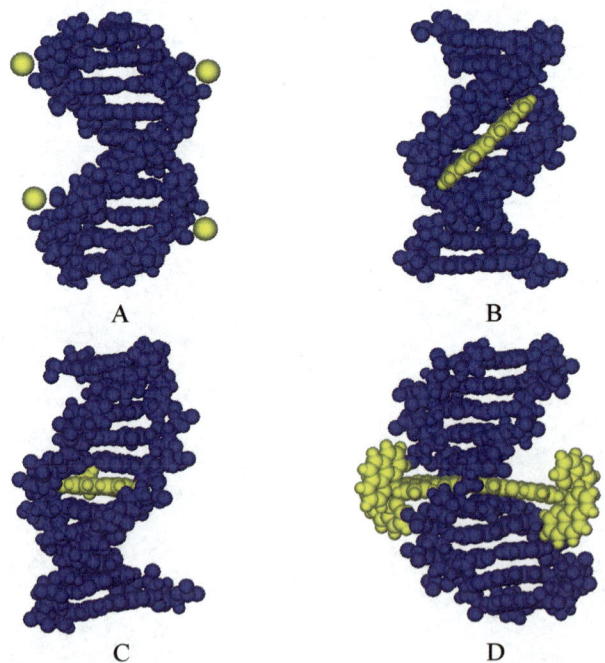

Figure 2 *Schematic of DNA binding modes. (A) External electrostatic binding of positive metal ions. (B) Groove binding of a minor groove binder such as DAPI. (C) Intercalation of ethidium. (D) Threading intercalation of a semirigid ruthenium dimer, $\Delta\Delta$-[μ-(11,11'-bidppz)(phen)$_4$Ru$_2$]$^{4+}$*

DNA structure, is characterized by fast binding kinetics approaching the diffusion limit ($k_a = 10^8 - 10^9$ M^{-1} s^{-1}).

4.2.2 Groove Binding

Groove binding agents associate with DNA in either the major or the minor groove. While a majority of DNA-binding proteins penetrate into the major groove most of the smaller molecules bind into the relatively narrow and deep minor groove (Figure 2B), which affords only a limited number of hydrogen bond donor or acceptors. Despite a smaller number of potential hydrogen bond interactions in the minor groove compared to its major counterpart, binding in the minor groove may in some cases be sequence selective mainly as a result of hydrogen bonding but also due to additional van der Waals and hydrophobic contacts as well as more specifically positioned electrostatic interactions. Small minor groove-binders like DAPI[13-16,3] (Figure 3A), Hoechst 33258[17,18] (Figure 3B), and distamycin[19-25] (Figure 3C) all preferentially bind to AT-rich regions of DNA. From the observation of a sign change in flow linear dichroism (LD) for pure GC contexts, where the exocyclic amino group prevents binding in the minor groove, however, it was concluded that intercalation may also occur with one of these ligands.[26,27] The groove binders normally display binding kinetics that, like external binding, approaches the diffusion limit ($k_a \approx 10^8$ M^{-1} s^{-1}), while their rate of dissociation can vary considerably ($k_d = 10^{-1} - 10^3$ s^{-1}). In spite of their size the somewhat larger polyamides,[28-33] which also bind in the minor groove, have binding kinetics in the same range as the smaller minor groove binders.[5] However, the rates of dissociation of hairpin polyamides are generally much slower ($k_d \approx 10^{-3}$ s^{-1})[5] than those of an average minor groove binder, thus leading to stronger binding.

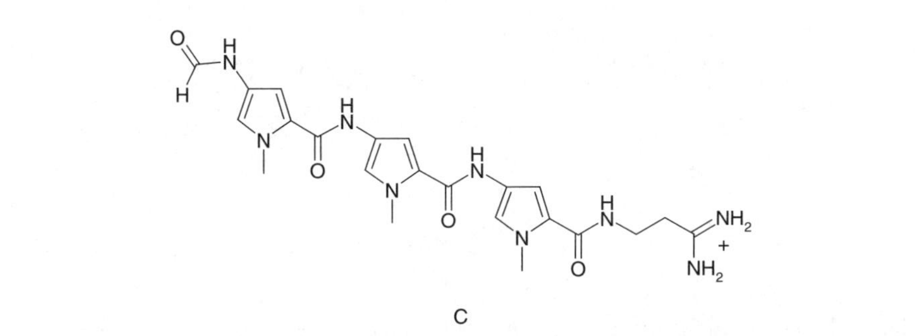

Figure 3 *DNA groove-binders DAPI (A), Hoechst 33258 (B), and distamycin (C)*

4.2.3   Intercalation

Intercalators associate with DNA by inserting themselves between adjacent base pairs (Figure 2C). In so doing they cause elongation of DNA by about the height of one DNA base (*i.e.* 3.4 Å) and a corresponding local unwinding of the helix. These local distortions are much more pronounced than occur with groove binders and can extend over several base pairs surrounding the actual site of intercalation. In addition to an attractive electrostatic contribution to the free energy of binding by normally cationic, classical intercalators like ethidium[34–38] (Figure 4A) a substantial part of their binding energy derives from hydrophobic interactions with the surrounding base pairs. In consequence, intercalative binding modes have a lower probability of being sequence specific. As a result of the requirement for larger and more complex distortions in the DNA structure, and also need for the intercalator to diffuse along DNA until it finds its (optimal) binding site, the kinetics of intercalation is generally slower than that for groove-binding. The associated helical distortions often make the mechanism of binding, and therefore also the kinetics, very complex. Consequently, intercalators like daunomycin[39–42] (Figure 4B) may follow bi- or multiphase association rate laws. Average overall rates of association for daunomycin and the other anthracycline antibiotics Adriamycin[43–46] (Figure 4C) and iremycin[41] (Figure 4D) are generally in the same order of magnitude as that of ethidium (k_a(anthracyclines) $= 10^6 – 10^7$ M^{-1} s^{-1} (refs. 47,48) and k_a(ethidium) $= 10^7$ M^{-1} s^{-1} (ref. 38)). Dissociation of these intercalators is

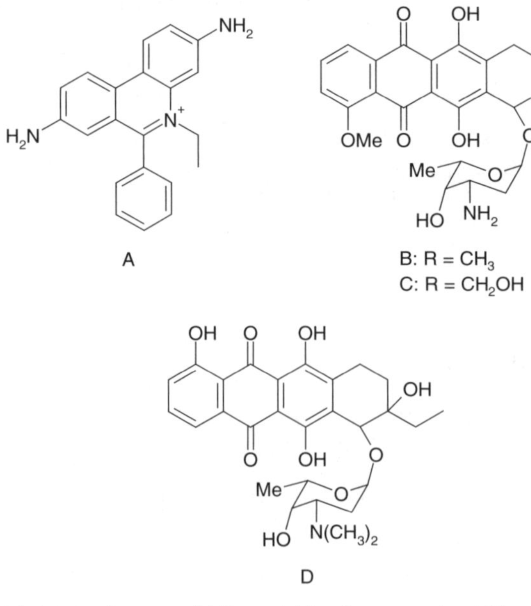

B: R = CH$_3$
C: R = CH$_2$OH

Figure 4   *DNA intercalators ethidium (A), daunomycin (B), Adriamycin (C), and iremycin (D)*

nucleic acids must also take into account the complexity associated with the energetic stability of the DNA targets.

The focus of this manuscript is to describe the interactions of a well-known sequence-specific DNA-binding agent, actinomycin D, and the evolution of our insight into the nature of the physico–chemical properties that govern the binding of this potent anticancer agent to a variety of nucleic acid structures and sequences. The researches that are described over a span of 60 years of intense biochemical and biophysical studies, focus on characterizing the nature of the actinomycin D-DNA complex. However, the actinomycin D story is not complete and indeed continues to provide exciting and novel insights into the unique properties associated with sequence-selective interactions of ligands with nucleic acids.

6.2 Introduction

6.2.1 Historical Perspectives

Certain small molecules that bind to DNA in a sequence-specific manner can be thought of as "hybrid" molecules, wherein a portion of the ligand structure is characterized by a planar heterocyclic chromophore that intercalates between adjacent base pairs of the DNA together with additional chemical element(s) that do not intercalate, but instead reside along the outside of DNA within the major or minor groove. The chromophore generally contributes to the overall stability of the ligand–DNA complex through stacking of the planar ring system with those of the DNA base pairs. However, quantitative footprinting studies by Waring, Dervan, Chaires, and others have provided pivotal evidence indicating that the portion of the compound that resides outside the intercalation site and within the major or minor groove of the DNA is the most likely determinant for sequence recognition.[5] Hence, the cyclic pentapeptide side chains of actinomycin D are predicted to play critical roles in both sequence-recognition as well as influencing the overall thermodynamic mechanism(s) driving the ligand–DNA complex formation.

Actinomycin has the distinction of being the first compound isolated as a soil-borne antibiotic to be used in the treatment of cancer.[6] The actinomycin family of "natural products" was first discovered in the 1940s by the Ukrainian born American scientist, Selman Waksman. After immigrating to the United States in 1910, Waksman studied soil-borne bacteria at Rutgers University and in doing so focused on the *Actinomyces* family. He received his Ph.D. degree from the University of California at Berkeley in 1918 and returned to Rutgers University to continue his work on bacteria and moulds in soils. In 1940, Waksman and H. Boyd Woodruff devised the growth inhibition zone method for screening natural products from soil microorganisms, particularly from the moulds *Streptomyces antibioticus* and *Streptomyces chrysomallus*. With their assay, they were able to distinguish compounds that exhibited antibacterial properties.[6] Using this method, Waksman identified the first true antibiotic isolated from *Streptomyces*, which he aptly named actinomycin C. Actinomycin C was found to be composed of several components including actinomycin C_1 (or actinomycin D), C_2, C_{2a}, and

approximately as fast ($k_d \approx 1$–$100 \, s^{-1}$)[38,47,48] as for normal groove binders like distamycin[4] and DAPI.[3]

4.2.4 Threading Intercalation

Like their classical counterparts, threading intercalators insert a part of the molecule between base pairs (Figure 2D) and interact attractively with DNA through electrostatic binding to the phosphates in the backbone as well as through π-stacking (hydrophobic interactions) with the surrounding base pairs. However, in addition, threading intercalators have bulky or polar substituents tethered to the part of the molecule that gets intercalated, which requires this substituent to be threaded through the DNA-base stack so that the resulting complex has one bulky substituent in each groove (Figure 2D). The substituents may or may not contribute to the binding of the threading molecule by hydrophobic interactions and hydrogen bonding to the floor of the grooves. During the process of binding, as the molecule contrives to insert itself between the base pairs, the substituents constitute major obstacles explaining the slow DNA-binding kinetics. In the case of naphthalene diimides[10] (Figure 5A) the obstacle to reaching the thermodynamic equilibrium geometry derives from the positively charged side chains whereas for nogalamycin[49–52] (Figure 5B) it comes from the two bulky sugar moieties attached to the aglycone anthracycline chromophore. Due to the even more extensive distortions of DNA structure compared to normal intercalators, and often sequential obstacles to the threading intercalator finding its energetically optimal site of intercalation, the kinetics of association of a threading intercalator is generally multiexponential.[53] The rate-limitations of reaching, and also leaving, the most stable

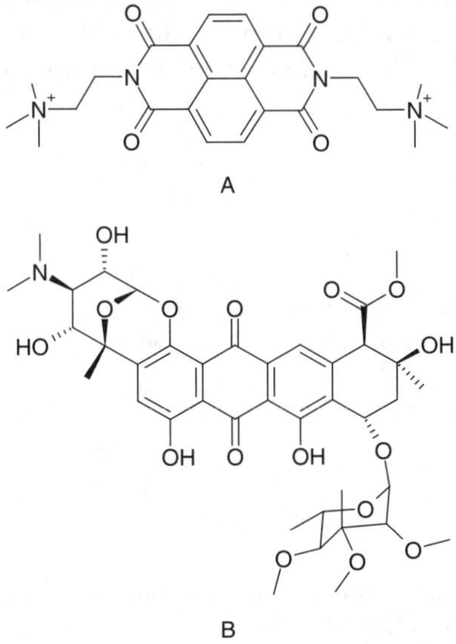

A

B

Figure 5 *DNA threading intercalators naphthalene diimide (A) and nogalamycin (B)*

binding geometry for a threading intercalator can be exemplified by the case of nogalamycin: the average overall rate of association of nogalamycin to DNA is very slow compared to that of the unhindered anthracycline intercalator daunomycin (k_a(nogalamycin) $\approx$ 10^3–10^5 M^{-1} s^{-1} (ref. 53)k_a(daunomycin) = 10^6–10^7 M^{-1} s^{-1})[47,48] as is the dissociation from DNA (k_d(nogalamycin) $\approx$ 10^{-3} s^{-1} at 0.01–0.5 M Na$^+$ and 40°C,[54] k_d(daunomycin) $\approx$ 10 s^{-1}).[47,48]

4.3   Common Slow DNA Binders

The focus throughout the remainder of this chapter will be on slow DNA-binding processes. The rates are slowed down either as a result of a requirement for large conformational change(s) in the DNA and/or of the drug molecule itself, or because of slow redistribution of the drug between several competing strong DNA binding sites, or both. Examples of these effects have been reported for several DNA–drug interactions in the literature and will be illustrated below, using the DNA binding of actinomycin D and nogalamycin as representative cases.

4.3.1   Actinomycin D

The antitumour antibiotic actinomycin D is composed of two identical cyclic pentadepsipeptides connected to a phenoxazone chromophore (Figure 6). It binds to DNA with a relatively high binding constant ($\sim 10^7$ M^{-1})[1] by intercalating its aromatic moiety between the base pairs[8] and extending its two pentadepsipeptides in opposite directions from the intercalation pocket along the valley of the minor groove. NMR,[55,56] X-ray crystallography,[57–59] and footprinting[23,24] studies have shown that actinomycin D is selective for intercalation at the dinucleotide GpC. The selectivity arises from specific interactions between the N-3 and exocyclic C-2 amino groups of guanine and

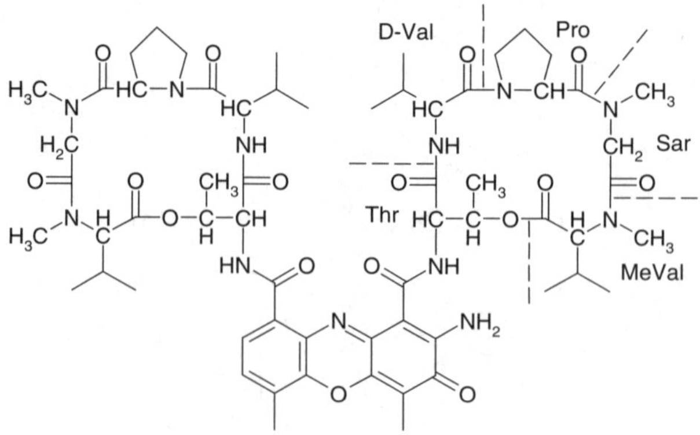

Figure 6   *The antitumour antibiotic actinomycin D. Names of the amino acids (Thr=threonine, MeVal=N-methylvaline, Sar=sarcosine, Pro=proline, and D-Val=D-valine) in the two cyclic pentadepsipeptides are given beside the right hand ring*

the threonine amide NH and carbonyl groups, respectively.[57–59] Besides a preference for the GpC dinucleotide the binding affinity is also highly dependent upon which other bases are flanking the intercalation site. The preferred sequence for actinomycin D binding is 5′-TGCA followed by CGCG, AGCT, and GGCC.[60] The differences in binding affinity among these sequences can be correlated to a large variation in the dissociation kinetics, where TGCA manifests a rate that is about four times slower than those from CGCG and AGCT and approximately two orders of magnitude slower than from GGCC.[2,60,61]

Compared to the DNA binding kinetics of normal intercalators mentioned above, both the association and dissociation rates for actinomycin D are very slow. The reported association rate constants are on the order of $10^{3-4} \, M^{-1} \, s^{-1}$ (refs. 1, 2) and the dissociation rate constants measured under the same conditions are roughly $10^{-3} \, s^{-1}$ (refs. 1, 2, 6–8,). Efforts to find the origins of the slow and multiexponential association kinetics have inspired many studies on the subject and various explanations have been proposed. In one study it was suggested that as many as five exponentials, of which three are unimolecular, are needed to describe the association and that they originate from slow intramolecular conversions of the two cyclic peptides and/or adjustments of the minor groove of the DNA to optimize the interaction area between the drug and the DNA.[8,62] On the other hand, NMR studies indicate that no significant conformational alterations of the cyclic peptides take place upon binding.[56] Another explanation that has been advanced suggests that the antibiotic molecule "shuffles" between low and high binding constant sites on the DNA.[23,63,64] However, it has been argued that this interpretation is not compatible with the three unimolecular exponentials in the association kinetics.[65] Instead, an attempt to explain the unimolecular processes has been suggested on the basis that in order to bind to certain sequences of DNA, actinomycin D has to wait for the thermal occurrence of an "improbable" distortion of the DNA.[2] This improbable event can be an inherent property of the DNA but can also result from other actinomycin D molecules binding to DNA sites close to the site of binding of the next drug molecule.[2]

Not unexpectedly the dissociation kinetics of actinomycin D from a heterogeneous DNA lattice also has to be described by more than a single exponential, where the different exponentials may reflect dissociation from binding sites of different strengths.[7,8] Lowering the drug concentration should result in fewer weak binding sites being occupied such that the influence of the fast dissociations decreases and instead the dissociation profile is dominated by a single slow component.[1,66] This result is in agreement with the single exponential processes observed when studying sodium dodecyl sulfate (SDS)-sequestered dissociation from sequences like poly(dG-dC)$_2$ and short oligonucleotides.[2,60,61] However, depending on which bases are surrounding the GpC intercalation site the dissociation rate constants can vary from $t_{1/2} = 3000 \, s$ to 100 times faster.[60] The slow dissociation process can be ascribed, like the association reaction, to the sluggishness of the intramolecular conversion of the two cyclic peptides and/or adjustments of the minor groove of the DNA as well as the specific constructive interactions between the ligand and the base functional groups exposed in the minor groove.[8,60] The rate of dissociation of actinomycin D has also been measured in an *in vitro* transcription assay that detects antibiotic-induced blockage to the elongation of an RNA transcript around an AGCT site, and been found to be slower than the average dissociation rate from the whole sequence measured by the SDS sequestration method.[6] This was not considered surprising since the SDS method gives an average dissociation rate constant whereas the transcriptional assay gives the actual dissociation rate from a certain site. Furthermore, recent results have shown that using the detergent sequestration technique can cause overestimation of the dissociation rate constants for slowly dissociating, positively charged DNA-binding molecules.[67,68] Also, preliminary results from a similar study with DNA and actinomycin D show that the dissociation rate constants are affected by the concentration of SDS and sodium decyl sulfate (Wilhelmsson, L.M., unpublished results).

4.3.2   Nogalamycin

The structurally interesting anthracycline antibiotic nogalamycin is built up of two non-aromatic ring systems, an uncharged nogalose sugar group, and a charged (at physiological pH) bicyclic aminoglucose moiety, connected to each end of a planar anthracycline chromophore (Figure 5B).[69] The various parts of this drug appear to have different functions to make binding robust. The "dumbbell" shaped molecule intercalates its aglycone group in the DNA base stack placing one bulky sugar moiety in each groove of the DNA double helix.[53,70,71] This threading intercalation is believed to explain how nogalamycin inhibits DNA-directed RNA synthesis *in vivo* as well as *in vitro* and interferes with topoisomerase activity.[72–75] However, unlike the other anthracyclines daunomycin and Adriamycin, higher cytotoxicity, and cardiotoxicity has limited the clinical use of nogalamycin.

Various techniques including viscometry,[70] NMR,[76,77] and X-ray crystallography[49–52,78–79] have verified that nogalamycin intercalates its anthracycline chromophore resulting in severe buckling of the surrounding base pairs. Furthermore, NMR and X-ray crystallography have provided evidence that the nogalose sugar is sandwiched between the walls of the minor groove and that the bicyclic aminoglucose moiety is positioned in the major groove of the distorted B-DNA double helix.[76–79,49–51] Numerous studies with different DNA sequences suggest that nogalamycin intercalates selectively at the 5' side of a purine and the 3' side of a pyrimidine, *e.g.* CpG and TpG, spanning three base pairs. In structures where nogalamycin binds to CpG sequences[76–80,49–51,53] it has been proposed that the antibiotic forms two hydrogen bonds in the major groove from its two -OH groups in the aminoglucose to the N7 of guanine and the N4 of the base paired C as well as one hydrogen bond in the minor groove from the keto oxygen of the nogalamycin methyl ester to the exocyclic N2 hydrogen of guanine.[76–79,49–51] These specific hydrogen bonds, together with several water-mediated hydrogen bonds and van der Waals interactions between the drug and the DNA, constitute the CpG sequence selective elements.[52]

Perhaps the most intriguing characteristic of the nogalamycin–DNA interaction is the sterically hindered threading of the approximately 1 nm wide bulky sugar moieties through the core of the DNA, which results in the slow association reaction ($k_a \approx 10^3 - 10^5$ M^{-1} s^{-1} at 0.01–0.5 M Na$^+$ and ambient temperature)[53] and the very slow dissociation kinetics ($k_d \approx 10^{-3}$ s^{-1} at 0.01–0.5 M Na$^+$ and 40 °C).[54] Two mechanism of threading have been proposed in which either a transient local melting of base pairs occurs[71] or else an extreme elongation of the base stack, without any disruption of hydrogen bonds, combined with severe buckling of surrounding base pairs to create a considerable opening in the duplex; in this way allowing the nogalamycin to enter its site of intercalation.[78] To furnish an adequate description of the kinetics of association with mixed sequence DNA an expression with three exponentials has to be used.[53] The fastest association process is approximately 10^5 M^{-1} s^{-1} (0.01–0.5 M Na$^+$ and ambient temperature), and the activation energies for the two slower rate constants are both about 90–100 kJ mol^{-1}, which would be consistent with a large conformational change of the DNA like, for example, the coherent opening of more than one base pair.[53] Binding to poly(dA-dT)$_2$ can be described by the sum of two exponentials, the fastest being on the order of 10^5 M^{-1} s^{-1} (0.01–0.5 M Na$^+$ and ambient temperature) whereas association with poly(dG-dC)$_2$ is mono-exponential with a rate constant of $2.5 \cdot 10^3$ M^{-1} s^{-1} (0.01–0.5 M Na$^+$ and ambient temperature).[53] Since AT sequences melt more easily and therefore are expected to experience more frequent transient openings than GC sequences, the results are in line with the first hypothesis of threading mentioned above. It has also been suggested that the three association constants observed for mixed sequence DNA most likely originates from heterogeneous binding of nogalamycin to regions with different stability, and even though it is tempting to suggest that the three exponentials are a result of intercalation into AT-rich (τ_1), GC-rich (τ_3), and mixed regions (τ_2) of the DNA this most certainly is an oversimplification.[53] Another proposal is that the complexity of the association with mixed-sequence DNA may arise from initial binding to easily accessible sites (AT-rich regions) followed by slow redistribution to thermodynamically more favoured sites (*vide infra*).[64]

Compared to other anthracyclines like Adriamycin, daunomycin, and iremycin ($k_d \approx 1$–100 s^{-1}),[47,48] nogalamycin has an extremely slow rate of dissociation from DNA ($k_d \approx 10^{-3}$ s^{-1} at 0.01–0.5 M Na$^+$ and 40 °C)[54]. These slow dissociation kinetics, in common with the association process, may be a result of the need to thread the bulky sugar moieties through the core of the DNA base stack. In synthetic polynucleotides like poly(dA-dT)$_2$ ($k_d = 4.0 \cdot 10^{-2}$ s^{-1}), poly(dG-dC)$_2$ ($k_d = 2.7 \cdot 10^{-4}$ s^{-1}), and poly(dA-dC) $\cdot$ poly(dG-dT) ($k_d = 3.4 \cdot 10^{-4}$ s^{-1}) the dissociation of nogalamycin can be described by a single exponential.[54] The more than two orders of magnitude slower dissociation from poly(dG-dC)$_2$ than from poly(dA-dT)$_2$ again suggests that nogalamycin has to wait longer for the more stable GC base pairs to transiently melt and enable the drug to escape from the DNA. When studying dissociation of nogalamycin from a mixture of poly(dG-dC)$_2$ and poly(dA-dT)$_2$ a single rate constant, close to the one observed for poly(dG-dC)$_2$, was found.[54] This indicates that nogalamycin in its final state of binding has a strong preference for GC compared to AT sequences. For mixed sequence DNA three exponentials ($k_{d1} = 3.3 \cdot 10^{-3}$ s^{-1}, $k_{d2} = 9.8 \cdot 10^{-4}$ s^{-1}, and $k_{d3} = 2.3 \cdot 10^{-4}$ s^{-1}) are needed to secure an adequate fit to the experimental data.[54] At higher drug-binding ratios it was found that the weighting of the two faster processes increases, indicating that these two binding sites are the weaker ones of three different sites. Another evidence for weak and strong sites comes from varying the equilibration time before starting the SDS sequestering.[54] In these experiments different rate constants were observed, which again indicates a redistribution ("shuffling") of the drug from a site that is easy to thread through to a thermodynamically more favourable site. The conclusion about mixed-sequence DNA, with regions of different stabilities mentioned for the association process above, is further supported since the three exponentials needed to describe the dissociation from the mixed DNA cannot be explained only by using the single exponentials of the synthetic polynucleotides.

4.4 Ruthenium Complexes Exhibiting Slow DNA Binding Kinetics

Octahedral ruthenium(II) complexes with aromatic ligands and their DNA binding properties have been the subject of extensive research during the last couple of decades.[81–93] The positively charged ruthenium centre and the planar aromatic hydrophobic ligands are well suited for interacting with the negatively charged backbone and the hydrophobic core of DNA, respectively. Furthermore, the inherent chirality of the complexes is of interest as the DNA duplex is chiral in itself. The rich photophysical repertoire of the complexes can be exploited in many ways to probe the binding to DNA; the techniques frequently used for this purpose are absorption, emission, circular dichroism (CD), and linear dichroism (LD) measurements.

Early experiments on the binding of ruthenium(II) complexes to DNA focused on [Ru(phen)$_3$]$^{2+}$ (phen = 1,10-phenanthroline, Figure 7A),[94–98] and before that the slowly inverting ferroin complex [Fe(phen)$_3$]$^{2+}$, which was found to show enantiopreferential binding as well as orientation with respect to the DNA.[99] Both groove binding and intercalation were suggested as the DNA binding mode(s). From flow-orientation studies it was inferred that both enantiomers of [Ru(phen)$_3$]$^{2+}$ bind by partly intercalating one phenanthroline edge into DNA, possibly by "quasi-intercalation" requiring only indentation of a nucleobase without separation of base pairs.[100] However, when one of the phen-ligands was enlarged to the larger dppz (dipyrido-[3,2-a:2',3'-c]-phenazine) this moiety was found to become deeply intercalated between separated base pairs, and the DNA locally unwound. The intercalation of the dppz moiety is associated with an astounding change in photophysical properties first observed for [Ru(bipy)$_2$dppz]$^{2+}$ (bipy = 2,2-bipyridine) and later also for

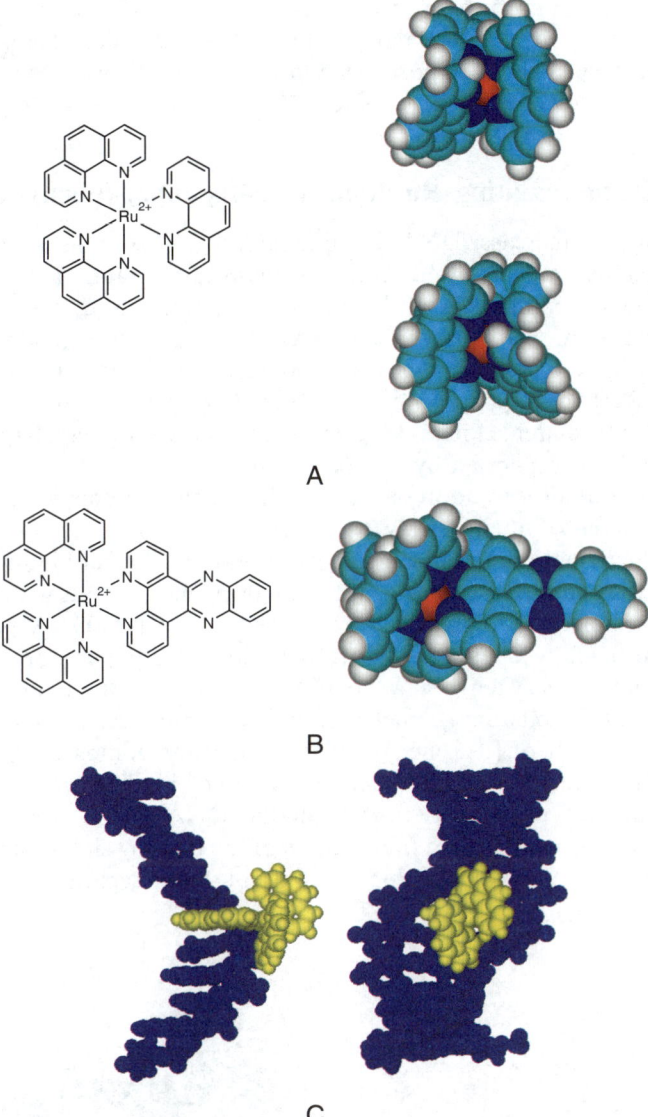

Figure 7 *Ruthenium conjugates. (A) [Ru(phen)₃]²⁺ (left) and its Δ- (top right) and Λ-enantiomers (bottom right). (B) [Ru(phen)₂dppz]²⁺ (left) and its Δ-enantiomer (right). (C) Δ-[Ru(phen)₂dppz]²⁺ intercalated into DNA with one of the strands omitted for clarity (left) and the arrangement of phen ligands of the conjugate in the DNA minor groove (right)*

the more extensively studied [Ru(phen)₂dppz]²⁺ (Figure 7B).[81,82] The drastic increase in luminescence quantum yield when these dppz complexes bind to DNA, the so-called "light-switch" effect, has been investigated in great detail and is generally believed to originate from the protection of the aza-nitrogens on the dppz moiety from hydrogen bonding to water. Upon binding to DNA the dppz moiety is firmly intercalated between the base pairs whereas the metal ion and ancillary ligands are positioned in one of the grooves (Figure 7C).[87,89] Which groove, however, is still debated.[84,87,92]

4.4.1 *Bis*-intercalating Ru-dimer [μ-c4(cpdppz)$_2$(phen)$_4$Ru$_2$]$^{4+}$

In an effort to increase DNA binding affinity as well as sequence- and enantioselectivity, and also to decrease the dissociation rates of Ru(II)polypyr-idyl compounds compared to the monomer [Ru(phen)$_2$dppz]$^{2+}$, the three stereoisomers (ΔΔ, ΔΛ = meso, and ΛΛ) of a dimer, [μ-c4(cpdppz)$_2$(phen)$_4$ Ru$_2$]$^{4+}$ (**1**, Figure 8), were synthesized and examined.[101,102] The dimer is built up of two [Ru(phen)$_2$dppz]$^{2+}$ moieties tethered *via* the dppz ligands through a flexible aliphatic diamide linker, long enough to allow *bis*-intercalation into two intercalation slots separated by two base pairs.

The initial question to address was whether each monomeric moiety of the dimer was actually interacting with DNA in the same way as [Ru(phen)$_2$ dppz]$^{2+}$, which ought to give rise to a correspondingly higher binding constant, or whether the combination of the bulky ancillary phen ligands with the flexible linker of the dimer would prevent it from intercalating into DNA. Remarkable similarity in the flow LD spectra of the enantiomerically pure monomeric and dimeric compounds when bound to mixed-sequence calf thymus (ct) DNA confirmed that the binding geometry of each monomeric unit of **1** was indeed very similar to that of [Ru(phen)$_2$dppz]$^{2+}$ (Figure 9; compare Figure 7C).[101] Moreover, the presumed *bis*-intercalative mode of binding was supported by the high luminescence quantum yield in the DNA-bound state.[101] Binding titration experiments indicated that the size of the binding site is four base pairs and that the two intercalated dppz ligands of **1** are separated by two base

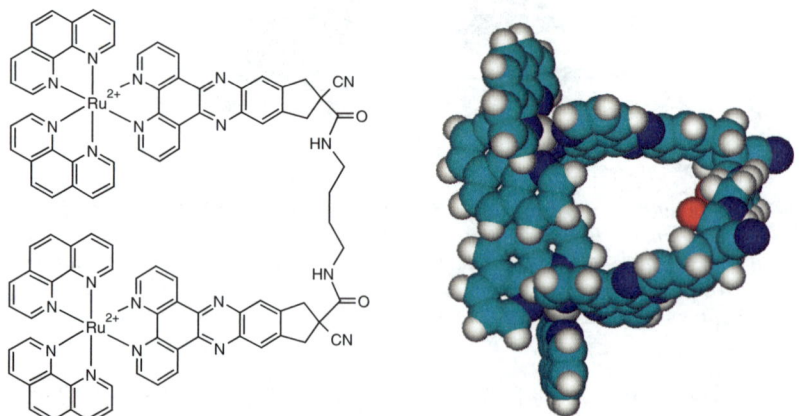

Figure 8 *Bis-intercalating ruthenium dimer, [μ-c4(cpdppz)$_2$(phen)$_4$Ru$_2$]$^{4+}$ (**1**) (left), and its ΔΔ-enantiomer (right)*

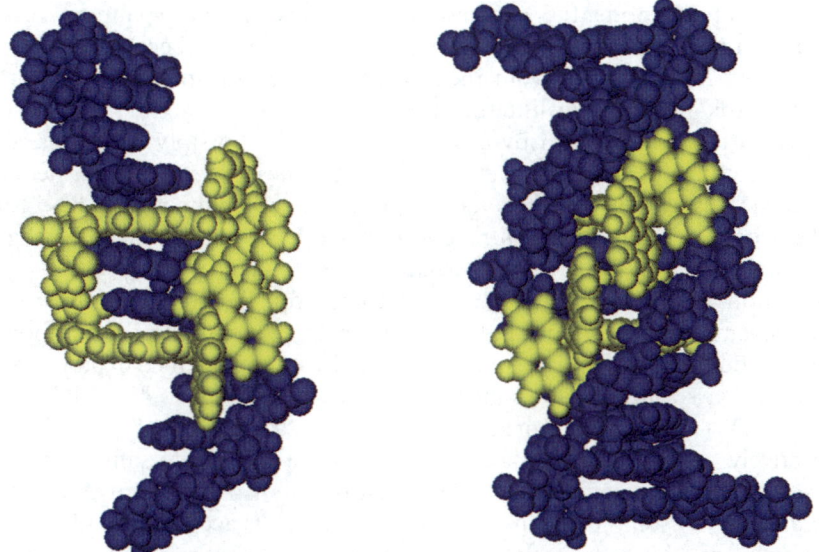

Figure 9 *DNA binding of ΔΔ-[μ-c4(cpdppz)$_2$(phen)$_4$Ru$_2$]$^{4+}$ (**1**), shown bis-intercalated into DNA with one of the strands omitted for clarity (left) and the arrangement of phenanthroline ligands of the conjugate in the DNA minor groove (right)*

pairs.[101] Finally, deconvolution of the LD and absorption spectra into differently polarized spectral components[89] showed that enantiomers of the dimer **1** both have a clockwise roll of their dppz ligands in the intercalation pocket (7° for ΔΔ-**1** and 19° for ΛΛ-**1**; Önfelt, Ph.D thesis). This small but significant rotation places the two phenanthroline ligands of ΔΔ-**1** directed along, and those of the ΛΛ-**1** roughly perpendicular to, a groove of the DNA (Figure 9).[102] These results are in excellent agreement with the DNA binding of the [Ru(phen)$_2$dppz]$^{2+}$ indicating that the aliphatic diamide tether has no significant influence upon the mode of binding for each subunit of the dimer.

It has been observed by calorimetric investigations that the binding of ΔΔ-**1** (Önfelt, unpublished data) and [Ru(phen)$_2$dppz]$^{2+}$ to DNA has a very small reaction enthalpy and therefore is mainly entropically driven. Furthermore, dilution has been used to estimate the DNA-binding constant of **1** to be approximately 10^{10} M^{-1} at 50 mM Na$^+$ (ref. 102) whereas the binding constant of [Ru(phen)$_2$dppz]$^{2+}$ under the same conditions is approximately 10^6 M^{-1}. Surprisingly, however, no significant enantioselectivity was observed. On the other hand, **1** has been found to bind more strongly to ct-DNA than to purely AT and GC sequences, thus indicating some selectivity.[102]

To reach the *bis*-intercalated state one of the ruthenium ions, together with its bulky phen ligands, has to be threaded back and forth through the core of the DNA. Alternatively, the tether could sling around two opened base pairs so as to place the two ruthenium moieties in the same groove. This topological obstacle causes the association process to be very complex.[102] To describe the binding a multiexponential expression and rate constants spanning a time-range from milliseconds to hours have to be used.[102] It has been suggested that this complexity originates from the existence of several intermediates on the way towards the final *bis*-intercalative mode of binding: first electrostatic external attraction followed by groove-binding, and possibly mono-intercala-tion. Also, the threading may involve several relatively stable intermediates, one being the single threaded geometry where one ruthenium moiety lies in each of the grooves. The association reaction, which has been shown to be slightly enantioselective, is faster for alternating AT sequences than GC.[102] This observation is analogous to what was found for the threading process of nogalamycin (*vide supra*) and indicates that base pair stability is an important factor for determining the rate of association. In a more detailed experiment the two fastest association rate constants for $\Delta\Delta$-**1** binding to 45 μM bases of poly(dA-dT)$_2$ were determined to be $k_{a1} = 29$ s^{-1} and $k_{a2} = 4.4$ s^{-1}, respectively.[102] However, it should be noted that under the same conditions the slowest process, which most likely involves redistribution of $\Delta\Delta$-**1**, takes several hours to complete. This slow reorganization is accompanied by a small increase in emission, a large increase in LD amplitude but insignificant changes in the LD spectral profile, thus suggesting that there is a minimal change in binding geometry of $\Delta\Delta$-**1** and a local, rather than global, increase in the orientation of the DNA. It has been proposed that this local increase of orientation arises as a result of differences in the binding cooperativity of **1** between early and later stages of the binding process.[102]

Dissociation of **1** from DNA is very slow and it has been suggested that the sluggishness of the process is a result of requirements for both base-pair opening and unstacking as the complex unthreads from the DNA. The slowness of the process can be exemplified with the two rate constants needed to explain the dissociation of meso-**1** from ct-DNA: $k_{d1} = 6.4 \cdot 10^{-3}$ s^{-1} and $k_{d2} = 1.5 \cdot 10^{-3}$ s^{-1}.[101] Under the same conditions dissociation from both alter-nating AT and GC sequences is faster than the corresponding process from ct-DNA. Furthermore, the dissociation of $\Delta\Delta$-**1** from AT sequences can be described by a single exponential expression with $k_d = 7.3 \cdot 10^{-3}$ s^{-1} [101] whereas for poly(dG-dC)$_2$ 95% of the dissociation can be assigned to a k_d of $4.5 \cdot 10^{-3}$ s^{-1} (ref. 102). The fact that the dissociation from ct-DNA is slower than those from purely alternating sequences indicates that it is not only the strength of the base pair hydrogen bonding that dictates the rate of the process but rather the local stability of the DNA duplex, as in the case of nogalamycin (*vide supra*). Experiments also demonstrated that the unthreading from poly(dA-dT)$_2$ is fastest for $\Lambda\Lambda$-**1** ($k_d = 4.2 \cdot 10^{-2}$ s^{-1}) followed by meso- and $\Delta\Delta$-**1** ($k_d = 1 \cdot 10^{-2}$ s^{-1} and $0.73 \cdot 10^{-2}$ s^{-1}, respectively).[101] This observation indicates that Λ moieties penetrate, and thus unthread, more easily than Δ moieties. The similarity between the rates of dissociation for meso- and $\Delta\Delta$-**1** also suggests that the Δ moiety of these stereoisomers serves as an anchor in the DNA during the rate limiting step from mono-intercalation to free complex in solution.[102] However, this enantioselectivity and also the accuracy of the estimated k_d:s for **1** as well as for actinomycin D and nogalamycin should probably be regarded with caution especially in the light of the recent finding that the anionic surfactants employed in the SDS-sequestration method can strongly affect the rate of dissociation depending on the nature of the leaving molecule and its interaction with DNA.[67]

4.4.2 Semirigid Ru-dimer $[\mu\text{-}(11,11'\text{-bidppz})(x)_4Ru_2]^{4+}$ (x = phen or bipy)

Another dimeric ruthenium conjugate that has recently received attention is the "dumbbell" shaped semirigid $[\mu\text{-}(11,11'\text{-bidppz})(x)_4Ru_2]^{4+}$, where x = bipy (**2**) or phen (**3**) (Figures 10A, B, and C).[103–107] This compound consists of two $[Ru(x)_2dppz]^{2+}$ moieties connected *via* one single bond between the outer rings of the two dppz ligands. The first report on DNA binding of this kind of compound, presented a decade ago, suggested from LD measurements and lack of luminescence that neither the ΔΔ- nor the ΛΛ-enantiomers can intercalate into DNA but instead bind in the grooves with a very high binding constant (K = 10^{12} M^{-1}, at 10 mM Na$^+$ and at room temperature).[103] Although the LD data for all four complexes were consistent with groove binding, the low energy part of the spectrum measured for ΔΔ-**3** binding to DNA contained a most unusual positive signal. This unique peculiarity, quite unexpected, prompted further investigation.

4.4.2.1 *Evidence of Extremely Slow Threading Intercalation of ΔΔ-3*

Upon investigation of the positive low energy part of the LD spectrum of ΔΔ-**3** in the presence of ct-DNA, a chance observation on a sample that had been left for two weeks (!) at room temperature revealed a change from positive LD amplitude to negative.[104] This result indicated a major change in the DNA binding geometry of ΔΔ-**3**. To facilitate more efficient studies of the binding process it was accelerated by adding salt and increasing temperature. An isosbestic point in the LD spectra collected at different times, supported by a singular value decomposition (SVD) analysis of the whole spectra, demonstrated that the change involved only two binding geometries: an initial mode corresponding to the positive LD spectrum originally observed and a final mode giving rise to the negative LD spectrum.[104] The final binding mode could be characterized by a binding angle of the bidppz ligand that was roughly perpendicular to the DNA helix axis, strong luminescence, and luminescence lifetimes that were comparable to those of Δ-$[Ru(phen)_2dppz]^{2+}$.[104] Furthermore, it was observed that dissociation from the initial binding mode was effectively instantaneous whereas dissociation from the final binding mode took several days even at elevated temperatures. It was suggested that the initial binding mode was a groove-bound metastable binding mode from which the ruthenium complex rearranged into the final threading-intercalated mode of binding by passing one of the Ru(phen)$_2$ moieties through the core of the

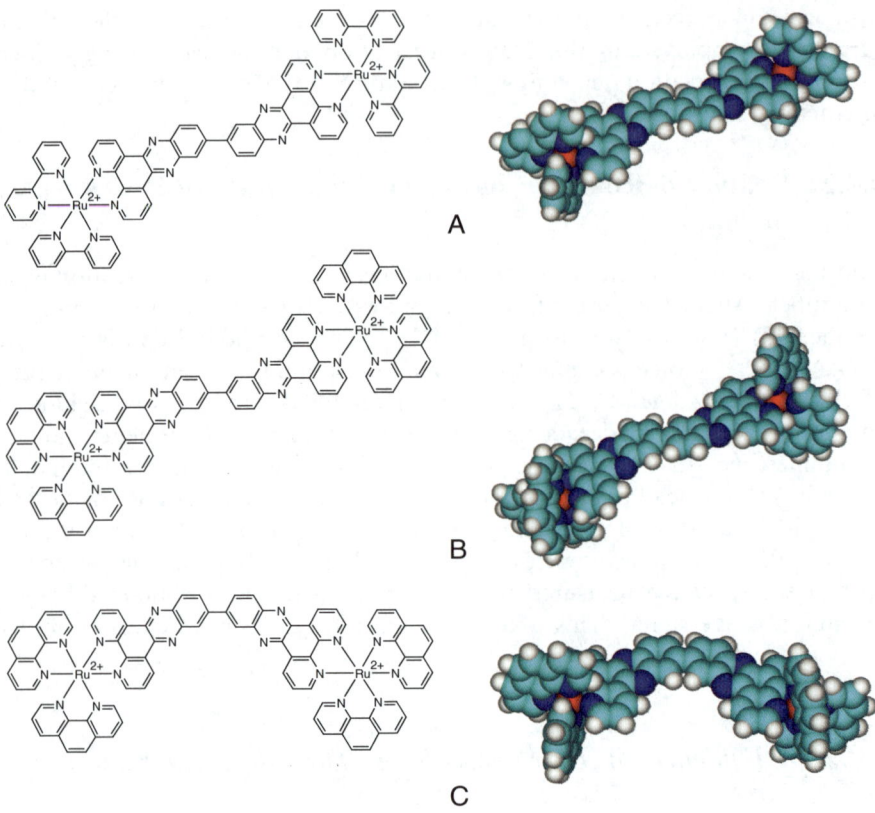

Figure 10 *Semirigid Ru-dimers [μ-(11,11'-bidppz)(x)₄Ru₂]⁴⁺ (x=phen or bipy). (A)*
*[μ-(11,11'-bidppz)(bipy)₄Ru₂]⁴⁺ (**2**) (left) and its ΛΛ-enantiomer (right)*
with the bidppz ligand in anti-conformation. (B) [μ-(11,11'-bidppz)(phen)₄-
*Ru₂]⁴⁺ (**3**) (left) and its ΔΔ-enantiomer (right) with the bidppz ligand in*
*anti-conformation. (C) [μ-(11,11'-bidppz)(phen)₄Ru₂]⁴⁺ (**3**) (left) and its*
ΛΛ-enantiomer (right) with the bidppz ligand in syn-conformation

DNA.[104] The kinetics of this rearrangement and threading process, measured
at 50 °C, 100 mM Na^+ and 160 μM ct-DNA (bases) required two exponentials
($k_{a1} = 1.2 \cdot 10^{-3}$ s^{-1} and $k_{a2} = 0.2 \cdot 10^{-3}$ s^{-1}) to fit the data. Measurements were
also performed at other temperatures and from these threading activation
energies of 94 and 83 kJ mol^{-1} were estimated for the processes corresponding
to the two exponentials.[104] It should be noted that these values are similar to
the threading activation energies of nogalamycin discussed above. The SDS-
sequestered dissociation process, under similar conditions, could also be
described by a two-exponential expression ($k_{d1} = 1.9 \cdot 10^{-3}$ s^{-1} and $k_{d2} =$
$0.8 \cdot 10^{-3}$ s^{-1}) but with activation barriers of 65 and 66 kJ mol^{-1}, respectively.[104] The surprising result that the dissociation rate constants provoked by
SDS were faster than the rearrangement rate constants triggered the discovery
of the surfactant-induced dissociation rate enhancement that was mentioned

earlier.[67,68] Thus, again the apparent values of the rate constants should be treated with caution and only be used for rough comparison of rates of dissociation of DNA-bound drugs. Recently, the non-catalysed rate of dissociation of ΔΔ-3 from ct-DNA has been estimated by studying the increase in luminescence as complexes are gradually transferred to poly(dA-dT)$_2$. Addition of poly(dA-dT)$_2$ to a pre-equilibrated sample of ΔΔ-3 and ct-DNA resulted in an emission trace that could be fitted to a single exponential with a rate constant of $2.4 \cdot 10^{-5} s^{-1}$ (150 mM NaCl, 50 °C; Nordell, P., unpublished results). This rate of dissociation is at least 30 times slower than that determined with the SDS sequestration technique.

4.4.2.2   Final and Metastable DNA Binding Modes of Stereoisomers of **2** and **3**

The initial metastable binding mode of ΔΔ-3 to ct-DNA characterized by the positive LD, low luminescence quantum yield and instantaneous SDS-driven dissociation was also observed for binding of ΔΔ-3 to poly(dG-dC)$_2$ and poly(dI-dC)$_2$.[105] The positive LD suggested that this metastable state is groove-bound, but the LD does not tell if it is major- or minor-groove binding. However, poly(dG-dC)$_2$ and poly(dI-dC)$_2$ are very different in the minor groove while these polynucleotides are chemically similar in the major groove. Therefore, the similar LD of ΔΔ-3 bound to poly(dG-dC)$_2$ and poly(dI-dC)$_2$ indicated that the primary, metastable binding mode is in the major groove.[105] By contrast, based on the immediate appearance of negative LD and extremely slow dissociation even after short equilibration times it was concluded that ΔΔ-3 threads directly through poly(dA-dT)$_2$ virtually without passing through an intermediate state.[104,105] This substantial "kinetic recognition" could be utilized for targeting AT-rich DNA sequences among other sequences, both for diagnostic and therapeutic purposes.

Both ΛΛ- and meso-**3** interacting with ct-DNA yield initial LD spectra that are negative and suggest a binding geometry where their bidppz ligand is inclined to the DNA helix axis by approximately 65°. This angle is too large to originate from minor-groove-binding of the bidppz ligand and is larger than for the initial binding mode of ΔΔ-3: it approaches that of the final intercalated form.[105] However, it was shown that both ΛΛ- and meso-**3** dissociate immediately from this binding mode, suggesting a non-threaded state of binding. It was concluded that the two enantiomers adopt a metastable binding mode in ct-DNA where the bidppz ligand might be quasi-intercalated in the *syn* conformation (Figure 10C) from the major groove side (Figure 11A).[105]

The general outline of the final binding geometry of all stereoisomers of **2** (Nordell, P *et al.* unpublished results) and **3** interacting with ct-DNA, poly(dA-dT)$_2$, poly(dG-dC)$_2$, or poly(dI-dC)$_2$ is that the conjugates all thread through DNA to end up with the bidppz ligand intercalated between the base pairs, which means that one Ru(phen)$_2$ moiety ends up in each groove of the DNA (Figure 11B).[105] This conclusion was drawn from the great similarity between the LD spectra of all stereoisomers of **2** and **3** bound to any of the DNAs.

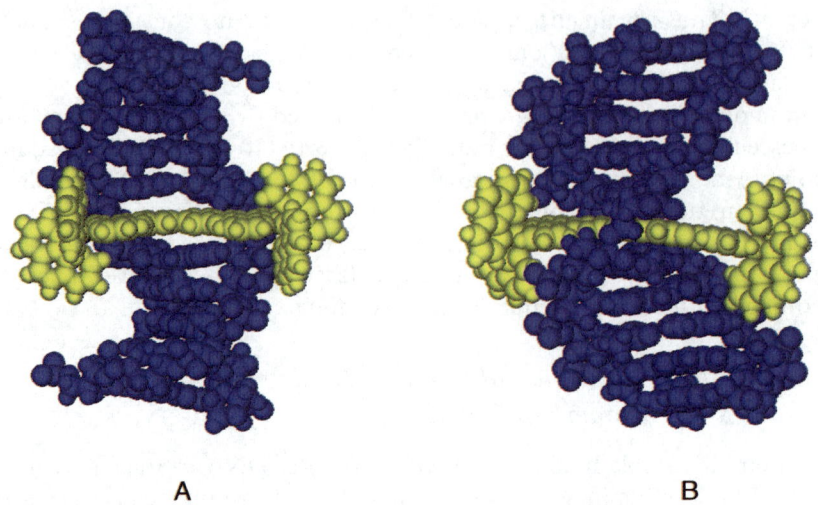

A   **B**

Figure 11   *DNA binding of ΔΔ-[μ-(11,11'-bidppz)(phen)₄Ru₂].⁴⁺ Initial quasi-intercalative major groove-bound form (A) and final threading-intercalated form (B)*

In addition, interaction of the three forms of **3** with all the DNAs except poly(dG-dC)$_2$ resulted in an increase in luminescence and lifetimes similar to those of the parent compound [Ru(phen)$_2$dppz]$^{2+}$, which further suggested an intercalative mode of binding.[105] Moreover, since dissociation of ΔΔ-**3** from the final binding mode in poly(dG-dC)$_2$ was found to be very slow, it was concluded that threading intercalation was the equilibrium binding mode for this system also.[105]

In a thorough investigation a detailed final binding geometry for **3** attached to poly(dA-dT)$_2$ has been suggested (Figure 11B).[105] From electrostatic arguments and luminescence properties it was concluded that **3** intercalates asymmetrically with one of the Ru(phen)$_2$ moieties deeply intercalated leaving the other moiety displaced from the centre of the DNA helix. Furthermore, it was concluded from LD that the central pivot bond of the bidppz ligand **3** is rotated so that both Ru(phen)$_2$ moieties are rotated clockwise, looking from the corresponding groove side. From CD and differential CD it was suggested that the bridging bidppz ligand is intercalated in its *anti*-conformation (Figure 10B). Finally, it was proposed from LD and luminescence that the deeply intercalated Ru(phen)$_2$ moiety resides in the minor groove and, with the help of meso-**3**, that there is an enantio-preference for a Λ moiety in the minor groove.

4.4.2.3   DNA Rearrangement Kinetics of ΛΛ-**2** and ΛΛ-**3**

In the vast majority of studies of the rearrangement of **2** and **3** binding to different DNAs, the kinetics has been found to be multiphasic. Quite unexpectedly, when studying the binding of ΛΛ-**2** to poly(dA-dT)$_2$ by luminescence it was found that the kinetics of the system, basically irrespective of basepair:conjugate ratio, could be described by a single exponential expression

($k_a = 3.4 \cdot 10^{-3}$ s^{-1} at 120 μM of base).[106] This suggested that ΛΛ-**2** was able to rearrange directly from its externally bound state to the threaded intercalated state. By contrast, the binding of ΛΛ-**3** to the same kind of alternating DNA at a ratio of 8 (basepair:complex) was found to follow a second order rate law: $k \times$ [non-threaded complex]^2.[106] When the binding density of that system was decreased a gradual transition towards pseudo-first-order kinetics was observed. To fit this behaviour of both bimolecular and pseudo-first-order kinetics a global fit employing conditional probabilities of a non-cooperative McGhee and von Hippel model was used. It was concluded that a first order rate constant of $1.2 \cdot 10^{-2}$ s^{-1} and second order rate constant of $5.0 \cdot 10^{-2}$ s^{-1} (at 120 μM of base, and [ΛΛ-**3**] = 7.5, 3.7, 1.9, and 0.95 μM) gave an excellent fit to the data.[106] It was suggested that the two different pathways originated from direct threading intercalation from a groove-bound form and indirect (bimolecular) threading intercalation from a groove that is catalysed by opening of the base stack from the opposite side of DNA by another ΛΛ-**3**.[106]

4.4.2.4 Slow Shuffling Rearrangement of ΔΔ-3

Another intriguing effect was found when studying the binding of ΔΔ-**3** to poly(dA-dT)$_2$. The association can in this case be described by a bimolecular expression with an additional extremely slow first-order process.[107] The bimolecular process ($k_{a1} \approx 10^{-3}$ s^{-1}, at 2 mM Mg^{2+} and 5 mM K$^+$, 50 °C, 80 μM base, and [ΔΔ-**3**] = 6, 4, and 2 μM), which gives rise to a large increase in luminescence quantum yield is, as mentioned above, a result of the conjugate becoming threading-intercalated.[107] What is surprising is that after the complex is threading-intercalated another extremely slow first-order process ($k_{a2} \approx 10^{-4}$ s^{-1}) follows. To investigate this rearrangement the DNA groove binder DAPI was used as an intrinsic probe. Using observations from CD, emission and energy transfer between DAPI and ΔΔ-**3** it was concluded that the reorganization results from a slow "shuffling" of ΔΔ-**3** from an initial slightly anti-cooperative bound form to its final thermodynamically most favourable distribution.[107] The slow shuffling process was suggested to be a result of dissociation, lateral diffusion along the DNA followed by reassociation as in the case of nogalamycin; alternatively, it could be explained by a mechanism in which the complex remains threaded and by sequential base pair opening diffuses laterally along the DNA.[107]

References

1. F.M. Chen, *Biochemistry*, 1988, **27**, 1843.
2. F.M. Chen, *Biochemistry*, 1990, **29**, 7684.
3. F.A. Tanious, J.M. Veal, H. Buczak, L.S. Ratmeyer and W.D. Wilson, *Biochemistry*, 1992, **31**, 3103.
4. R. Baliga and D.M. Crothers, *Proc. Natl. Acad. Sci. U.S.A.*, 2000, **97**, 7814.

5. R. Baliga, E.E. Baird, D.M. Herman, C. Melander, P.B. Dervan and D.M. Crothers, *Biochemistry*, 2001, **40**, 3.
6. D.R. Phillips and D.M. Crothers, *Biochemistry*, 1986, **25**, 7355.
7. M.C. Fletcher and K.R. Fox, *Nucleic Acids Res.*, 1993, **21**, 1339.
8. W. Müller and D.M. Crothers, *J. Mol. Biol.*, 1968, **35**, 251.
9. V.A. Bloomfield, D.M. Crothers and I. Tinoco Jr., *Nucleic Acids – Structures, Properties, and Functions. University Science Books*, Sausalito, 2000.
10. D.S. Johnson and D.L. Boger, in *Comprehensive Supramolecular Chemistry*. Elsevier Science Ltd, Oxford, 1996, 73.
11. H. Deng, V.A. Bloomfield, J.M. Benevides and G.J. Thomas, *Nucleic Acids Res.*, 2000, **28**, 3379.
12. T.J. Thomas and V.A. Bloomfield, *Biopolymers*, 1984, **23**, 1295.
13. S. Eriksson, S.K. Kim, M. Kubista and B. Nordén, *Biochemistry*, 1993, **32**, 2987.
14. J. Kapuscinski and K. Yanagi, *Nucleic Acids Res.*, 1979, **6**, 3535.
15. M. Kubista, B. Åkerman and B. Nordén, *Biochemistry*, 1987, **26**, 4545.
16. T.A. Larsen, D.S. Goodsell, D. Cascio, K. Grzeskowiak and R.E. Dickerson, *J. Biomol. Struct. Dyn.*, 1989, **7**, 477.
17. C. Bailly, P. Colson, J.P. Henichart and C. Houssier, *Nucleic Acids Res.*, 1993, **21**, 3705.
18. N. Spink, D.G. Brown, J.V. Skelly and S. Neidle, *Nucleic Acids Res.*, 1994, **22**, 1607.
19. M.L. Kopka, C. Yoon, D. Goodsell, P. Pjura and R.E. Dickerson, *Proc. Natl. Acad. Sci. U.S.A.*, 1985, **82**, 1376.
20. A. Abu-Daya, P.M. Brown and K.R. Fox, *Nucleic Acids Res.*, 1995, **23**, 3385.
21. J.G. Pelton and D.E. Wemmer, *Biochemistry*, 1988, **27**, 8088.
22. J.G. Pelton and D.E. Wemmer, *J. Am. Chem. Soc.*, 1990, **112**, 1393.
23. K.R. Fox and M.J. Waring, *Nucleic Acids Res.*, 1984, **12**, 9271.
24. M.W. van Dyke, R.P. Hertzberg and P.B. Dervan, *Proc. Natl. Acad. Sci. U.S.A.*, 1982, **79**, 5470.
25. M. Coll, C.A. Frederick, A.H.J. Wang and A. Rich, *Proc. Natl. Acad. Sci. U.S.A.*, 1987, **84**, 8385.
26. U. Sehlstedt, S.K. Kim and B. Nordén, *J. Am. Chem. Soc.*, 1993, **115**, 12258.
27. W.D. Wilson, F.A. Tanious, H.J. Barton, R.L. Jones, K. Fox, R.L. Wydra and L. Strekowski, *Biochemistry*, 1990, **29**, 8452.
28. M. Mrksich and P.B. Dervan, *J. Am. Chem. Soc.*, 1993, **115**, 9892.
29. M. Mrksich, M.E. Parks and P.B. Dervan, *J. Am. Chem. Soc.*, 1994, **116**, 7983.
30. M.E. Parks, E.E. Baird and P.B. Dervan, *J. Am. Chem. Soc.*, 1996, **118**, 6147.
31. S. White, J.W. Szewczyk, J.M. Turner, E.E. Baird and P.B. Dervan, *Nature*, 1998, **391**, 468.

32. L.A. Dickinson, R.J. Gulizia, J.W. Trauger, E.E. Baird, D.E. Mosier, J.M. Gottesfeld and P.B. Dervan, *Proc. Natl. Acad. Sci. U.S.A.*, 1998, **95**, 12890.
33. P.B. Dervan, *Bioorg. Med. Chem.*, 2001, **9**, 2215.
34. C.G. Reinhardt and T.R. Krugh, *Biochemistry*, 1978, **17**, 4845.
35. W. Fuller and M.J. Waring, *Berichte der Bunsengesellschaft für Physikalische Chemie*, 1964, **68**, 805.
36. J.-B. LePecq, *Methods of Biochemical Analysis*, 1971, **20**, 41.
37. M.J. Waring, *J. Mol. Biol.*, 1965, **13**, 269.
38. R.B. Macgregor, R.M. Clegg and T.M. Jovin, *Biochemistry*, 1985, **24**, 5503.
39. J.B. Chaires, N. Dattagupta and D.M. Crothers, *Biochemistry*, 1982, **21**, 3927.
40. J.B. Chaires, N. Dattagupta and D.M. Crothers, *Biochemistry*, 1982, **21**, 3933.
41. H. Fritzsche, H. Triebel, J.B. Chaires, N. Dattagupta and D.M. Crothers, *Biochemistry*, 1982, **21**, 3940.
42. L. Gianni, B.J. Corden and C.E. Myers, *Rev. Biochem. Toxicol.*, 1983, **5**, 1.
43. R.L. Momparler, M. Karon, S.E. Siegel and F. Avila, *Cancer Res.*, 1976, **36**, 2891.
44. M. Eriksson, B. Nordén and S. Eriksson, *Biochemistry*, 1988, **27**, 8144.
45. D.E. Graves and T.R. Krugh, *Biochemistry*, 1983, **22**, 3941.
46. K.M. Tewey, T.C. Rowe, L. Yang, B.D. Halligan and L.F. Liu, *Science*, 1984, **226**, 466.
47. W. Förster and E. Stutter, *Int. J. Biol. Macromol.*, 1984, **6**, 114.
48. J.B. Chaires, N. Dattagupta and D.M. Crothers, *Biochemistry*, 1985, **24**, 260.
49. Y.C. Liaw, Y.G. Gao, H. Robinson, G.A. van der Marel, J.H. van Boom and A.H.J. Wang, *Biochemistry*, 1989, **28**, 9913.
50. Y.G. Gao, Y.C. Liaw, H. Robinson and A.H.J. Wang, *Biochemistry*, 1990, **29**, 10307.
51. M. Egli, L.D. Williams, C.A. Frederick and A. Rich, *Biochemistry*, 1991, **30**, 1364.
52. C.K. Smith, G.J. Davies, E.J. Dodson and M.H. Moore, *Biochemistry*, 1995, **34**, 415.
53. K.R. Fox and M.J. Waring, *Biochim. Biophys. Acta*, 1984, **802**, 162.
54. K.R. Fox, C. Brassett and M.J. Waring, *Biochim. Biophys. Acta*, 1985, **840**, 383.
55. T.R. Krugh, E.S. Mooberry and Y.C.C. Chiao, *Biochemistry*, 1977, **16**, 740.
56. S.C. Brown, K. Mullis, C. Levenson and R.H. Shafer, *Biochemistry*, 1984, **23**, 403.
57. H.M. Sobell and S.C. Jain, *J. Mol. Biol.*, 1972, **68**, 21.
58. S. Kamitori and F. Takusagawa, *J. Mol. Biol.*, 1992, **225**, 445.
59. H. Robinson, Y.G. Gao, X.L. Yang, R. Sanishvili, A. Joachimiak and A.H.J. Wang, *Biochemistry*, 2001, **40**, 5587.

60. F.M. Chen, *Biochemistry*, 1988, **27**, 6393.
61. F.M. Chen, *Biochemistry*, 1992, **31**, 6223.
62. R. Bittman and L. Blau, *Biochemistry*, 1975, **14**, 2138.
63. K.R. Fox and M.J. Waring, *Eur. J. Biochem.*, 1984, **145**, 579.
64. K.R. Fox and M.J. Waring, *Nucleic Acids Res.*, 1986, **14**, 2001.
65. S.C. Brown and R.H. Shafer, *Biochemistry*, 1987, **26**, 277.
66. J.J. Duffy and T.J. Lindell, *Biochem. Pharmacol.*, 1985, **34**, 1854.
67. F. Westerlund, L.M. Wilhelmsson, B. Nordén and P. Lincoln, *J. Am. Chem. Soc.*, 2003, **125**, 3773.
68. Marcus and R.A. Micelle-Enhanced, Dissociation of a Ru Cation /DNA Complex, *J. Phys. Chem. B*, 2005, **109**(45), 21419–21424.
69. B.K. Bhuyan and C.G. Smith, *Proc. Natl. Acad. Sci. U.S.A.*, 1965, **54**, 566.
70. W. Kersten, H. Kersten and W. Szybalski, *Biochemistry*, 1966, **5**, 236.
71. D.A. Collier, S. Neidle and J.R. Brown, *Biochem. Pharmacol.*, 1984, **33**, 2877.
72. J. Fok and M.J. Waring, *Mol. Pharmacol.*, 1972, **8**, 65.
73. L.H. Li, S.L. Kuentzel, L.L. Murch, L.M. Pschigoda and W.C. Krueger, *Cancer Res.*, 1979, **39**, 4816.
74. H.L. Ennis, *Antimicrob. Agents Chemother.*, 1981, **19**, 657.
75. F. Zunino and G. Capranico, *Anti-Cancer Drug Design*, 1990, **5**, 307.
76. X.L. Zhang and D.J. Patel, *Biochemistry*, 1990, **29**, 9451.
77. M.S. Searle, J.G. Hall, W.A. Denny and L.P.G. Wakelin, *Biochemistry*, 1988, **27**, 4340.
78. L.D. Williams, M. Egli, G. Qi, P. Bash, G.A. Van der Marel, J.H. Van Boom, A. Rich and C.A. Frederick, *Proc. Natl. Acad. Sci. U.S.A.*, 1990, **87**, 2225.
79. C.K. Smith, J.A. Brannigan and M.H. Moore, *J. Mol. Biol.*, 1996, **263**, 237.
80. K.R. Fox and M.J. Waring, *Biochemistry*, 1986, **25**, 4349.
81. A.E. Friedman, J.-C. Chambron, J.-P. Sauvage, N.J. Turro and J.K. Barton, *J. Am. Chem. Soc.*, 1990, **112**, 4960.
82. A.E. Friedman, C.V. Kumar, N.J. Turro and J.K. Barton, *Nucleic Acids Res.*, 1991, **19**, 2595.
83. A.M. Pyle and J.K. Barton, *Prog. Inorganic Chem.*, 1990, **38**, 413.
84. Y. Jenkins, A.E. Friedman, N.J. Turro and J.K. Barton, *Biochemistry*, 1992, **31**, 10809.
85. C.S. Chow and J.K. Barton, *Methods Enzymol.*, 1992, **212**, 219.
86. I. Haq, P. Lincoln, D.C. Suh, B. Nordén, B.Z. Chowdhry and J.B. Chaires, *J. Am. Chem. Soc.*, 1995, **117**, 4788.
87. C. Hiort, P. Lincoln and B. Nordén, *J. Am. Chem. Soc.*, 1993, **115**, 3448.
88. B. Nordén, P. Lincoln, B. Åkerman and E. Tuite, in *Metal Ions in Biological Systems*, A. Siegel and H. Siegel (eds), vol. 33, Marcel Dekker, Inc, New York, Basel, Hong Kong, 1996, 177.
89. P. Lincoln, A. Broo and B. Nordén, *J. Am. Chem. Soc.*, 1996, **118**, 2644.
90. K.E. Erikkila, D.T. Odom and J.K. Barton, *Chem. Rev.*, 1999, **99**, 2777.

91. E. Tuite, P. Lincoln and B. Nordén, *J. Am. Chem. Soc.*, 1997, **119**, 239.
92. C.M. Dupureur and J.K. Barton, *J. Am. Chem. Soc.*, 1994, **116**, 10286.
93. C. Moucheron, A. Kirsch DeMesmaeker and S. Choua, *Inorg. Chem.*, 1997, **36**, 584.
94. S. Satyanarayana, J.C. Dabrowiak and J.B. Chaires, *Biochemistry*, 1992, **31**, 9319.
95. S. Satyanarayana, J.C. Dabrowiak and J.B. Chaires, *Biochemistry*, 1993, **32**, 2573.
96. J.P. Rehmann and J.K. Barton, *Biochemistry*, 1990, **29**, 1701.
97. J.P. Rehmann and J.K. Barton, *Biochemistry*, 1990, **29**, 1710.
98. M. Eriksson, M. Leijon, C. Hiort, B. Nordén and A. Gräslund, *J. Am. Chem. Soc.*, 1992, **114**, 4933.
99. B. Nordén and F. Tjerneld, *FEBS Lett.*, 1976, **67**, 368.
100. P. Lincoln and B. Nordén, *J. Phys. Chem. B*, 1998, **102**, 9583.
101. B. Önfelt, P. Lincoln and B. Nordén, *J. Am. Chem. Soc.*, 1999, **121**, 10846.
102. B. Önfelt, P. Lincoln and B. Nordén, *J. Am. Chem. Soc.*, 2001, **123**, 3630.
103. P. Lincoln and B. Nordén, *Chem. Commun.*, 1996, **18**, 2145.
104. L.M. Wilhelmsson, F. Westerlund, P. Lincoln and B. Nordén, *J. Am. Chem. Soc.*, 2002, **124**, 12092.
105. L.M. Wilhelmsson, E.K. Esbjörner, F. Westerlund, B. Nordén and P. Lincoln, *J. Phys. Chem. B*, 2003, **107**, 11784.
106. P. Nordell and P. Lincoln, *J. Am. Chem. Soc.*, 2005, **127**, 9670.
107. F. Westerlund, L.M. Wilhelmsson, B. Nordén and P. Lincoln, *J. Phys. Chem. B*, 2005, **109**, 21140–21144.

CHAPTER 5

DNA Gene Targeting using Peptide Nucleic Acid (PNA)

PETER E. NIELSEN

University of Copenhagen, Department of Medical Biochemistry and Genetics, The Panum Institute, Blegdamsvej 3c, Copenhagen N DK-2200, Denmark

5.1 Introduction

Peptide nucleic acids (PNA) were introduced in 1991 as a new class of pseudo peptide DNA mimics (Figure 1). PNA was originally conceived as a ligand that could recognize double-stranded DNA sequence specifically in the major groove, mimicking the DNA recognition of triplex-forming oligonucleotides.[1,2] Surprisingly, however, it was discovered that homopyrimidine PNA oligomers are able to invade the DNA helix by forming very stable triplex invasion

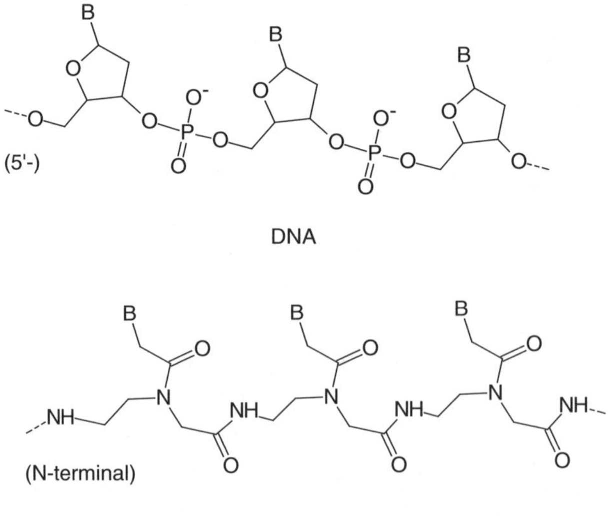

Figure 1   *Chemical structures of PNA compared to DNA*

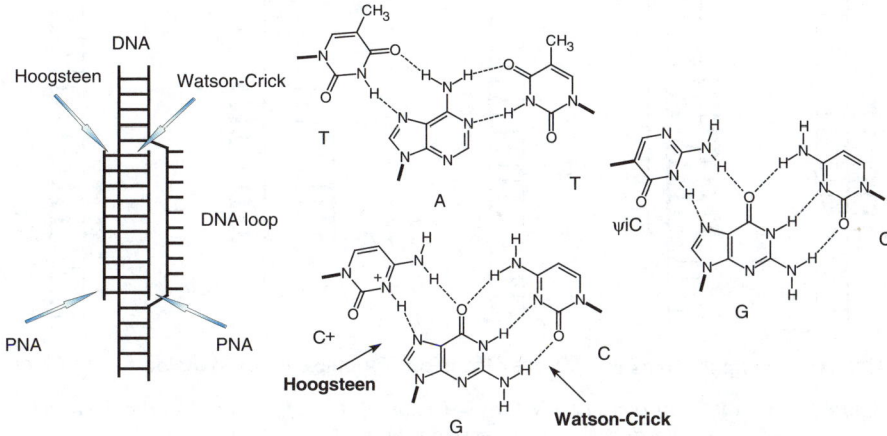

Figure 2 *Triplex invasion by homopyrimidine PNA oligomers. One PNA-strand binds via Watson–Crick base pairing (preferably in the anti-parallel orientation), while the other binds via Hoogsteen base pairing (preferably in the parallel orientation). It is usually advantageous to connect the two PNA strands covalently via a flexible linker into a bis-PNA, and to substitute all cytosines in the Hoogsteen strand with pseudoisocytosines (ΨiC), which do not require low pH for protonation at N^3*

complexes in which a PNA_2–DNA triplex is formed on the sequence-complementary purine DNA strand while the pyrimidine strand is extruded as a single-stranded loop (P-loop)[1,3,4] (Figure 2).

It was also immediately realized that PNA is a very potent structural mimic of DNA and RNA capable of forming both PNA–DNA and PNA–RNA (as well as PNA–PNA) duplexes which are generally of higher stability and are endowed with greater sequence discrimination than the corresponding DNA–DNA or DNA–RNA duplexes.[2,5] Subsequent studies have also revealed that more complex structures, such as triplexes, quadruplexes and hairpins can be adopted by PNA oligomers of appropriate sequence.[6–10] The properties of PNA have inspired and attracted the attention of many disciplines of science ranging from organic and physical chemistry, over molecular biology to genetic diagnostics, nanotechnology and drug discovery – even so far as to include aspects of the origin of life.[11–14]

The present chapter will focus on the efforts and progress made towards understanding sequence-specific targeting of duplex DNA using PNA, and the prospects of exploiting PNA for targeted control of gene expression.

5.2 Duplex DNA Recognition *in vitro*

Four distinct modes of recognizing duplex DNA have been discovered for PNA oligomers (Figure 3) depending upon the target and PNA sequence as well as the conditions. Triplex invasion is (so far) restricted to homopurine DNA targets, and is thermodynamically and kinetically stabilized by the formation of

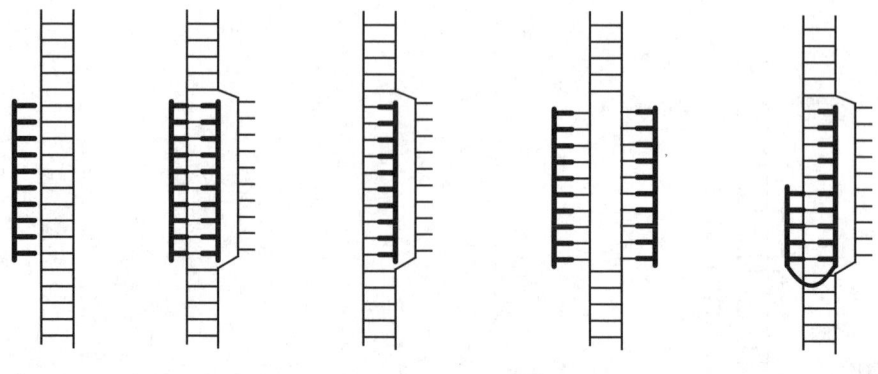

Triplex   Triplex Invasion   Duplex Invasion   Double Duplex Invasion   Tail-Clamp

Figure 3   *Five different types of PNA–dsDNA complexes. DNA is schematically drawn as
a ladder, and the PNA oligomers are in bold*

a Watson–Crick–Hoogsteen PNA_2–DNA triplex.[3,4] Simple duplex invasion
relies on Watson–Crick base pairing and in principle suffers no sequence
constraints, but under most conditions these complexes are not very stable *in
vitro*. However, binding has been demonstrated with homopurine PNA oligomers, which form exceptionally stable PNA–DNA duplexes,[15] and also for
PNAs conjugated to cationic peptides when binding to targets under negative
superhelical stress.[16] The situation *in vivo* may be more complex (*vide infra*).

Duplex invasion complexes can be stabilized through binding of a second
PNA oligomer to the single-stranded loop thereby forming a double-duplex
invasion complex. However, this requires the employment of sterically or
otherwise compromised pseudo-complementary PNA molecules because such
a pair of PNAs would otherwise quench each other by hybridization[17] (Figure
4). Furthermore, combinations of these binding modes, for example, using
PNA "tail-clamps" are possible.[18,19] Finally, conventional triplex binding of
homopyrimidine PNAs in the major groove of the DNA helix at sequence-
complementary homopurine DNA targets may also take place by Hoogsteen
base pairing. All types of helix invasion complexes require opening ("breathing") of the DNA double helix, and the kinetics of formation of such complexes
are therefore very slow and sensitive to increasing ionic strength and other
factors which stabilize the double helix. Thus in order to obtain stable complexes, the dissociation rate must also be very slow. Accordingly, in most cases
that have been examined the binding, including sequence discrimination, is
kinetically controlled – a phenomenon that has been exploited.[4,20,21] By contrast, conventional-triplex formation has much faster kinetics and is much less
sensitive to ionic strength.[6] This mode of binding has not yet been extensively
investigated.

Although PNA binding does not show the almost all-or-none orientation
dependence observed with DNA oligonucleotides, Watson–Crick duplexes are
most stable in the antiparallel configuration (amino-end of the PNA facing the

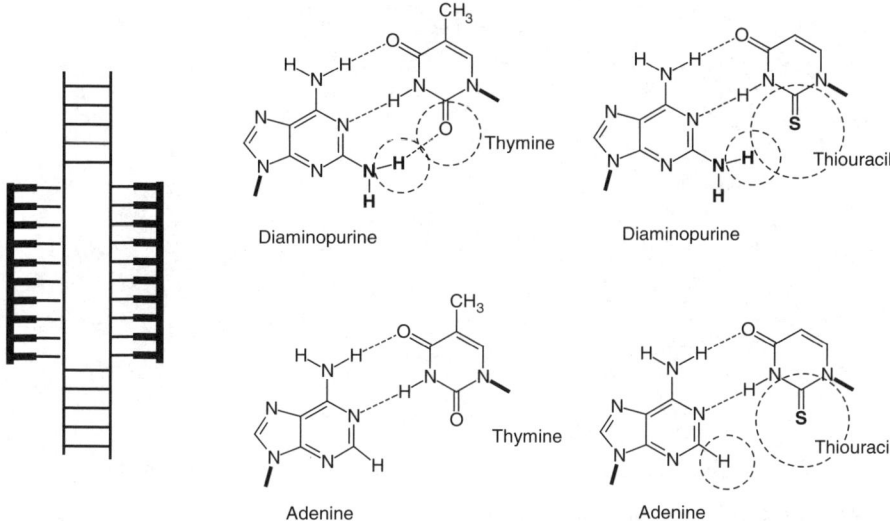

Figure 4 *Double duplex invasion of pseudo-complementary PNAs. In order to obtain efficient binding the target (and thus the PNAs) should contain at least 50% AT, though no other sequence constraints apply, and in the PNA oligomers all adenines and thymines are substituted with 2,6-diaminopurine or 2-thiouracil, respectively. Pairing between these base analogues is very unstable due to steric hindrance. Therefore the two sequence-complementary PNAs will not be able to bind to each other, but they bind to their complementary DNA sequences very well*

3'-end of the oligonucleotide), whereas the parallel orientation is preferred for Hoogsteen binding to DNA.[22]

Based on these observations, *bis*-PNAs (or PNA clamps) for triplex invasion have been constructed in which the Watson–Crick and Hoogsteen domains are connected *via* a flexible chemical linker.[22] Furthermore, cytosine can optimally and advantageously be replaced by pseudoisocytosine in the Hoogsteen domain as this virtually eliminates the requirement for an acidic pH otherwise needed to ensure N^3-protonation of cytosine.[22]

5.3 PNA Conjugates

A variety of other DNA-interactive ligands have been conjugated to (*bis*) PNAs in order to improve or to modify DNA recognition and binding. The rate of binding can be significantly improved by making cationic PNAs,[23] for example, *via* lysine "conjugation" which results in a rate enhancement of almost 10-fold for each charge (up to +4). This effect is ascribed to an increase in local PNA concentration proximal to the DNA thereby increasing the probability of the PNA "catching" a dynamic DNA opening.

An analogous effect is obtained using PNAs conjugated to the DNA intercalator 9-aminoacridine (Figure 5), but additionally the binding of such conjugates to dsDNA is much less sensitive to inhibition by elevated ionic strength.

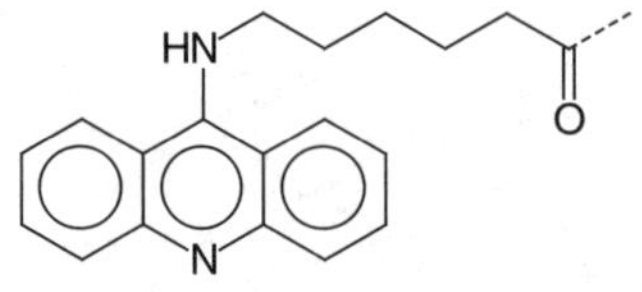

9-aminoacridine

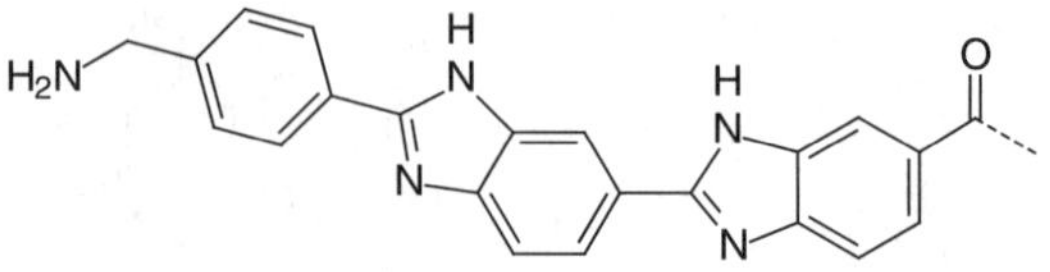

"Hoechst"

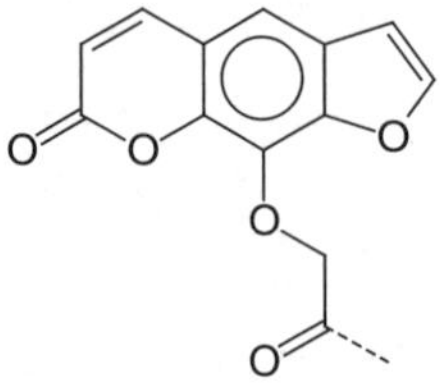

8-alkoxypsoralen 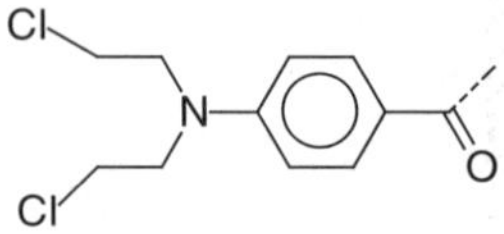   Phenylacetyl nitrogen mustard

Figure 5   *Ligands used for conjugation to PNAs*

In fact *bis*-PNA-acridine conjugates bind to relaxed duplex DNA targets at submicromolar concentration in buffer containing 150 mM K^+, 2 mM Mg^{++}.[24]

Although it has been demonstrated that decameric (*bis*)PNAs bind to their sequence-complementary target with very high kinetically controlled discrimination,[20] it has (not surprisingly) been observed that this kinetic discrimination is dramatically reduced with increasing target length. Therefore, 15–16-mer targets which would be required for uniqueness in the human genome cannot be selectively targeted by PNA oligomers. However, the kinetic control can be exploited using PNAs conjugated to another DNA-recognizing ligand that relies on thermodynamically controlled recognition with a fast on-rate, thereby poising the PNA in close proximity to its recognition site on the DNA. This principle has been exemplified using a *bis*-PNA conjugated to a minor groove-binding Hoechst analogue.[25] It was demonstrated that this PNA was able to recognize a PNA target proximal to an AT-rich Hoechst binding site up to 30 times more effectively than an "isolated" PNA target. Therefore, by using other DNA ligands such as hairpin polyamides that by themselves exhibit exquisite sequence-specificity[26] PNA conjugates that effectively recognize 16-mer DNA sequences should be feasible.

PNA conjugates may also be exploited in order to direct covalent modification to specific DNA targets as exemplified by conjugates attached to nitrogen mustard[27] or psoralen.[28] With such conjugates interstrand DNA cross-linking is specifically introduced proximal to a PNA target.

5.4   Effect of PNA Binding on DNA Structure

Obviously PNA invasion complexes have dramatic effects on the DNA structure by the mere fact that the helix is opened. However, quite dramatic decreases in the mobility of PNA–dsDNA complexes upon gel electrophoretic analysis (EMSA) have indicated that the overall properties of the DNA helix such as its flexibility and/or (directional) bending can also be affected. More detailed studies of this phenomenon have shown that triplex invasion unwinds the DNA helix by 12° per base, essentially reflecting the larger pitch of a PNA2DNA triplex (*ca.* 16 bases per turn) compared to B-DNA (10.5 bp). It also introduces an additional unwinding of about 60° at the dsDNA–PNA2DNA triplex junctions.[29] Similar changes in DNA mobility are observed for PNAds-DNA double duplex invasion complexes,[17] and in this case directional bending appears to be involved.[30] Indeed it has been shown that such complexes may mimic the function of proteins evolved by nature to induce kinks or bends in the otherwise stiff DNA helix and thereby bring distant DNA regions into proximity so as to facilitate functional interaction.[30]

5.5   Cellular Gene Targeting

Specific binding of PNA to duplex DNA targets *in vivo* or *ex vivo* in cells in culture has so far not been directly demonstrated, but has been inferred from biological effects. Using homopyrimidine PNAs Glazer and co-workers[31] have found that PNAs delivered to cells in culture can produce a 2–3 fold increase in mutation rates within or proximal to the PNA target indicating that PNA triplex invasion complexes are at least to some extent mutagenic, probably by activating the repair machinery.

Most surprisingly Cutrona *et al.*[32,33] have reported that treatment of cancer cells with mixed purine–pyrimidine sequence PNAs targeted to the sense strand of the *c-myc* (and *mycn*) gene results in a decreased level of c-myc mRNA transcript as well as cell death by apoptosis.The authors ascribe these effects to anti-gene binding of the PNA to the DNA of the *c-myc* gene, but have no evidence supporting this assumption. Indeed the accumulated knowledge from experiments *in vitro* shows that complexes between mixed purine–pyrimidine sequence PNAs and double-stranded DNA are neither thermodynamically nor kinetically stable. However, such complexes may form at DNA ends[34] or with cationic peptide-PNA conjugates directed to negatively supercoiled targets,[16] which in general can increase PNA binding by up to two orders of magnitude.[35] Thus the cellular effects of these anti-*myc* PNAs are not mechanistically understood at present, and other possible targets such as sense RNA transcripts or micro RNAs should also be considered.

It is well established that PNA-triplex invasion complexes can arrest elongating RNA polymerases *in vitro*, in particular, when positioned on the template DNA strand.[36–38] Most interestingly, Corey *et al.* recently demonstrated that mixed-sequence PNA oligomers targeted to the open transcription initiation complex of RNA polymerase can cause specific inhibition of transcription.[39] They ascribed the observed effect to binding of the PNA to the DNA loop in the open complex. This interpretation is consistent with earlier results reported by Sigman *et al.*[40–42] who showed that oligonucleotides can specifically bind to isolated *Escherichia coli* RNA polymerase open promoter complexes and thereby block transcription initiation. Moreover, the results are compatible with earlier findings showing that active transcription catalyses PNA invasion.[43]

5.6   Activation of Gene Transcription

The PNA helix invasion P-loop is reminiscent of an open transcription initiation complex with RNA polymerase, and indeed it was found that *E. coli* as well as mammalian RNA polymerase recognizes this DNA structure as a starting point for transcription.[44] There is even evidence that transcription in cells may be activated by PNA-triplex invasion complexes.[45] Thus PNA targets may function as artificial "promoters" where cognate PNA oligomers can act after the manner of transcription factors. Following a more traditional approach, conjugates between PNA oligomers and peptide transcriptional activator domains have been shown to activate transcription *in vitro*.[46–47]

5.7   Gene-Targeted Repair

The possibility of being able to correct genes in somatic cells by sequence-targeted repair has great appeal in basic science and not least in drug discovery. Although results of initial reports exploiting DNA–RNA chimeras for the purpose could not be reproduced, it was recently found that activation of homologous recombination through site-specific introduction of double-strand DNA breaks by a genetically expressed engineered nuclease can result in up to 20% gene repair in mammalian cells.[48] It is noteworthy that DNA adducts formed by PNA-psoralen[49] or PNA-nitrogen mustard[50] conjugates would induce double-strand breaks upon action by the cellular repair machinery, and it may well be that PNA conjugates of these types may analogously be employed to catalyse gene-targeted repair. Interestingly, it has been reported that gene correction of limited efficiency can be induced by a PNA–DNA chimera.[51] Thus, this exciting new area deserves further attention.

5.8   Cellular Delivery and Bioavailability *in vivo*

PNA oligomers are large, hydrophilic molecules and – like oligonucleotides and most peptides – do not diffuse passively through the lipid membrane. Furthermore, since no cellular transporters appear to be present, simple PNA oligomers are very poorly taken up by prokaryotic and eukaryotic cells, and modified PNAs and/or delivery agents are required for effective penetration into cells.[52] With cells in culture, these methods include cationic liposome-assisted delivery of PNA–DNA hybrids[53] in which the negatively charged oligonucleotide functions as a carrier of the uncharged PNA. Alternatively, association of PNA with the liposomes may be accomplished through conjugation to lipids[54] or other lipophilic entities such as polyaromatics, of which 9-aminoacridine[55] and a Hoechst analogue[56] (Figure 5) have proved particularly effective.

A variety of cationic "cell penetrating peptides" (CPPs) have also been successfully employed.[42,57] However, recent studies have shown that most of these peptides are taken up by endocytotic pathways,[58,59] and therefore the PNA conjugates have to escape from the endosomal compartment in order to reach their target within the nucleus (or cytoplasm). Endosomal release is triggered by a variety of viral peptides,[60] which could be conjugated to PNA, but which unfortunately are rather large (20-mers or larger). Alternatively, a variety of auxiliary agents such as chloroquine, Ca^{2+} ions[61] and photosensitizing dyes[62] have been found to very significantly augment the biological effects of PNA through induction of endosomal release. Thus, sufficiently effective methods are available for studies of biological effects of PNA in cell culture.

PNAs also exhibit very poor bioavailability *in vivo* (studied in mice, rats and baboons) due to very fast renal excretion ($t_{1/2}$ of $1/2$ h), and consequently good activity in whole animals requires the use of chemically modified PNAs. Simple conjugation to peptides such as oligolysine[63] or an alternating KFF peptide[64] has resulted in improved bioavailability, as has conjugation to receptor-targeted ligands such as oligo N-acetylgalactosamine[65] which has high affinity for the hepatic asialoglycoprotein receptor (ASGP-R). However, significantly improved methods for *in vivo* delivery of PNA oligomers are needed before these molecules can be effectively used for applications involving whole animals.

5.9 Prospects

Clearly sequence-specific gene targeting with PNA oligomers at the level of DNA offers great opportunities due to the variety and versatility associated with different PNA–dsDNA binding modes as well as the availability of a wide selection of chemical modifications of PNA oligomers. Nonetheless, much still needs to be learned about DNA targeting in cells and its consequences for gene expression. Also, cellular delivery and not least bioavailability of these large molecules and conjugates *in vivo* is still very much a challenge that demands attention. However, the reward – if successful – will be the possibility of manipulating gene function at the sequence level, not only in animals but also in humans; perhaps even with organ and cell specificity. PNAs are now recognized as highly promising agents for the discovery and development of novel tools for biological research as well as specifically designed gene-targeted drugs in molecular medicine.

References

1. P.E. Nielsen, M. Egholm, R.H. Berg and O. Buchardt, Sequence-selective recognition of DNA by strand displacement with a thymine-substituted polyamide, *Science*, 1991, **254**, 1497–1500.
2. M. Egholm, O. Buchardt, L. Christensen, C. Behrens, S.M. Freier, D.A. Driver, R.H. Berg, S.K. Kim, B. Nordén and P.E. Nielsen, PNA hybridizes to complementary oligonucleotides obeying the Watson–Crick hydrogen-bonding rules, *Nature*, 1993, **365**, 566–568.

3. P.E. Nielsen, M. Egholm and O. Buchardt, Evidence for (PNA)2/DNA triplex structure upon binding of PNA to dsDNA by strand displacement, *J. Mol. Recogn.*, 1994, **7**, 165–170.

4. V.V. Demidov, E. Protozanova, K.I. Izvolsky, C. Price, P.E. Nielsen and M.D. Frank-Kamenetskii, Kinetics and mechanism of the DNA double helix invasion by pseudocomplementary peptide nucleic acids, *Proc. Natl. Acad. Sci. USA*, 2002, **99**, 5953–5958.

5. P. Wittung, P.E. Nielsen, O. Buchardt, M. Egholm and B. Nordén, DNA-like double helix formed by peptide nucleic acid, *Nature*, 1994, **368**, 561–563.

6. P. Wittung, P. Nielsen and B. Nordén, Extended DNA-recognition repertoire of peptide nucleic acid (PNA): PNA–dsDNA triplex formed with cytosine-rich homopyrimidine PNA, *Biochemistry*, 1997, **36**, 7973–7979.

7. S.A. Kushon, J.P. Jordan, J.L. Seifert, H. Nielsen, P.E. Nielsen and B.A. Armitage, Effect of secondary structure on the thermodynamics and kinetics of PNA hybridization to DNA hairpins, *J. Am. Chem. Soc.*, 2001, **123**, 10805–10813.

8. B. Datta, C. Schmitt and B.A. Armitage, Formation of a PNA2-DNA2 hybrid quadruplex, *J. Am. Chem. Soc.*, 2003, **125**, 4111–4118.

9. Y. Krishnan-Ghosh, E. Stephens and S. Balasubramanian, A PNA(4) quadruplex, *J. Am. Chem. Soc.*, 2004, **126**, 5944–5945.

10. B. Petersson, B.B. Nielsen, H. Rasmussen, I.K. Larsen, M. Gajhede, P.E. Nielsen and J.S. Kastrup, Crystal structure of a partly self-complementary peptide nucleic acid (PNA) oligomer showing a duplex–triplex Network, *J. Am. Chem. Soc.*, 2005, **127**, 1424–1430.

11. H. Stender, PNA FISH: An intelligent stain for rapid diagnosis of infectious diseases, *Expert Rev. Mol. Diagn.*, 2003, **3**, 649–655.

12. G.L. Igloi, Single-nucleotide polymorphism detection using peptide nucleic acids, *Expert Rev. Mol. Diagn.*, 2003, **3**, 17–26.

13. V.L. Marin, S. Roy and B.A. Armitage, Recent advances in the development of peptide nucleic acid as a gene-targeted drug, *Expert Opin. Biol. Ther.*, 2004, **4**, 337–348.

14. O. Brandt and J.D. Hoheisel, Peptide nucleic acids on microarrays and other biosensors 50, *Trends Biotechnol.*, 2004, **22**, 617–622.

15. P.E. Nielsen and Christensen, Strand displacement binding of a duplex-forming homopurine PNA to a homopyrimidine duplex DNA target, *J. Am. Chem. Soc.*, 1996, **118**, 2287–2288.

16. X. Zhang, T. Ishihara and D.R. Corey, Strand invation by mixed base PNAs and a PNA-peptide chimera, *Nucleic Acids Res.*, 2000, **28**, 3332–3338.

17. J. Lohse, O. Dahl and P.E. Nielsen, Double duplex invasion by peptide nucleic acid: A general principle for sequence-specific targeting of double-stranded DNA, *Proc. Natl. Acad. Sci. USA*, 1999, **96**, 11804–11808.

18. T. Bentin, H.J. Larsen and P.E. Nielsen, Combined triplex/duplex invasion of double-stranded DNA by "tail-clamp" peptide nucleic acid, *Biochemistry*, 2003, **42**, 13987–13995.

19. K. Kaihatsu, R.H. Shah, X. Zhao and D.R. Corey, Extending recognition by peptide nucleic acids (PNAs): Binding to duplex DNA and inhibition of transcription by tail-clamp PNA-peptide conjugates, *Biochemistry*, 2003, **42**, 13996–14003.

20. H. Kuhn, V.V. Demidov, M.D. Frank-Kamenetskii and P.E. Nielsen, Kinetic sequence discrimination of cationic bis-PNAs upon targeting of double-stranded DNA, *Nucleic Acids Res.*, 1998, **26**, 582–587.

21. H. Kuhn, V.V. Demidov, P.E. Nielsen and M.D. Frank-Kamenetskii, An experimental study of mechanism and specificity of peptide nucleic acid (PNA) binding to duplex DNA, *J. Mol. Biol.*, 1999, **286**, 1337–1345.

22. M. Egholm, L. Christensen, K.L. Dueholm, O. Buchardt, J. Coull and P.E. Nielsen, Efficient pH-independent sequence-specific DNA binding by pseudoisocytosine-containing bis-PNA, *Nucleic Acids Res.*, 1995, **23**, 217–222.

23. M.C. Griffith, L.M. Risen, M.J. Greig, E.A. Lesnik, K.G. Sprankle, R.H. Griffey, J.S. Kiely and S.M. Freier, Single and bis peptide nucleic acids as triplexing agents: binding and stoichiometry, *J. Am. Chem. Soc.*, 1995, **117**, 831–832.

24. T. Bentin and P.E. Nielsen, Superior duplex DNA strand invasion by acridine conjugated peptide nucleic acids, *J. Am. Chem. Soc.*, 2003, **125**, 6378–6379.

25. P.E. Nielsen, K. Frederiksen and C. Behrens, Extended target sequence specificity of PNA-minor-groove binder conjugates, *Chembiochem*, 2005, **6**, 66–68.

26. P.B. Dervan and B.S. Edelson, Recognition of the DNA minor groove by pyrrole-imidazole polyamides, *Curr. Opin. Struct. Biol.*, 2003, **13**, 284–299.

27. Z.V. Zhilina, A.J. Ziemba, P.E. Nielsen and S.W. Ebbinghaus, PNA-nitrogen mustard conjugates are effective suppressors of HER-2/neu and biological tools for recognition of PNA/DNA interactions (2006) (submitted).

28. K.H. Kim, P.E. Nielsen and P.M. Glazer, Site-specific gene modification by PNAs conjugated to psoralen, *Biochemistry*, 2006, **45**, 314–323.

29. J.H. Kim, K.H. Kim, N.E. Møllegaard, P.E. Nielsen and H.S. Koo, Helical periodicity of $(PNA)_2(DNA)$ triplexes in strand displacement complexes, *Nucleic Acids Res.*, 1999, **27**, 2842–2847.

30. E. Protozanova, V.V. Demidov, V. Soldatenkov, S. Chasovskikh and M.D. Frank-Kamenetskii, Tailoring the activity of restriction endonuclease *Ple*I by PNA-induced DNA looping, *EMBO Reports*, 2002, **3**, 956–961.

31. A.F. Faruqi, M. Egholm and P.M. Glazer, Peptide nucleic acid-targeted mutagenesis of a chromosomal gene in mouse cells, *Proc. Natl. Acad. Sci. USA*, 1998, **95**, 1398–1403.

32. G. Cutrona, E.M. Carpaneto, M. Ulivi, S. Roncella, O. Landt, M. Ferrarini and L.C. Boffa, Effects in live cells of a c-*myc* anti-gene PNA linked to a nuclear localization signal, *Nat. Biotechnol.*, 2000, **18**, 300–303.

33. R. Tonelli, S. Purgato, C. Camerin, R. Fronza, F. Bologna, S. Alboresi, M. Franzoni, R. Corradini, S. Sforza, A. Faccini, J.M. Shohet, R. Marchelli and A. Pession, Anti-gene peptide nucleic acid specifically inhibits MYCN expression in human neuroblastoma cells leading to cell growth inhibition and apoptosis, *Mol. Cancer Ther.*, 2005, **4**, 779–786.

34. I.V. Smolina, V.V. Demidov, V.A. Soldatenkov, S.G. Chasovskikh and M.D. Frank-Kamenetskii, End invasion of peptide nucleic acids (PNAs) with mixed-base composition into linear DNA duplexes, *Nucleic Acids Res.*, 2005, **33**, e146.

35. T. Bentin and P.E. Nielsen, Enhanced peptide nucleic acid binding to supercoiled DNA: Possible implications for DNA "Breathing" dynamics, *Biochemistry*, 1996, **35**, 8863–8869.

36. N.J. Peffer, J.C. Hanvey, J.E. Bisi, S.A. Thomson, C.F. Hassman, S.A. Noble and and L.E. Babiss, Strand-invasion of duplex DNA by peptide nucleic acid oligomers, Proc. Nat. Acad. Sci. U. S. Am., 1993, **90**, 10648–10652.

37. P.E. Nielsen, M. Egholm and O. Buchardt, Sequence-specific transcription arrest by peptide nucleic acid bound to the DNA template strand, *Gene*, 1994, **149**, 139–145.

38. E. Guffanti, R. Corradini, S. Ottonello and G. Dieci, Functional dissection of RNA polymerase III termination using a peptide nucleic acid as a transcriptional roadblock, *J. Biol. Chem.*, 2004, **279**, 20708–20716.

39. B.A. Janowski, K. Kaihatsu, K.E. Huffman, J.C. Schwartz, R. Ram, D. Hardy, C.R. Mendelson and D.R. Corey, Inhibiting transcription of chromosomal DNA with antigene peptide nucleic acids, *Nat. Chem. Biol.*, 2005, **1**, 210–215.

40. L. Milne, Y. Xu, D.M. Perrin and D.S. Sigman, An approach to gene-specific transcription inhibition using oligonucleotides complementary to the template strand of the open complex, *Proc. Natl. Acad. Sci. USA*, 2000, **97**, 3136–3141.

41. L. Milne, D.M. Perrin and D.S. Sigman, Oligoribonucleotide-based gene-specific transcription inhibitors that target the open complex, *Methods*, 2001, **23**, 160–168.

42. J.T. Hwang, F.E. Baltasar, D.L. Cole, D.S. Sigman, C.H. Chen and M.M. Greenberg, Transcription inhibition using modified pentanucleotides, *Bioorg. Med. Chem.*, 2003, **11**, 2321–2328.

43. H.J. Larsen and P.E. Nielsen, Transcription-mediated binding of peptide nucleic acid (PNA) to double-stranded DNA: Sequence-specific suicide transcription, *Nucleic Acids Res.*, 1996, **24**, 458–463.

44. N.E. Møllegaard, O. Buchardt, M. Egholm and P.E. Nielsen, Peptide nucleic acid-DNA strand displacement loops as artificial transcription promoters, *Proc. Natl. Acad. Sci. USA*, 1994, **91**, 3892–3895.

45. G. Wang, X. Xu, B. Pace, D.A. Dean, P.M. Glazer, P. Chan, S.R. Goodman and I. Shokolenko, Peptide nucleic acid (PNA) binding-mediated induction of human g-globin gene expression, *Nucl. Res.*, 1999, **27**, 2806–2813.

46. B. Liu, Y. Han, D.R. Corey and T. Kodadek, Toward synthetic transcription activators: Recruitment of transcription factors to DNA by a PNA-peptide chimera, *J. Am. Chem. Soc.*, 2002, **124**, 1838–1839.
47. B. Liu, Y. Han, A. Ferdous, D.R. Corey and T. Kodadek, Transcription activation by a PNA-peptide chimera in a mammalian cell extract, *Chem. Biol.*, 2003, **10**, 909–916.
48. F.D. Urnov, J.C. Miller, Y.L. Lee, C.M. Beausejour, J.M. Rock, S. Augustus, A.C. Jamieson, M.H. Porteus, P.D. Gregory and M.C. Holmes, Highly efficient endogenous human gene correction using designed zinc-finger nucleases, *Nature*, 2005, **435**, 646–651.
49. K.H. Kim, P.E. Nielsen and P.M. Glazer, Site-specific gene modification by PNAs conjugated to psoralen, *Biochemistry*, 2006, **45**, 314–323.
50. Z.V. Zhilina, A.J. Ziemba, P.E. Nielsen and S.W. Ebbinghaus, PNA-nitrogen mustard conjugates are effective suppressors of HER-2/neu and biological tools for recognition of PNA/DNA interactions, *Bioconjugate Chem.*, 2006, **17**, 214–222.
51. F.A. Rogers, K.M. Vasquez, M. Egholm and P.M. Glazer, Site-directed recombination via bifunctional PNA–DNA conjugates, *Proc. Natl. Acad. Sci. USA*, 2002, **99**, 16695–16700.
52. U. Koppelhus and P.E. Nielsen, Cellular delivery of peptide nucleic acid (PNA), *Adv. Drug Deliver. Rev.*, 2003, **55**, 267–280.
53. S.E. Hamilton, C.G. Simmons, I.S. Kathiriya and D.R. Corey, Cellular delivery of peptide nucleic acids and inhibition of human telomerase, *Chem. Biol.*, 1999, **6**, 343–351.
54. T. Ljungstrøm, H. Knudsen and P.E. Nielsen, Cellular uptake of adamantyl conjugated peptide nucleic acids, *Bioconjugate Chem.*, 1999, **10**, 965–972.
55. T. Shiraishi and P.E. Nielsen, Down-regulation of MDM2 and activation of p53 in human cancer cells by antisense 9-aminoacridine-PNA (peptide nucleic acid) conjugates, *Nucleic Acids Res.*, 2004, **32**, 4893–4902.
56. T. Shiraishi, N. Nadia Bendifallah and P.E. Nielsen, Cellular delivery of polyheteroaromate-peptide nucleic acid (PNA) conjugates mediated by cationic lipids, *Bioconjugate Chem.*, 2006, **17**, 189–194.
57. K. Kilk and U. Langel, Cellular delivery of peptide nucleic acid by cell-penetrating peptides, *Methods Mol. Biol.*, 2005, **298**, 131–141.
58. U. Koppelhus, S.K. Awasthi, V. Zachar, H.U. Holst, P. Ebbesen and P.E. Nielsen, Cell-dependent differential cellular uptake of PNA, peptides, and PNA-peptide conjugates, *Antisense Nucleic Acid Drug Devel.*, 2002, **12**, 51–63.
59. M. Fotin-Mleczek, R. Fischer and R. Brock, Endocytosis and cationic cell-penetrating peptides – a merger of concepts and methods, *Curr. Pharm. Des.*, 2005, **11**, 3613–3628.
60. J.S. Wadia, R.V. Stan and S.F. Dowdy, Transducible TAT-HA fusogenic peptide enhances escape of TAT-fusion proteins after lipid raft macropinocytosis, *Nat. Med.*, 2004, **10**, 310–315.

61. T. Shiraishi, S. Pankratova and P.E. Nielsen, Calcium ions effectively enhance the effect of antisense Peptide nucleic acids conjugated to cationic tat and oligoarginine peptides, *Chem. Biol.*, 2005, **12**, 923–929.
62. T. Takehashi and P.E. Nielsen, Photochemically enhanced cellular delivery of cell penetrating peptide-PNA conjugates *FEBS Lett.*, 2006 (in press).
63. P. Sazani, F. Gemignani, S.-H. Kang, M.A. Maier, M. Manoharan, M. Persmark, D. Bortner and R. Kole, Systemically delivered antisense oligomers upregulate gene expression in mouse tissues, *Nat. Biotechnol.*, 2002, **20**, 1228–1233.
64. E. Kristensen, In *vitro and in vivo studies on pharmacokinetics and metabolism of PNA constructs in rodents*, in *Peptide Nucleic Acids: Methods and Protocols*, P.E. Nielsen (ed), Humana Press, Copenhagen, Totowa, NJ, United States, 2002, 259–269.
65. R. Hamzavi, F. Dolle, B. Tavitian, O. Dahl and P.E. Nielsen, Modulation of the pharmacokinetic properties of PNA: Preparation of galactosyl, mannosyl, fucosyl, N-acetylgalactosaminyl, and N-acetylglucosaminyl derivatives of aminoethylglycine peptide nucleic acid monomers and their incorporation into PNA oligomers, *Bioconjugate Chem.*, 2003, **14**, 941–954.

CHAPTER 6

Actinomycin D: Sixty Years of Progress in Characterizing a Sequence-Selective DNA-Binding Agent

DAVID E. GRAVES

Department of Chemistry, University of Alabama at Birmingham, 901 14th Street South, Birmingham AL 35294, USA

6.1 Summary

The interactions of small molecules with nucleic acids have provoked considerable interest in the biophysical research community for more than half a century. By virtue of their critical cellular functions, nucleic acids have long been a preferred target for therapeutic agents such as antibiotics and anticancer agents. Modifications to the informational content of nucleic acids through the DNA base sequence, structure, and/or stability have been shown to exert profound influences on nucleic acid function as evidenced by mutations, cessation of DNA replication or transcription, or selected targeting of single- and double-strand breaks modulated by the nuclear enzymes topoisomerase I and II.[1] Additionally, the interactions of small molecules with nucleic acids can be viewed as model systems for gaining insights into the subtle but profound nature of sequence-selective interactions. The sequence-dependent interactions of proteins and small molecules were long thought to be strictly driven by the structural nature of the ligand-DNA complex, wherein specific non-covalent contacts must be made between the ligand and the target DNA for sequence-selectivity to be observed. This was chiefly the case for protein-DNA interactions and is exemplified by the interactions of restriction endonucleases with their target palindromic DNA sequences.[2–4] However, more detailed studies that link both the structural and energetic properties of nucleic acids are now beginning to reveal the multifaceted nature through which the DNA base sequence directs subtle but significant changes in the energetic stabilities as well as structural features of nucleic acids. Hence, the sequence-selective interactions of small molecules or proteins with

C_3, with the numerical subscript referring to increasing R_f values. Actinomycin C was shown to exhibit potent bacteriostatic and bactericidal properties.[6]

Waksman and Tishler characterized the chemical properties of actinomycin as a quinine-like orange pigment that was highly active against Gram-positive bacteria. Although very potent, further development of the actinomycins for treatment of diseases in humans was halted because of their extreme toxicity in animals; Waksman[7] went on to successfully identify over 20 new biologically active compounds including streptomycin and neomycin, and proposed the term "antibiotic" as a class of natural compounds that act as growth inhibitors. His discovery of streptomycin led to his receiving the Nobel Prize in Physiology or Medicine in 1952. Tishler went on to become the first president of Merck Sharpe and Dohme Research Laboratory, a division of Merck & Co and remained in that position until 1970.[8]

Almost 12 years after Waksman's discovery of actinomycin, a German scientist, Christian Hackmann, exploited the therapeutic value of actinomycin C as a cytotoxic agent against lymphocytes for its potential use in treating cancer.[9,10] Upon initial animal screening at the National Cancer Institute, actinomycin was found ineffective against adult cancers. However, on closer inspection by Dr. Sidney Farber[11] at the Boston Children's Hospital in 1955, actinomycin was shown to product remissions in childhood cancers, most notably Wilms' tumor.[12] Subsequently, the use of actinomycin D as an effective treatment of cancer evolved through the 1950s is routinely used in combination with vincristine and cyclophosphamide (VAC combination chemotherapy) as an effective treatment targeted toward several cancers including choriocarcinoma, Wilms' tumor, rhabdomyosarcoma, and Kaposi's sarcoma.[13–15] Both the effectiveness and selectivity of actinomycin D against specific solid tumors has stimulated considerable interest in the mechanism(s) associated with the anticancer activity of this drug and prompted considerable research efforts to determine the structural, biophysical, and biological properties of the compound with respect to its cellular target.

The literature reports over 400 different analogues of actinomycin that have been isolated or synthesized.[16–18] Most have been characterized with respect to their biological activities and reveal key structural and/or chemical determinants that are required for activity.[19–22] The structural features of actinomycin D (shown in Figures 1 and 2) are almost unique, in that as a natural product the drug has a highly substituted planar heterocyclic phenoxazone chromophore with two cyclic pentapeptide side chains.[23–26] Interestingly, the amino acids making up the peptide side chains are extensively modified and include the sequence (L)Thr-(D)Val-(L)Pro-Sar-(L)N-methylVal. The makeup of the oligopeptide side chains differs for the various actinomycin compounds (C_1, C_2, C_{2A}, and C_3) within the family; however, the threonine at residue 1 and sarcosine at residue 4 are invariant for all of the actinomycins.[16,18,27,28] The heterocyclic phenoxazone ring carries methyl groups at the 4 and 6 positions as well as amino and keto substituents at positions 2 and 3, respectively.

Modifications to the amino acid residues result in altered bacteriostatic activity. For example, opening of the cyclic pentapeptide ring results in

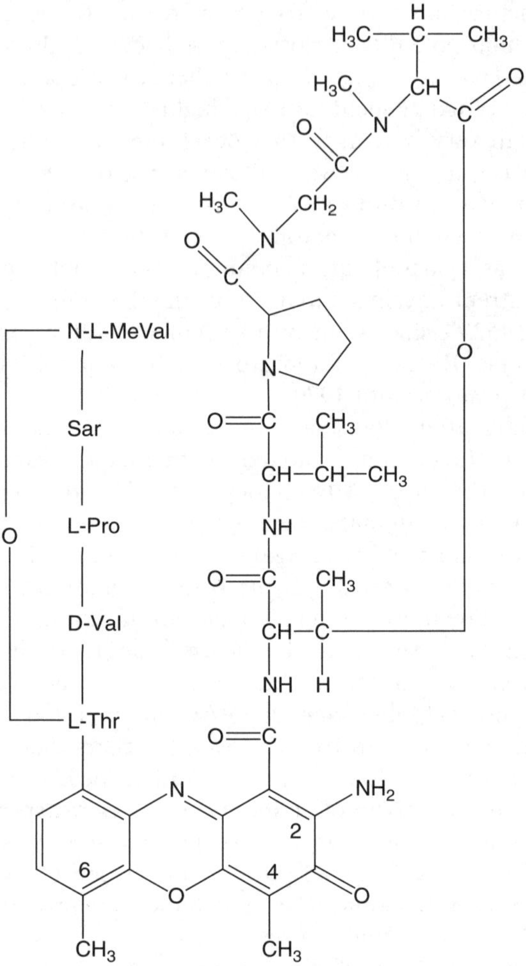

Figure 1 *Structure of actinomycin D showing the amino acid residues making up the cyclic pentapeptide side chain (right) and names of the amino acid residues (left)*

complete inactivation of the drug. Substitution of the 4′-proline with the 4′-oxoproline results in enhanced antimicrobial activity. By contrast replacement of the amino group at the 2-position on the phenoxazone ring yields low or limited activity. Replacement or removal of the methyl groups at the 4 and/ or 6 positions also results in low or limited activity. Modifications to the phenoxazone chromophore at the 7-position by substitution with Cl, Br, or NO₂ result in enhanced activity, while addition of OH and NH₂ substituents results in decreased activity.[29] Hence, the integrity of the pentapeptide ring and the presence of the 2-amino substituent were deemed essential for biological activity. Actinomycin D is moderately water soluble despite the predominantly hydrophobic nature of the amino acid residues composing the pentapeptide side chains. The solubility of the drug is, unusually, enhanced with decreasing

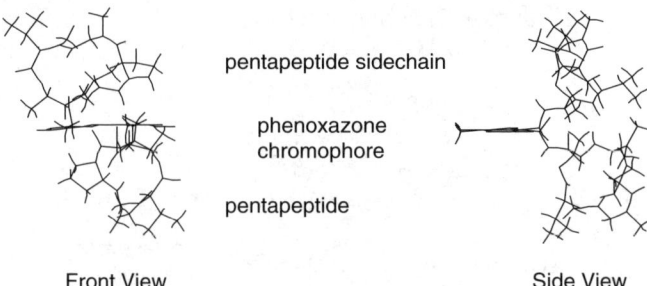

Front View   Side View

Figure 2 *Three-dimensional structure of actinomycin D as characterized by X-ray crystallography. Note the planar heterocyclic phenoxazone chromophore located in the center portion of the molecule and the two cyclic pentapeptide side chains located above and below the phenoxazone ring*

temperature. The cyclic pentapeptide side chains are relatively rigid and do not undergo significant structural change upon complex formation.

6.3 DNA-Binding Studies: The Early Years

6.3.1 The Intercalation Model

Numerous studies have provided a wealth of structural and biophysical information characterizing the interactions of actinomycin D with nucleic acids and specifically with the 5′-GpC-3′ sequence.[30–37] The actinomycin D-DNA complex is formed through intercalation of the planar phenoxazone ring between the 5′-GpC-3′ base pairs from the direction of the minor groove. The two cyclic pentapeptide side chains reside above and below the intercalated chromophore and cover four base pairs of the DNA.[38,39] Hence, the actinomycin D-DNA complex has bimodal features integrating both intercalation as well as groove binding interactions. The sequence-selectivity of actinomycin D for 5′-GpC-3′ is thought to arise through four critical hydrogen bonds that are formed between the threonine carbonyls and guanine-2-amino substituents, and the threonine amide-NH and guanine-N3 atoms. These hydrogen bonds are stabilized due to the shielding from water by the two cyclic pentapeptide chains residing in the minor groove above and below the intercalation site.[31,40] Detailed structural analyses by both X-ray diffraction and NMR methods demonstrate the effects of actinomycin D binding on structural features of the B-DNA helix that are required to accommodate both the phenoxazone chromophore and the pentapeptide side chains within the minor groove.[41–47]

Figure 3 shows an energy-minimized structure for actinomycin D intercalated into the central 5′-GpC-3′ step of an 11-mer DNA duplex containing a central – TGCA – binding site. This view depicts the intercalative stacking of the planar phenoxazone ring and the DNA bases around the drug. Also, the minor groove is widened by the presence of the cyclic pentapeptide side chains

duplex DNA sequence
(5'-CTAT**TGCA** TAC-3')

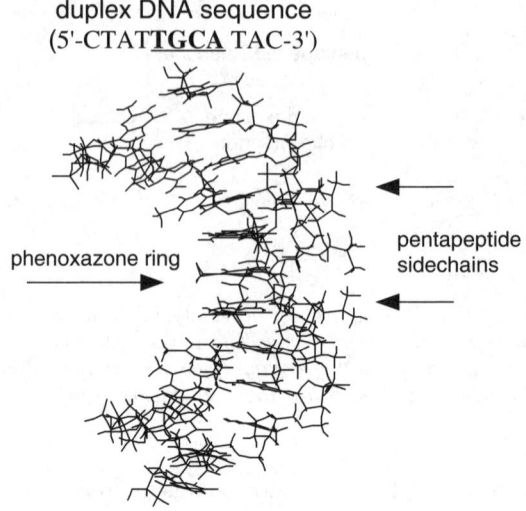

phenoxazone ring

pentapeptide
sidechains

Figure 3 *Energy-minimized structure showing the complex formed between actinomycin*
*D and the 5'-CTAT**TGCA**TAC-3'/5'-GTATGCAA-TAG-3' duplex. The view is*
from the minor groove of the DNA illustrating the intercalative geometry of the
phenoxazone ring at the GpC step and the pentapeptide side chains residing
tightly within the minor groove of the DNA. The structure was generated using
Discover and Insight II (Accelrys, Inc.)

and the DNA double helix, which is bent so that the helical axis is no longer
vertical. Notice that the side chains span a distance of two base pairs in both
directions from the intercalation site. From this view it is easy to see the
influence that the sequence, both at the intercalation site and flanking the
site, has on actinomycin D binding. The base sequence dictates the make-up of
the chemical substituents residing on the floor of the minor groove; their
positioning serves to either hinder or favor the formation of contacts with the
ligand.

For the past 44 years, actinomycin D has served as a paradigm for sequence-
selective DNA-binding ligands. In 1962, studies by Reich and Goldberg demonstrated the linkage between the binding of actinomycin D to DNA and the
guanine content of the DNA.[48] Using a variety of nucleic acids they demonstrated that actinomycin D binds to double-stranded DNA quite well; however,
the drug was shown to bind poorly if at all to double-stranded RNA, a DNA/
RNA hybrid duplex, or to a single-stranded DNA or RNA.[49] The pioneering
work by Kersten in 1960 and Felsenfeld and co-workers in 1965 laid much of
the groundwork for our understanding of the binding properties of actinomycin D with DNA.[11,25] Their studies revealed that the greater the GC
content, the higher the affinity of actinomycin D for DNA.[31,50] Of particular
interest was the observation by Felsenfeld and co-workers that the binding of
actinomycin D to calf thymus DNA appeared to be entropy-driven with a ΔS of
+31 (eu) and a ΔH near zero.[31]

Throughout the late 1960s and into the 1970s, extensive biophysical studies characterizing the binding of actinomycin D to various DNAs of heterogenous base sequence were performed to determine the hydrodynamic, kinetic, and thermodynamic properties associated with actinomycin D-DNA complex formation, as well as the linkage with anticancer activities. From these studies it was concluded that actinomycin D has a general requirement for deoxyguanosine within the duplex DNA structure, and in particular the 2-amino substituent of guanine.[30,31], [50–55] Early studies confirmed that actinomycin D binds strongly to double-strand DNAs (typically K_{int} of $1–5 \times 10^6\ M^{-1}$) with binding strength directly correlated with its GC-content, and thus the stringency of actinomycin D for deoxyguanosine-containing DNA sequences was firmly established.[30] However, in an elegant study reported by Wells and Larson (1970) on the relationship between actinomycin D-binding affinity, percent GC content, and sequence variation in synthetic polynucleotides, key insights were revealed into the promiscuity of actinomycin D for sequences other than the 5′-dGpC-3′ paradigm that would be the subject of numerous studies more than two decades later.[56] In contrast to all other known intercalating drugs, Gellert and coworkers demonstrated that the thermodynamic driving force associated with the interaction between actinomycin D and duplex DNA was a favorable entropy term, with a binding enthalpy estimated to be near zero.[31]

6.3.2 Sequence-Selectivity of Actinomycin D

Throughout the 1960s, investigators examined the complicated nature of actinomycin D-DNA interaction. The lack of actinomycin D binding at sites rich in A-T base pairs prompted speculation as to its sequence selectivity.[30] In the 1970s, UV–visible and NMR spectroscopic studies by Krugh and coworkers characterized the binding propensities of actinomycin D to a series of duplexes formed by self-complementary dinucleotides and demonstrated a clear preference for intercalation of the phenoxazone chromophore between the 5′-G and 3′-C of a 5′-GpC-3′ duplex. Other self-complementary and non-self complementary sequences had dramatically decreased affinity for the antibiotic.[33–35] Characterization of the first crystal structure of actinomycin D was accomplished by Sobell and co-workers in 1972, who proposed an intercalative binding model, wherein the actinomycin D chromophore inserts between the 5′-GpC-3′ sequence leaving the cyclic pentapeptide side chains forming hydrogen bonding contacts through the carbonyl oxygen of a threonine residue to the guanine-2-amino in the minor groove of the DNA.[32,33] The sequence-selectivity for the 5′-GpC-3′ sequence was postulated to arise from the formation of these hydrogen bonds. The interactions of the pentapeptide side chains with the minor groove of the double helix above and below the intercalation site served to shield and stabilize these hydrogen bonds from solvent exposure. Hence, a rather large drug-DNA complex was formed covering a minimum of 4 base pairs that resulted in a conformational accommodation of the bulky side chains. The major breakthrough in determining the mechanism that drives the sequence-selectivity of actinomycin D for the 5′-GpC-3′ sequence was reported by Bailly and Waring in 1993, who demonstrated through footprinting studies that the 2-amino substituent of guanine was indeed the key recognition element associated with this sequence recognition. Through specific replacement of adenine residues within a well-defined *tyr*T-DNA fragment with 2,6-diaminopurine (DAP), actinomycin D was shown to effectively target the former AT sites now bearing an additional purine 2-amino group.[57–59]

6.4   Characterization of the Actinomycin D-DNA Complex

Throughout the 1990s, further X-ray crystallography and high-resolution NMR studies were devoted to characterizing the interaction of actinomycin D with a variety of DNA sequences.[41–46] In 1994, Takusagawa and co-workers reported the structure of actinomycin D complexed with the self-complementary deoxyoctanucleotide d(GAAGCTTC)$_2$.[41] Their study revealed that upon binding actinomycin D the DNA becomes distorted and, interestingly, the resulting structure embodied cyclic pentapeptide side chains of actinomycin D residing in the minor groove that were asymmetric. Additional crystal structures and NMR structures provided key insights regarding the critical contacts between the DNA and both the drug chromophore and side chains, including hydrophobic interactions between the surface of the minor groove and the proline, sarcosine and methylvaline residues of the cyclic pentapeptides.[42] These detailed structural studies led to predictions that the bases, which flank the intercalation site would be highly influential in directing actinomycin D binding, and suggested that the 5′- flanking base adjacent to the GpC intercalation site would be restricted to either A or T, owing to hydrophobic and steric interactions.

6.4.1   Role of Bases Flanking the Actinomycin D-Binding Site

With the emergence of DNase I footprinting methods in the 1980s, highly accurate sequence preferences for DNA-binding agents could be determined. These provided clear evidence for the sequence-specific interaction of actinomycin D with DNA. Using this method, not only could the actual intercalation site be deduced but the influences of bases that flank the intercalation site could also be ascertained. This method relies on the ability of the ligand to protect bases associated with the DNA-binding site from a DNA cleaving agent, such as DNase I.[60–66] non-covalent interactions of actinomycin D with DNA were stringently examined using this technique and shown to exhibit clear preferential binding to specific base sequence motifs. Footprinting methodologies were extended to furnish *quantitative footprinting* that allowed not only sequence preferences to be determined, but also provided a reliable means for measuring the thermodynamic properties associated with the interaction of a drug at specific sites along the DNA lattice.[67–70] Through these quantitative footprinting studies, considerable insights were derived refining the sequence-selectivity of actinomycin D and revealing a diversity of intrinsic binding constants at discrete sequence locations. These studies demonstrated clear preference of actinomycin D for the sequences 5'-TGCA-3' followed by 5'-CGCG-3' > 5'-AGCT-3' > 5'-GGCC-3'. In addition, this method provided key information about additional "non-traditional" [*i.e.* non-d(GpC)] sequences that were found to furnish high-affinity-binding sites for actinomycin D.[64,66,67,71] Subsequently, more rigorous investigations examining the role of flanking base sequences on the binding affinity of actinomycin D were undertaken by Chen and co-workers. They characterized the binding interactions of actinomycin D with self-complementary deoxyoligonucleotide duplexes containing all possible 5'-XGCY-3' sequences.[73] Their studies confirmed earlier footprinting results showing that the 5'-TGCA-3' tetranucleotide was the sequence of choice, and found it to exhibit the highest actinomycin D-binding affinity followed by 5'-CGCG-3' > 5'-AGCT-3' > 5'-GGCC-3'.[72,73] Chen also showed that the rate of dissociation of actinomycin D from the 5'-TGCA-3' duplex was more than two orders of magnitude slower than from the 5'-GGCC-3' duplex, indicating that not only does the drug bind preferentially to this site, but the complex that is formed is very stable.[74] Studies reported by Graves have demonstrated that the bases surrounding the preferred 5'-GpC-3' actinomycin D-binding site significantly influence the binding energy that governs complex formation.[75–80] The isothermal calorimetric studies from the Graves laboratory reveal significant differences in the binding enthalpy associated with actinomycin D-complex formation as a result of the nature of 5'-base immediately flanking the d(GpC) sequence. Using non-self complementary 11-mer duplexes, Graves and co-workers observed similar trends to those reported by Chen as regards to relative-binding affinities and dissociation rates, with 5'-TGCA-3' being highly favored followed by 5'-CGCA-3' > 5'-AGCA-3' > 5'-GGCA-3'.[77,78] However, upon parsing the binding-free energies into enthalpic and entropic contributions, key insights were gained regarding the thermodynamic-binding mechanisms. These studies revealed that actinomycin D binding to the 5'-TGCA-3' duplex was predominately *enthalpy* driven with a ΔH of −5.9 kcal mol^{-1}. By contrast, binding of actinomycin D to the 5'-CGCA-3' and 5'-AGCA-3' duplexes showed markedly less favorable binding enthalpies (−3.4 and −3.9 kcal mol^{-1}, respectively) and more favorable entropic contributions toward complex formation.[77] The interaction of actinomycin D with the 5'-GGCA-3' duplex was too weak for accurate measurements to be obtained by isothermal calorimetry. Although the binding enthalpy values are relatively modest in magnitude as compared with other intercalating agents such as ethidium and acridines, where binding enthalpy ranges from –8 to –12 kcal mol^{-1}, the observation of a binding enthalpy of −5.9 kcal mol^{-1} for actinomycin D binding to the -5'-TGCA-3' duplex is highly significant because historically, the thermodynamic-binding mechanism ascribed to actinomycin D-DNA interaction has always been reported as an entropy-driven process.[80] Hence, there appears to be a significant influence exerted by the 5'-flanking base adjacent to the 5'-GpC-3' intercalation site governing both the magnitude of the binding affinity as well as the thermodynamic mechanism that drives complex formation.[77]

6.4.2   Promiscuity in the Sequence Selectivity of Actinomycin D

An alternative method to 'protection' footprinting by DNase I was employed by Rill and co-workers that utilized a photoreactive analogue of actinomycin D (7-azidoactinomycin D), developed by the Graves laboratory, to probe the sequence-selective interactions of actinomycin with nucleic acids.[81] Using this method, 7-azidoactinomycin D was added to the target DNA and allowed to equilibrate in the absence of light. After a time for equilibration, the ligand–DNA complex was subjected to photolysis using visible light to activate the photoreactive azido moiety to the highly reactive nitrene resulting in covalent attachment within the ligand–DNA-binding site. The drug-DNA adduct that was formed was susceptible to piperidine cleavage because of the covalent attachment of actinomycin D, and the footprint of the ligand-binding site was easily accessible through electrophoretic footprinting methods.[75,76] There are several advantages of the photoaffinity labeling footprinting methodology over that of equilibrium binding of the ligand followed by DNase I strand cleavage. First, because the ligand is covalently attached rather than reversibly bound to the DNA, much lower ligand concentrations are required to generate an accurate footprint; hence, not only are the strongest sequence-selective sites observed, but alternative sites with lower-binding affinities can also be identified. Second, the cleavage of the DNA occurs with single-base precision at the site of covalent attachment of the ligand rather than indirectly *via* protection from DNase I cleavage.[75,76] Hence, using this method Rill, Graves, and co-workers[75,76] demonstrated the accuracy of the photoaffinity footprinting method for actinomycin D and found additional high-affinity-binding sites containing both the traditional – XGCY – sequence motif as well as non-traditional "non d(GpC)" sites.

One of the most interesting outcomes of the photoaffinity footprinting experiments was the direct observation of sequence-selective binding of act-inomycin D to unorthodox binding sequences.[76] These studies revealed a novel non-traditional actinomycin D-binding site, the 5'-T(G)nT-3' sequence motif.[78] Further work on this unexpected binding motif was conducted in the Graves laboratory and revealed that the affinity of actinomycin D binding to these sites was comparable to that observed for the traditional 5'-TGCA-3' duplex.[77,78] These photoaffinity footprinting studies established that 5'-XGG-3' (where X = pyrimidine) was actually a preferred actinomycin D-binding site and prompted more extensive biophysical characterization of interactions of act-inomycin D with the T(G)nT-3' motif.[76,77] Using the non self-complementary oligonucleotide sequence 5'-CTA-T(G)nT–AC-3' where $n = 1$, 2, or 3 guanine residues, binding isotherms were constructed by spectroscopic and isothermal titration calorimetry. The results of these studies revealed that the number of G residues in the T(G)nT motif was a key determinant of the binding affinity of actinomycin D for the duplexes, with 5'-TGGGT-3' having the strongest binding affinity followed by 5'-TGGT-3' > 5'-TGGGGT-3'.[78] No binding was detectable with the 5'-TGT-3' duplex. The interactions were characterized by a relatively strong DNA-binding affinity (~10^6 M^{-1}), with enthalpy and entropy of approximately -3 kcal mol^{-1} and $+16$ eu, respectively for the TGGGGT and TGGGT duplexes. These thermodynamic values are comparable to those observed for the interaction of actinomycin D with the 5'-CGCA-3' and 5'-AGCA-3' duplexes.[77,78] By contrast, the binding enthalpy determined for the interaction of actinomycin D with the 5'-TGGT-3' duplex was markedly different. The affinity was slightly lower compared to the other sequences; however, the binding enthalpy was found to be $+3.0$ kcal mol^{-1} (a difference of more than 6 kcal mol^{-1} compared to the other sequences), indicating that the enthalpic contribution toward complex formation was decidedly unfavorable. The adverse-binding enthalpy was compensated by a markedly more favorable entropic contribution ($+41$ eu). Hence, the deletion from the DNA sequence of a single G residue (5'-TGGGT-3'–5'-TGGT-3') results in a complete reversal of the thermodynamic-binding mechanism driving complex formation between the two duplex species.[78] These observed variations in the binding enthalpy and entropy clearly demonstrate strong, though perhaps subtle, influences of base sequence on the thermodynamic mechanism(s) through which actinomycin D (and other ligands and proteins) may interact with DNA that in turn may influence functional properties of the complexes.

These observations led us to speculate whether this unexpected binding behavior could be attributed to the interaction of actinomycin D with single-stranded DNA or some other structural motif. They were supported by examination of the thermal stability in the presence of actinomycin D for the 5'-T(G)nT-3' containing duplexes. Our studies revealed that the presence of actinomycin at a 1:1 (ligand/duplex) ratio stabilized the -5'-TGGGGT-3' and 5'-TGGGT-3' duplexes by approximately 9 °C, similar to the stability enhancement observed for the 5'-TGCA-3' and 5'-CGCA-3' sequences. By contrast, the 5'-TGGT-3' duplex showed no appreciable change in thermal stability in the presence of a 1:1 ligand/duplex ratio. Accordingly, these observations led us to begin an investigation of possible interactions between actinomycin D and single-stranded DNAs.[77,78]

6.5 Global *Vs.* Microscopic Sequence-Recognition

In 1968, Müller and Crothers[30] published an elegant manuscript characterizing the biophysical properties of actinomycin D-DNA interaction. Of considerable importance in this paper were their observations of the association and dissociation kinetics of the actinomycin D-DNA complex. Complex formation with heterogeneous duplex DNA was described as a complex multi-rate process. Similarly, the dissociation of actinomycin D from heterogeneous DNA was demonstrated to be very slow and to require a minimum of three-rate constants to approximate the whole dissociation process. In 1986, the "shuffling hypothesis" was adumbrated by Fox and Waring[82] as an explanation for the time-dependent DNase I footprinting patterns that they had observed for actinomycin D-DNA recognition.[82] According to the "shuffling hypothesis", the interactions of actinomycin D with nucleic acids initially are relatively non-specific, with the drug forming many weak interactions along the DNA lattice.

These interactions rapidly dissociate and the ligand moves along the lattice from site to site until a preferred site is reached where the drug forms a high-affinity complex characterized by a markedly slower dissociation time.

6.5.1 The Shuffling Hypothesis Revisited

Collaborative efforts of Waring, Bailly, and Graves[83,84] provided the opportunity for a direct test of the shuffling hypothesis using the photoreactive analogue 7-azido-actinomycin D. In this study, the photoreactive actinomycin D analogue was allowed to interact with a 178 base pair [^{32}P]-end labeled restriction fragment for specified lengths of time ranging from 20 s to 45 min. After incubation (in the dark), the drug-DNA complex was subjected to photolysis, rendering the drug covalently bound to the DNA. The modified DNA was then treated with hot piperidine and analyzed on a sequencing gel. The results were quantitated by densitometric analysis and revealed time-dependent migration from selected sites on the DNA lattice, with classically low affinity sites decreasing in intensity (*i.e.* actinomycin D binding) and high affinity sites increasing in reactivity as shown in Figure 4. These studies provide strong support for the "shuffling hypothesis" by demonstrating the locations of the actinomycin D on the DNA lattice change significantly with time. Interestingly, even at the shortest times, the DNA sequencing patterns revealed actinomycin to be excluded from DNA sequences having a high AT content, indicating that the initial event in the sequence-recognition process requires recognition of a global structural feature such as minor groove geometry.[83,84]

The minor grooves of DNA lattices having high AT content within localized sites are narrow in comparison to other regions of DNA with high GC

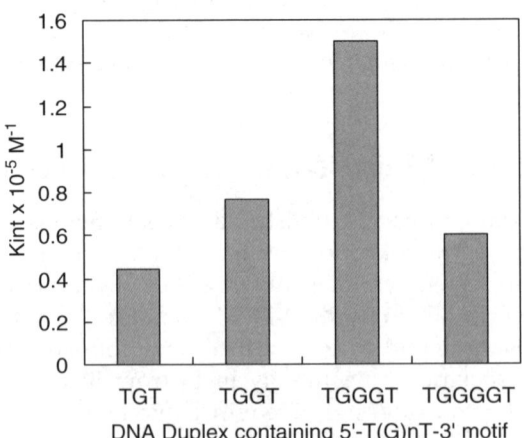

Figure 4   *Time-course of 7-azidoactinomycin D-related strand breaks for selected nucleotides within the DNA sequence 5'-GTGAGC$_{46}$GG$_{44}$AT-3'. As observed from the plot, position 44 (left panel) shows a decrease over time in the amount of actinomycin bound to the G residue there. By contrast, the C residue at position 46 shows a marked increase in actinomycin concentration as a function of time (see ref. 83 for details)*

or mixed sequences. The lack of actinomycin D binding to the AT-rich sites even at early equilibration times may indicate an initial recognition of some sequence-dependent structural determinant necessary for actinomycin D binding. By the same analogy, DNA sequences with high GC or mixed content seemed to accommodate actinomycin D quite readily so that early in the reaction time the ligand could be observed at many such sequences. As time went on the ligand was observed to accumulate steadily at specific sequences (*i.e.* 5'-GpC-3' and 5'-GGG-3'), while many of the initially reactive sites lost their reactivity (Figure 4). The concomitant decreases and increases in band intensity observed at neighboring nucleotide positions on the sequencing gels are best interpreted as being due to the "shuffling" of single-ligand molecules from nonspecific-binding sites to preferred 5'-GpC-3' containing sites best visualized as a redistribution of ligand molecules between local sites within the sequence (Figure 5).[83]

The ultimate determinants of actinomycin D sequence-recognition reside in the specific contacts made between the ligand (*i.e.* phenoxazone ring and pentapeptide side chains) and functional groups of the DNA both at the site of chromophore insertion and within the minor groove at bases flanking the intercalation site.[57–59] Optimal binding of actinomycin D relies on the availability and formation of specific hydrogen bonds, van der Waals contacts, and hydrophobic interactions between the ligand and DNA, all of which translate into unique thermodynamic components that contribute to the "thermodynamic origin" of sequence-recognition.

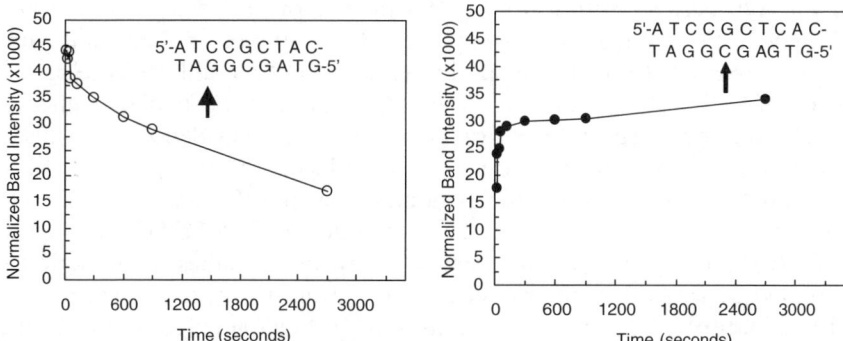

Figure 5 *Histogram showing the influence of the number of centrally located G residues in a DNA duplex having the sequence 5'-CTA T(G)nT TAC-3' (where n is equal to 1, 2, 3, or 4 guanine residues) on the intrinsic binding affinity of actinomycin D. Absorbance titrations were carried out in 10 cm cylindrical quartz cells at a constant temperature of 10 ⚥C. DNA-binding isotherms were obtained by adding aliquots of actinomycin D to fixed concentrations of DNA. To minimize aggregation, the level of free actinomycin D was maintained below 5 μM throughout all binding studies. Concentrations of bound actinomycin D were determined by monitoring changes in the drug absorbance. Theoretical fits of the binding data were calculated using a nonlinear least-squares analysis (Origin, Microcal Inc.)*

6.6   Structural Motifs as Actinomycin D Targets

6.6.1   The Era of Single-Strand DNA Binding

In the early 1990s, the once stringent sequence-selectivity that had so fittingly defined actinomycin D-DNA interactions was being eroded into a plethora of sequence promiscuity. Numerous manuscripts from different laboratories were reporting the binding of actinomycin D to a variety of non-GpC sequences.[85–87] At the same time that the Graves laboratory began to probe the binding of actinomycin D to single-stranded DNA, Wadkins Jovin, Rill, Chen, and co-workers[88–94] were also reporting the binding of actinomycin D to a number of non-GC sequences and indeed, to non-duplex DNA structures. The mechanistic ramifications of such actinomycin D interactions are broad with demonstrated inhibitory activity toward viral ligase, helicase, and inhibition of (–) strand transfer in the replication of HIV and other retroviral reverse transcriptases.[95–97] Early on, these single-strand binding studies revealed interaction with both single-strands as well as 5'-GpC-3' sites in hairpin stems in a manner analogous to interactions with duplex DNAs. In 1997, Jovin and co-workers[98] demonstrated that actinomycin D would bind to single-strand DNA that was lacking 5'-GpC-3' sequences.[98–100] Further investigations led Wadkins and later Chen to demonstrate binding of actinomycin D to DNA hairpins and it appeared that the drug could be involved in inducing hairpin formation.[90] In detailed studies, Chen examined the binding of actinomycin D to the sequence 5'-CGTCGTCG-3' and made single base changes to identify the critical base required for drug binding. His studies revealed very tight binding (10^7 M^{-1}), and through base substitution the 3'-G was demonstrated to be essential for actinomycin D binding. Conclusions from these studies implied that a monomeric hairpin was formed wherein the actinomycin D would prefer to have the 3'-sides of both G bases stacking on the opposite faces of its planar phenoxazone chromophore. This monomeric hairpin was not in evidence in the absence of actinomycin D, and appeared to be induced by the presence of the drug. Further evidence of such induced hairpin formation came from Chen in his characterization of the interactions of 5'-TGTCT(n)G-3' and 5'-TGT(n)GTCT-3' with actinomycin D.[101] In these studies monomeric hairpins could be formed in the presence of the drug in which the 3'-end of the DNA strand folded back to form a hairpin, with actinomycin D displacing a T juxtaposing two G bases thereby allowing stacking against both faces of the chromophore.[101] Rill and co-workers[93] described the binding of actinomycin D to an unstructured, single-stranded DNA of sequence 5'-AACCATAG-3'. The binding to this single-strand octanucleotide was very strong ($~10^7$ M^{-1} strand) and did not require any apparent preexisting hairpin or secondary structure prior to binding. However, the 5'-TAG-3' sequence was demonstrated to be essential for actinomycin D binding to occur. In addition to this sequence specificity, Rill and co-workers noted that a minimum of eight nucleotides were required for tight binding. These studies indicate that the initial binding of actinomycin D to single-strand DNA may result in substantial conformational changes to the flanking sequences in order to accommodate the drug and achieve a tight actinomycin D-single-strand DNA complex.[93]

Studies from the Graves laboratory that focused on the single-strand deca-deoxynucleotide 5'-GTAACCATAG-3' revealed significant interactions with actinomycin D. Although we initially started our studies with a working hypothesis, which supposed that the central – ACCA – was directing the binding, we subsequently determined that the 3'-G is in fact the driving force for the actinomycin D interaction. Removal of the 5'-G from the other end did not adversely affect the binding of actinomycin D to the 5'- TAACCATAG-3' sequence. By contrast, removal of the 3'-G resulted in total loss of actinomycin D binding (unpublished results). Hence, the role of the 3'G in the binding of actinomycin D may reside in stabilizing the stacking of actinomycin D on the face of the G base paired with one of the central C's in the DNA strand, resulting in the formation of a stable pseudo-hairpin structure as postulated by Chen and co-workers.

The most recent evidence to date comes from a 2005 manuscript wherein Jovin and co-workers reported the detailed crystallographic, NMR, and molecular dynamics characterization of unique structures that are formed between 7-amino-actinomycin D and single-stranded DNAs of sequence 5'-TTAGTT-3', 5'-TTAG^{Br}UT-3' and 5'-TTTAGTTT-3'.[102] They found that actinomycin D forms two crystalline complexes (orthorhombic and trigonal) in a 2:4 drug/ DNA ratio. In both cases the phenoxazone chromophore of actinomycin D appears stacked between the central G residues of two 5'-TTAG^{Br}UT-3' strands in a manner schematically shown below

$$5'- T \ T \ A \ G \ ^{Br}U \ T\text{-}3'$$

$$| \ | \ X \ | \ |$$

$$3'\text{-}T \ ^{Br}U \ G \ A \ T \ T\text{-}5'$$

The complex formed between actinomycin D and the non-self complementary d(TTAG ^{Br}UT), d(TTAGTT), and d(TTTAGTTT) sequences appears to display features more closely associated with duplex DNA than with purely single-strand or hairpin DNA structures, in which the actinomycin D phenoxazone chromophore is stacked between the guanine residues of two antiparallel DNA strands. The two guanines are paired to the ^{Br}U bases on the opposite strand, and additional A-T base pairs are formed between the strands providing additional stability to the complex. Even though neither of the DNA strands would form any energetically favored duplex or hairpin conformation by itself, in the presence of actinomycin D the phenoxazone chromophore serves as an anchor and the two pentapeptide side chains appear to lock the strands into an ordered conformation (Figure 6).[98]

The four DNA strands are conformationally distinct and form two antiparallel partially base paired double strands with extensive base stacking and some base pairing. The actinomycin D chromophore is sandwiched between the two guanines of two antiparallel strands. The bases on each strand appear to easily

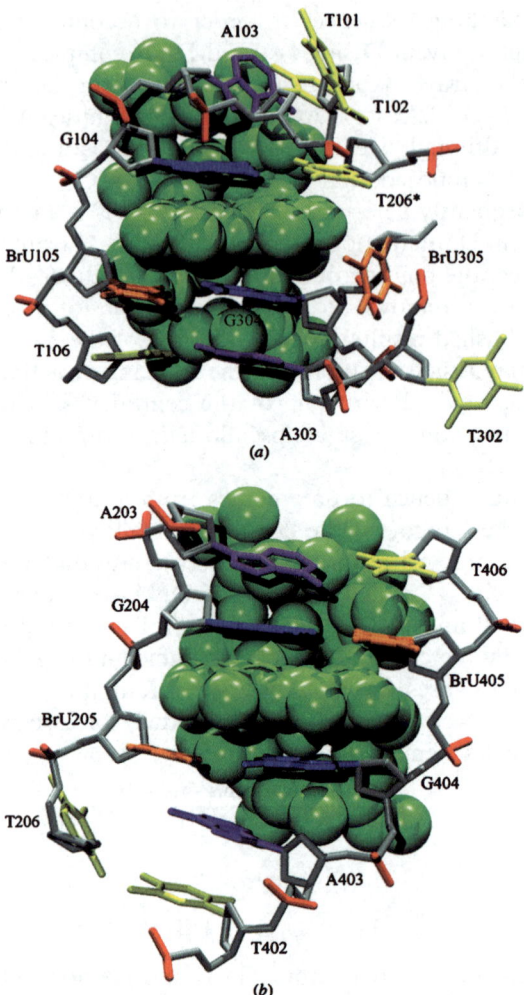

Figure 6 *Structure derived from X-ray crystallography of actinomycin D bound to single-stranded DNA d(TTAG^{Br}UT) reproduced from E. Alexopoulos, E.A. Jares-Erijman, T.M. Jovin, R. Klement, R. Machinek, G.M. Sheldrick, and I. Uson, Acta Crystallographica D, 2005, **61**, 407–415 by kind permission of the author and the journal. The upper panel shows the complex (unit 1) that consists of strands 1 and 3 interacting with actinomycin molecule 1. The lower panel shows strands 2 and 4 interacting with a second molecule of actinomycin D. Color coding guanine, blue; thymine, yellow; adenine, pink; bromouracil, orange; sugar, grey; and phosphate, red*

accommodate looped-out structures to facilitate both base stacking and base pairing with the opposite strand.

  These results reveal new insights into the binding properties of actinomycin D with single-stranded DNA. In the previous hemi-intercalation model proposed by Jovin, the phenoxazone chromophore was stacked between a 5′-ApG-3′

sequence. Based upon the structural results derived from crystallography and NMR, the model is revised so that the actinomycin D chromophore now lies on the 3'-side of the guanine residue. The molecular dynamics simulations reveal that the thymine residues flanking the 5'-AGT-3' core are quite mobile. Additionally, the stability of the G-^{Br}U base pairing is tenuous and therefore does not contribute significantly to the stability of the complex that is formed. Instead, the principal driving force for the drug interaction with this DNA sequence arises from the tight stacking and hydrogen bonding interactions between the phenoxazone chromophore and the guanine residue. This apparent tight stacking and hydrogen bonding serves as an anchor for additional DNA strands to then assimilate and add to the ligand–DNA complex. From these data, Jovin and co-workers propose a model for the assembly process wherein the actinomycin D and DNA strand begin to associate *via* a single-stacking interaction between the G residue and the phenoxazone chromaphore. This 1:1 complex then acquires a second (antiparallel) DNA strand with concomitant stacking between the phenoxazone chromophore and the guanine of the second strand, forming a 1:2 complex. Finally, the complex aggregates into a 2:4 drug/strand complex with quasi-antiparallel double-stranded helical structure.

6.7 Conclusions

From the initial characterizations of the interaction of actinomycin D with nucleic acids, the drug was purported to bind only to duplex DNA. However, unlike other intercalating drugs, actinomycin D exhibited a unique mechanism of action through its targeting of RNA polymerase and inhibition of transcription. More than 40 years of intense biophysical studies have led to a realization that in fact the binding of actinomycin D to DNA is much more promiscuous that once believed, both as regards its base sequence specificity as well as its DNA structural selectivity. Although the bulk of the 60 years of studies of actinomycin D have primarily focused on complexes formed with duplex DNA molecules, the past decade has dealt primarily with the binding of actinomycin to non-duplex structures including hairpin and single-stranded DNAs. Considerable advances have been made toward our understanding of mechanism(s) that drive both sequence-selectivity as well as complexes formed with non-duplex DNA structures. However, with the latest evidence reported by Jovin's 2005 manuscript, it appears that actinomycin D will find a way to form stable complexes with many forms of DNA – whether it be duplex, hairpin, or single-strand in nature. Close scrutiny of this most recent work indicates that actinomycin D behaves much like molecular Velcro when it comes into interactions with nucleic acids, specifically targeting the guanine residue (or more accurately, 2-amino purine). With the stacking and hydrogen bonding interactions anchoring the phenoxazone chromophore to the guanine and added protection from solvent provided by the pentapeptide side chains, the DNA strands are mobile enough to find optimal geometries to assemble into stable stacked complexes around the actinomycin D anchor. It is salutary that after 60 years of investigations, we continue to be amazed by the novel properties that make actinomycin D such a fascinating product of Nature.

Acknowledgments

Supported by NSF-MCB 0334785 from the National Science Foundation. I thank Dr. Nichola Garbett and Jessica Record for their comments and work. I also thank Dr. Isabel Usón (author) and Dr. Louise Jones (editor) of *Acta Crystallographica D* for permission to include a figure from their manuscript in this work.

References

1. M.J. Waring, *Ann. Rev. Biochem.*, 1981, **50**, 159–192.
2. G.A. Thomas, W.L. Kubasek, W.L. Peticolas, P. Greene, J. Grable and J.M. Rosenberg, *Biochemistry*, 1989, **28**, 2001–2009.
3. J.A. McClarin, C.A. Fredrick, B.C. Wang, P. Greene, H.W. Boyer, J. Grable and J.M. Rosenberg, *Science*, 1986, **234**, 1526–1541.
4. M.C. Needels, S.R. Fried, R. Love, J.M. Rosenberg, H.W. Boyer and J.P. Green, *Proc. Natl. Acad. Sci. USA*, 1989, **86**, 3579–3583.
5. K. Fox and M.J. Waring, *Method. Enzymol.*, 2001, **340**, 412–430.
6. S.A. Waksman and H.B. Woodruff, *J. Bacteriol.*, 1940, **40**, 581–600.
7. S.A. Waksman and H.B. Woodruff, *J. Bacteriol.*, 1941, **42**, 231–249.
8. N. Kresge, R.D. Simoni and R.L. Hill, *J. Biol. Chem.*, 2004, **279**, 7–8.
9. C. Hackmann, *Z. Krebsforsch*, 1952, **58**, 607–613.
10. C. Hackmann, *Z. Krebsforsch*, 1954, **60**, 250–255.
11. S. Farber, G. D'Angilo, A. Evans and A. Mitus, *Ann. N.Y. Acad. Sci.*, 1960, **89**, 421–425.
12. S. Farber, G. D'Angilo, A. Evans and A. Mitus, *J. Urology*, 2002, **168**, 2560–2562.
13. S. Farber, *J. Am. Med. Assoc.*, 1966, **198**, 826–836.
14. S.L. Lewis, *Cancer*, 1972, **30**, 1517–1521.
15. E. Frei, *Cancer Chemoth. Rep.*, 1974, **58**, 49–54.
16. H. Brockmann, *Cancer Chemoth. Rep.*, 1974, **58**, 9–20.
17. S.K. Sengupta, Y. Kogan, C. Kelly and J. Szabo, *J. Med. Chem.*, 1988, **31**, 768–774.
18. R.H. Shafer, R.R. Burnette and P.A. Mirau, *Nucleic Acids Res.*, 1980, **8**, 1121–1132.
19. W. Kersten, H. Kersten and H.M. Rauer, *Nature*, 1960, **187**, 60–61.
20. E. Reich, R.M. Franklin and A.J. Shatkin, *Science*, 1961, **134**, 556–557.
21. E. Reich and I.H. Goldberg, *Prog. Nucleic Acid Res. Mol. Biol.*, 1964, **3**, 183–234.
22. I.H. Goldberg, *Am. J. Med.*, 1965, **39**, 722–752.
23. E. Katz, in *Antibiotics II*, D. Gottlieb and P.D. Shaw (eds), Springer, New York, 1967, 276.
24. U. Hollstein, *J. Am. Chem. Soc.*, 1974, **96**, 8036–8040.
25. E. Bullock and A.W. Johnson, *J. Chem. Soc. UK*, 1957, **51**, 3280–3285.

26. S.S. Danyluk and T.A. Victor, in *Quantum Aspects of Heterocyclic Compounds Chem. Biochem. Proc., Int. Symp.*, 2nd edn, E.D. Bergmann (ed), Israel Acad. Sci. Hum., Jerusalem, Israel, 1970, 394–411.
27. P.A. Mirau and R.H. Shafer, *Biochemistry*, 1982, **21**, 2622–2626.
28. P.A. Mirau and R.H. Shafer, *Biochemistry*, 1982, **21**, 2626–2631.
29. J. Meienhofer, *Cancer Chemoth. Rep.*, 1974, **58**, 21–34.
30. W. Muller and D.M. Crothers, *J. Mol. Biol.*, 1968, **35**, 251–290.
31. M. Gellert, C.E. Smith, D. Neville and G. Felsenfeld, *J. Mol. Biol.*, 1965, **11**, 445–457.
32. R.D. Wells and J.E. Larson, *J. Mol. Biol.*, 1970, **49**, 319–342.
33. T.R. Krugh, *Proc. Natl. Acad. Sci. USA*, 1972, **69**, 1911–1914.
34. T.R. Krugh and Y.C. Chem, *Biochemistry*, 1975, **14**, 4912–4922.
35. T.R. Krugh and J.W. Neely, *Biochemistry*, 1973, **12**, 1775–1782.
36. D.J. Patel, *Biochemistry*, 1974, **13**, 1476–1482.
37. D.J. Patel, *Biochemistry*, 1974, **14**, 2396–2402.
38. H.M. Sobell, *Prog. Nucleic Acid Res. Mol. Biol.*, 1973, **13**, 153–190.
39. S.C. Jain and H.M. Sobell, *J. Mol. Biol.*, 1972, **68**, 1–20.
40. D.M. Crothers and D.I. Ratner, *Biochemistry*, 1968, **7**, 1823–1827.
41. S. Kamitori and F. Takusagawa, *J. Am. Chem. Soc.*, 1994, **116**, 4154–4165.
42. H. Robinson, Y.-G. Gao, R. Sanishvili, A. Joachimiak and A.H.-J. Wang, *Biochemistry*, 2001, **40**, 5587–5592.
43. C. Lian, H. Robinson and A.H.J. Wang, *J. Am. Chem. Soc.*, 1996, **118**, 8791–8801.
44. D.R. Brown, M. Kurz, D.R. Kearns and V.L. Hsu, *Biochemistry*, 1994, **33**, 651–664.
45. H. Chen, X. Liu and D.J. Patel, *J. Mol. Biol.*, 1996, **258**, 457–479.
46. S.H. Chou, K.H. Chin and F.M. Chen, *Proc. Natl. Acad. Sci. USA*, 2002, **99**, 6625–6630.
47. P.A. Mirau and R.H. Shafer, *Biochemistry*, 1982, **21**, 2626–2631.
48. E. Reich, I.H. Goldberg and M. Rabinowitz, *Nature*, 1962, **196**, 743–748.
49. R. Haselkorn, *Science*, 1964, **143**, 682–684.
50. W. Kersten, *Biochim. Biophys. Acta*, 1961, **47**, 610–611.
51. M.J. Waring, *J. Mol. Biol.*, 1970, **54**, 247–279.
52. M. Hogan, N. Dattagupta and D.M. Crothers, *Biochemistry*, 1979, **18**, 280–288.
53. J.C. Wang, *Biochim. Biophys. Acta*, 1971, **232**, 246–251.
54. D.J. Patel, *Acc. Chem. Res.*, 1979, **12**, 118–125.
55. H.M. Sobell, C.C. Tsai, S.C. Jain and S.G. Gilbert, *J. Mol. Biol.*, 1977, **114**, 333–345.
56. R.D. Wells, *Science*, 1969, **165**, 75–76.
57. C. Marchand, C. Bailly, M.L. McLean, S.E. Moroney and M.J. Waring, *Nucleic Acids Res.*, 1992, **20**, 5601–5606.
58. M.J. Waring and C. Bailly, *Gene*, 1994, **49**, 69–79.
59. C. Bailly, C. Marchand and M.J. Waring, *J. Am. Chem. Soc.*, 1993, **115**, 3784–3785.

60. M.J. Lane, J.C. Dabrowiak and J.N. Vournakis, *Proc. Natl. Acad. Sci. USA*, 1983, **80**, 3260–3264.
61. Scamrov, A.V. and Beabealashvilli, R. S. (1983) FEBS Lett. 164, 97–101.
62. A.V. Van Dyke and P.B. Dervan, *Biochemistry*, 1983, **22**, 2373–2377.
63. M.W. Van Dyke, R.P. Hertzberg and P.B. Dervan, *Proc. Natl. Acad. Sci. USA*, 1983, **79**, 5470–5474.
64. K.R. Fox and M.J. Waring, *Nuc. Acids Res.*, 1984, **12**, 9271–9285.
65. R.J. White and D.R. Phillips, *Biochemistry*, 1989, **28**, 6259–6269.
66. R. Rehfuss, J. Goodisman and J.C. Dabrowiak, *Biochemistry*, 1990, **29**, 777–781.
67. M. Shudsba, H. Kishikawa, J. Goodisman and J. Dabrowiak, *J. Mol. Recognit.*, 1994, **7**, 133–139.
68. J.B. Chaires, J.E. Herrera and M.J. Waring, *Biochemistry*, 1990, **29**, 6145–6153.
69. J.B. Chaires, J.E. Herrera, M. Britt, K.R. Fox and M.J. Waring, *Biochem. Pharmacol.*, 1988, **37**, 1785–1786.
70. J.B. Chaires, K.R. Fox, J.E. Herrera, M. Britt and M.J. Waring, *Biochemistry*, 1987, **26**, 8227–8237.
71. K. Waterloh and K.R. Fox, *Biochim. Biophys. Acta*, 1992, **131**, 300–306.
72. F.M. Chen, *Biochemistry*, 1992, **31**, 6223–6228.
73. C. Liu and F.M. Chen, *Biochemistry*, 1996, **35**, 16346–16353.
74. F.M. Chen, C.M. Jones and Q.L. Johnson, *Biochemistry*, 32, 5554–5559.
75. G.A. Marsch, D.E. Graves and R.L. Rill, *Nucleic Acids Res.*, 1995, **23**, 1252–1259.
76. R.L. Rill, G.A. Marsch and D.E. Graves, *J. Biomol. Struct. Dyn.*, 1989, **7**, 591–605.
77. S.A. Bailey, D.E. Graves, R.L. Rill and G. Marsch, *Biochemistry*, 1993, **32**, 5881–5887.
78. S.A. Bailey, D.E. Graves and R.L. Rill, *Biochemistry*, 1994, **33**, 11493–11500.
79. S.A. Bailey, D.E. Graves and M.R. Eftink, *Proceeding of SPIE – The International Society for Optical Engineering*, 1994, **2137**, 462–468.
80. J. Ren, T.C. Jenkins and J.B. Chaires, *Biochemistry*, 2000, **39**, 8439–8447.
81. D.E. Graves and R.M. Wadkins, *J. Biol. Chem.*, 1989, **264**, 7262–7266.
82. K.R. Fox and M.J. Waring, *Nucleic Acids Res.*, 1986, **14**, 2001–2014.
83. C. Bailly, D.E. Graves, G. Ridge and M.J. Waring, *Biochemistry*, 1994, **33**, 8736–8745.
84. G.S. Ridge, C. Bailly, D.E. Graves and M.J. Waring, *Nucleic Acids Res.*, 1994, **22**, 5241–5246.
85. E. Scott, G. Zon, L.G. Marzilli and W.D. Wilson, *Biochemistry*, 1988, **27**, 7940–7951.
86. J.G. Snyder, N.G. Hartman, B.L. D'Estantoit, O. Kennard, D.P. Remeta and K.J. Breslauer, *Proc. Natl. Acad. Sci. USA*, 1989, **86**, 3968–3972.
87. F.M. Chen, *Biochemistry*, 1998, **37**, 3955–3964.
88. R.M. Wadkins and T.M. Jovin, *Biochemistry*, 1991, **30**, 9469–9478.

89. R.M. Wadkins, E.A. Jares-Erijman, R. Klement, A. Ruediger and T.M. Jovin, *J. Mol. Biol.*, 1996, **262**, 53–68.
90. R.M. Wadkins, B. Vladu and C.S. Tung, *Biochemistry*, 1998, **37**, 11915–11923.
91. F. Sha and F.M. Chen, *Biophys. J.*, 2000, **79**, 2095–2104.
92. F.M. Chen and F. Sha, *Biochemistry*, 2001, **40**, 5218–5225.
93. H. Yoo and R.L. Rill, *J. Mol. Recog.*, 2001, **14**, 145–150.
94. F.M. Chen and F. Sha, *Biochemistry*, 2002, **41**, 5043–5049.
95. W.R. Davis, S. Gabbara, D. Hupe and J.A. Peliska, *Biochemistry*, 1998, **37**, 14213–14221.
96. J. Guo, T. Wu, J. Bess, L.E. Henderson and J.G. Levin, *J. Virology*, 1998, **72**, 6716–6724.
97. T. Imamichi, T.P. Conrads, M. Zhou, Y. Liu, J.W. Adlesberger, T.D. Veenstra and H.C. Lane, *J. Acq. Immn. Def. Synd.*, 2005, **40**, 388–397.
98. E.A. Jares-Erijman, R. Klement, R. Machinek, R.M. Wadkins, B.I. Kankia, L. Marky and T.M. Jovin, *Nucleos.Nucleot.*, 1997, **16**, 661–667.
99. F.M. Chen, *Biochemistry*, 2001, **41**, 5043–5049.
100. F.M. Chen, F. Sha, K.H. Chin and S.H. Chou, *Nucleic Acids Res.*, 2003, **31**, 4238–4246.
101. F.M. Chen, F. Sha, K.H. Chin and S.H. Chou, *Biophys. J.*, 2003, **84**, 432–439.
102. E. Alexopoulos, E.A. Jares-Erijman, T.M. Jovin, R. Klement, R. Machinek, G.M. Sheldrick and I. Uson, *Acta Cryst. D*, 2005, **61**, 407–415.

Thermal Denaturation of Drug–DNA Complexes: Tools and Tricks

JONATHAN B. CHAIRES AND XIAOCHUN SHI

James Graham Brown Cancer Center, University of Louisville, 529 S. Jackson St., Louisville KY 40202, USA

7.1 Introduction

Only a few months after Watson and Crick presented their model for the DNA double helix,[1] the alkaline and thermal denaturation of DNA, as monitored by changes in UV absorbance, was reported.[2] This seminal work explicitly recognized that the observed hyperchromism resulted from " . . . the destruction of a secondary molecular structure constituted by labile bonds involving the puric and pyrimidic rings", and specifically referred to the Watson–Crick structure. Soon after, the Doty laboratory published a remarkable and prescient series of studies on the acid, alkaline, and thermal denaturation of DNA, employing a wide array of biophysical tools to show that the duplex strands separated upon denaturation.[3–10] That line of research led directly to the well-known paper by Marmur and Doty[11] that related the DNA melting temperature (T_m) to the GC content of the DNA. Concurrently, statistical mechanical theories for the effect of preferential ligand binding on the helix-to-coil transition were developed,[12,13] albeit with explicit reference to protein transitions rather than to DNA melting. These theories were general, however, and were equally applicable to DNA transitions. Only five years after the appearance of the Watson–Crick model, calorimetry was used to study the enthalpy of the acid denaturation of DNA.[14]

Against this background, thermal denaturation became a commonly used tool for the study of drug–DNA complexes from the early 1960s onward. The attraction of melting studies lies in their simplicity and readily available, inexpensive, instrumentation. Melting provides a simple and unambiguous demonstration of drug binding to DNA. Small molecules that bind preferentially to the DNA duplex stabilize the structure and elevate its T_m. Intercalators and groove binders both recognize particular features of duplex DNA, and consequently raise the T_m. Apart from a simple qualitative demonstration of binding to DNA, melting studies can be analyzed to obtain quantitative information about the binding interaction. A variety of approaches for such quantitative analysis have appeared over the years,[15–19] and new approaches continue to be developed.[20] One purpose of this chapter is to illustrate the utility of such methods in the quantitative analysis of drug–DNA-melting curves to obtain thermodynamic data.

Practical procedures and detailed protocols for properly conducting DNA melting experiments have been published,[21–23] and will not be repeated here. The focus, instead, will be on the analysis and interpretation of such data once the experiment is completed.

Current interest in drug–DNA interactions lies in the specificity of the binding process. Do DNA binders recognize particular sequences or structures? While the question of binding specificity is usually addressed by chemical and enzymatic footprinting methods, or by competition dialysis, the potential of thermal denaturation methods in this area should not be overlooked. Some recently developed tricks to use UV-melting methods to explore sequence and structural selectivity of drug binding will be described and illustrated to conclude this chapter.

7.2   Thermal Denaturation Tools

7.2.1   Analysis of T_m Shifts in the Presence of Drug

Figure 1 shows sample melting curves for the polydeoxynucleotide poly dA:poly dT in the presence of increasing concentrations of the groove binder netropsin.[24,25] These experiments show that the melting temperature of the duplex DNA is dramatically increased upon addition of netropsin. At the

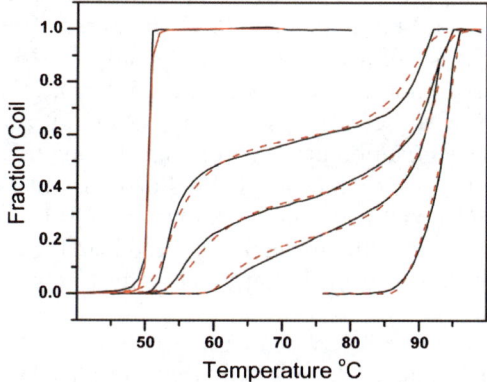

Figure 1   *Thermal denaturation of poly dA:poly dT in the presence of netropsin. The polynucleotide concentration was 45 μM bp. The molar ratio of added netropsin (drug/bp) was, from left to right: 0; 0.08; 0.14; 0.18; 0.28*

highest molar ratio of added netropsin shown in Figure 1, the T_m is elevated from 50.5 to 93.2°C, a difference of 42.7 degrees. In the absence of netropsin, or at saturating concentrations, the melting transition is sharp, with the transition complete over a span of just a few degrees. At lower molar ratios, below saturation, melting curves are seen to be broad and multiphasic. As will be discussed in more detail later, such behavior is a consequence of a complex underlying mechanism that involves ligand redistribution over the course of duplex melting. Qualitatively, the melting curves shown in Figure 1 show unambiguously that netropsin binds preferentially to the duplex form, stabilizing it against denaturation. Were netropsin to bind preferentially to single strands, a decrease in the melting temperature would be expected. The experimental behavior in Figure 1 is simply an illustration of Le Chatelier's principle at work. Perturbation of the helix-to-coil equilibrium (by drug binding) shifts the reaction in the direction of the favored species, the duplex in this case. Apart from such qualitative conclusions, what quantitative information might be extracted from the data in Figure 1?

This is where theories of DNA melting in the presence of ligands are needed. Among the various treatments, those presented by Crothers[15] and by McGhee[17] have proven to be the most accessible and useful. Both McGhee and Crothers provided closed-form equations that permit quantitative interpretation of T_m shifts under certain limiting conditions. Under conditions where the DNA duplex is saturated with ligand, and where there is assumed to be no binding to single-stranded forms, the shift in melting temperature can be described by relatively simple equations. McGhee provided the equation

$$\frac{1}{T_m^0} - \frac{1}{T_m} = \frac{R}{\Delta H_{DNA}} \ln\left[(1 + K_h L)^{1/n}\right] \tag{1}$$

where T_m^0 and T_m are the melting temperatures in the absence and presence of saturating concentrations of drug, respectively, expressed in degrees Kelvin. In Equation (1), R is the gas constant, ΔH_{DNA} is the enthalpy of DNA melting (per bp), K_h is the ligand-binding constant to the duplex form, L is the ligand concentration, and n is (approximately) the ligand site size in bp. For a full discussion of the origin and limitations of Equation (1), the original reference should be studied.[17] Crothers provided an equation with a slightly different algebraic form but which embodies the same underlying mechanism.

By using Equation (1), estimates of the binding constant K_h can be obtained *if* the enthalpy of DNA melting is known, along with a value for the site size (n). The ligand concentration can normally be assumed to be the total added. Binding constants obtained from Equation (1) refer to the temperature for the melting of the complex.

Figures 2 and 3 map the behavior of Equation (1) as a function of the key variables K_h and n. Figure 2 shows that the interplay between the binding constant and the site size results in a complex surface. The contour representation of that surface (Figure 3, right) shows that K_h and n are tightly coupled, such that several combinations of these variables can elevate T_m to the same

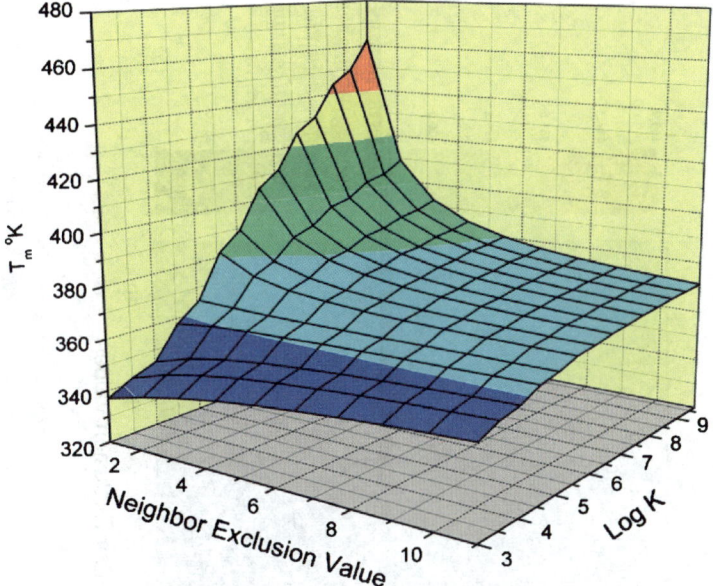

Figure 2 *Surface map computed for values of T_m from Equation (1). Values for K and n were systematically varied, with the remaining parameters held constant, with $T_m^0 = 333.15$ K, $\Delta H_{DNA} = 8.0$ kcal mol^{-1}, and $L = 20$ µM*

extent. ΔT_m values thus are not direct indications of binding affinity, as is often erroneously asserted, since the shift in T_m is influenced just as strongly by the site size. Figures 3A and B show slices of the surface in Figure 2 to emphasize the isolated effects of K_h and n. With a constant binding constant (Figure 3A) the magnitude of the elevation of T_m is seen to decrease with increasing site size. That behavior is somewhat counterintuitive. When the site size is constant (Figure 3B), the magnitude of the elevation of T_m increases with increasing binding affinity. For a series of compounds, ΔT_m values provide a measure of the relative binding affinity *only* if the binding site size is identical for all compounds.

For netropsin, with $n = 4$, a ΔT_m value of 53.1°C was measured for a total ligand concentration of 75 µM for the melting of poly dA:poly dT at 0.3 mM bp. By using Equation (1), a binding constant of 1.5×10^8 M^{-1} was measured at 103.6°C. Correction of this binding constant to lower temperatures requires knowledge of the binding enthalpy, which can be obtained by differential scanning calorimetry (DSC).

7.2.2 Obtaining Binding Enthalpy Values by DSC

Figure 4 shows the results of DSC experiments for the melting of poly dA:poly dT in the presence and absence of netropsin. Higher polynucleotide concentrations are required for the DSC experiment compared to the optical melting

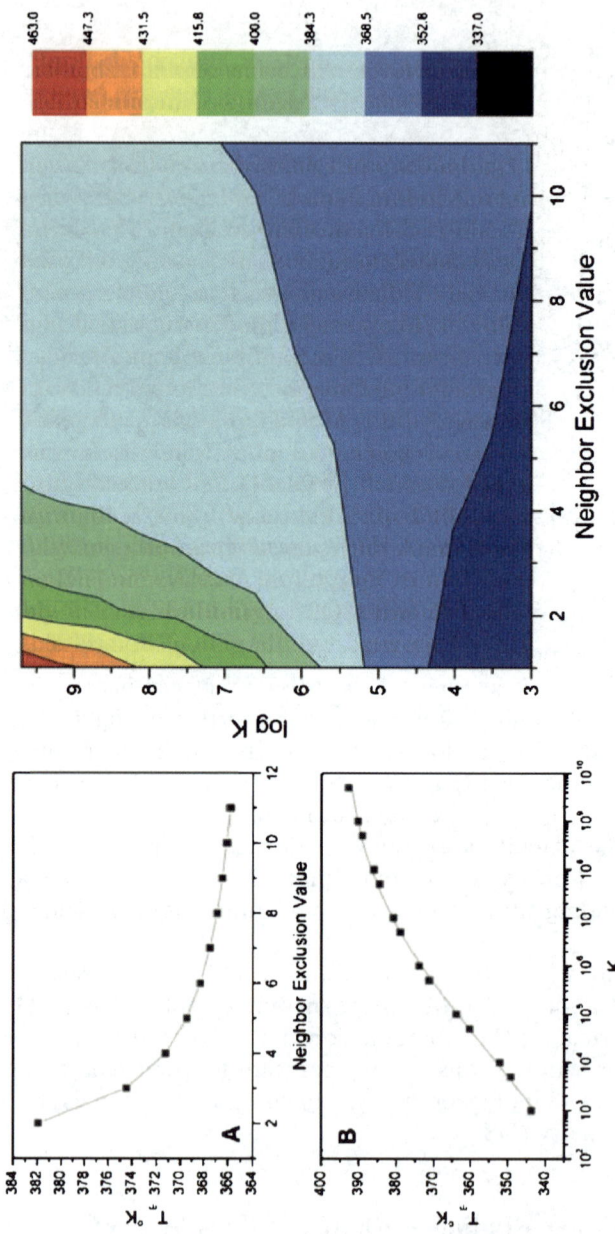

Figure 3 *Isolated effects of K and n on the elevation of T_m. The right side shows a contour representation of the surface shown in Figure 2. Panel A shows a slice across the surface with $K = 5 \times 10^5$ M^{-1} and n varied. Panel B shows a slice with n = 4 and K varied*

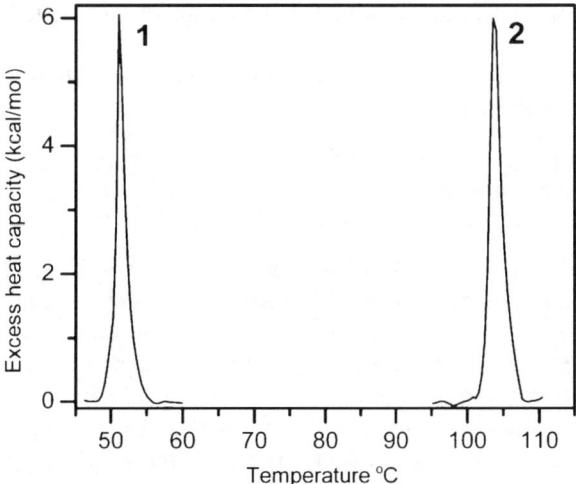

Figure 4 *Differential scanning calorimetry experiment with poly dA:poly dT in the absence (1) and presence (2) of netropsin*

experiment. With saturating amounts of netropsin, a dramatic elevation in ΔT_m is again evident, but DSC has the added value of providing enthalpy values directly from the peak areas seen in Figure 4.

Analysis of DSC data to obtain ligand-binding enthalpies uses Hess's law.[26] The following equilibria are considered:

1. duplex $\leftrightarrow$ 2(single strands) $\qquad\qquad\qquad\qquad$ ΔH_1
2. duplex–ligand $\leftrightarrow$ 2(single strands) + ligand $\qquad\quad$ ΔH_2
3. duplex–ligand $\leftrightarrow$ duplex + ligand $\qquad\qquad\quad$ $\Delta H_3 = \Delta H_2 - \Delta H_1$

The derived reaction 3 is the desired binding enthalpy, and can be obtained from the areas of the two peaks shown in Figure 4. The enthalpy ΔH_3 needs to be corrected for amount bound (r_b), and its sign changed to correspond to binding

$$\Delta H_b = -(\Delta H_3/r_b)$$

For the data shown in Figure 4, ΔH_1 was determined to be 10.6 kcal mol^{-1} and ΔH_2 was determined to be 13.6 kcal mol^{-1}. The binding ratio r_b was determined to be 0.25, leading to an estimate of $\Delta H_b = -12.0$ kcal mol^{-1}.

Once ΔH_b is known, the binding constant can be calculated for any desired temperature from the van't Hoff equation

$$\ln\left(\frac{K_2}{K_1}\right) = -\frac{\Delta H_b}{R}\left(\frac{1}{T_2} - \frac{1}{T_1}\right) \qquad\qquad (2)$$

For netropsin, this yields a value for the equilibrium constant of 1.4×10^{10} M^{-1} at 20°C.

The complete thermodynamic profile for netropsin can then be easily derived from standard relationships:

$$\Delta G_{20} = -RT\ln K = -13.6 \text{ kcal mol}^{-1}$$
$$\Delta H_b = -12.0 \text{ kcal mol}^{-1}$$
$$-T\Delta S = \Delta G_{20} - \Delta H_b = -1.6 \text{ kcal mol}^{-1}$$

Netropsin binding to DNA is thus driven by favorable contributions from both enthalpy and entropy. While this approach provides valuable quantitative data, it must be stressed that application of the methods and theory require conditions where the DNA lattice is saturated with ligand. In addition, a rather long extrapolation is required to correct values to lower standard temperatures. Possible heat capacity changes are, of necessity, neglected, perhaps introducing serious error into the estimates of parameters at the lower temperatures. Little can be done to alleviate that drawback, unless the particular drug–DNA system of interest is amenable to study by isothermal titration calorimetry.[27]

7.2.3 Modeling Melting Curves by McGhee's Algorithm

It is possible to analyze melting curves under conditions where the DNA lattice is not fully saturated. The whole family of melting curves shown in Figure 1 can be analyzed using a statistical mechanical model developed by McGhee and implemented into a Fortran program that contains some 1000 lines of code.[17,18] The essence of the model is portrayed in Figure 5. DNA can exist as a duplex helix or single-stranded coil. The helix–coil transition is described by a nearest-neighbor Ising model with nucleation parameter σ and a propagation constant s. Ligand binding to both helix or coil regions is allowed, with each described by neighbor-exclusion binding with a binding constant K, a neighbor-exclusion parameter n, and a cooperativity parameter ω. Both the helix propagation constant (s) and ligand binding constants (K) are temperature-dependent as governed by melting and binding enthalpy values. For a selected set of parameter values, McGhee's program will compute a melting curve that may

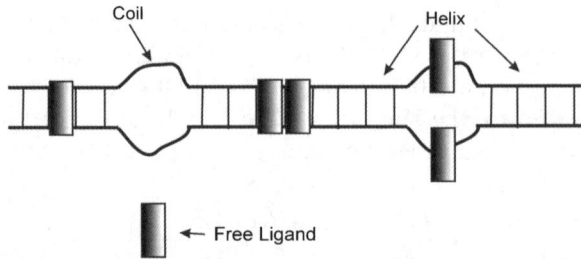

Figure 5 *Schematic representation of McGhee's model for the melting of DNA in the presence of ligand*

Table 1 *Binding constants obtained from the melting curves shown in Figure 1 for the interaction of netropsin with poly dA poly dT[a]*

[Netropsin] μM	Molar Ratio Nt/DNA	$K/10^{10}$ M^{-1}	n bp	σ	ω
0	0	0	0	5×10^{-5}	1
3.7	0.08	1.3	4	6×10^{-4}	1
6.5	0.14	1.3	4	5×10^{-4}	1
8.0	0.18	1.3	4	1×10^{-5}	1
12.5	0.28	1.3	4	1×10^{-5}	1

[a] Binding parameters were obtained by successive approximation "fits" to the experimental data using McGhee's statistical mechanical mode.[17,18] The concentration of $poly(dA) \cdot poly(dT)$ was fixed at 45 μM bp. [Netropsin] is the concentration of netropsin used in the experiments. The parameters K (the binding constant at 20°C), n (the binding site size), σ (the nucleation parameter), and ω (the ligand cooperativity constant) were systematically varied to produce the optimal fit to the experimental data. The following parameters were constrained in the simulation: $T_m = 50.5°C$, $\Delta H_m = -10.6$ kcal mol^{-1}, and $\Delta H_b = -12$ kcal mol^{-1}.

be compared to a normalized experimental curve. By successive approximation and with judicious constraint of some parameters, it is possible to estimate a set of parameters that describe an entire family of melting curves. The procedure is described in detail in a useful review by Spink and Wellman.[18] Figure 1 shows optimized "fits" to the experimental data for the netropsin system, with parameter values shown in Table 1. In this case, binding to single-stranded DNA was found to be negligible. Inspection of Table 1 shows, remarkably but as expected, that a single set of binding and melting parameters can (within error) accurately predict the entire family of melting curves. Only the changes in total and free netropsin concentrations dictate the complex shapes and positions of melting curves below saturation.

The complex shapes of melting curves evident in Figure 1 deserve comment and explanation. Their complexity, as revealed by McGhee's model, arises from redistribution of ligand over the course of helix denaturation. Binding sites are, by the model, assumed to be homogeneous along the lattice. At less than saturating ligand concentrations, as one region of the helix melts, ligand is transiently released, but then rebinds to any remaining duplex regions. These duplex regions thus become increasingly saturated over the course of the melt, leading to broad, multiphasic transition curves for which there is no well defined or simple T_m value.

7.2.4 Case Studies: Bisintercalating Anthracyclines and Echinomycin

The thermodynamic characterization of netropsin binding as described above is fully consistent with previous studies of the system,[28–30] providing validation of the experimental methods. Additional case studies will now be reviewed that demonstrate the further applicability of these methods, and which also reveal some significant advantages of the approach. These examples are drawn from studies of two types of *bis*intercalating compounds, first, a new class of rationally designed *bis*anthracyclines.[31–33] and, second, the natural product echinomycin.[34] Each of these cases posed particular difficulties that rendered traditional spectrophotometric titration methods useless, making the melting methods described here the only viable alternative available for quantitative studies.

The anthracycline antibiotics daunorubicin and doxorubicin have been mainstays of cancer chemotherapy.[35–37] A new class of *bis*anthracyclines (Figure 6) was rationally designed from considerations of available high-resolution crystal structures of anthracycline–DNA complexes. In high-resolution structures of daunorubicin–DNA complexes, two drug molecules are bound to a DNA hexanucleotide.[38] The two drug molecules are generally intercalated near the ends of the oligonucleotide in a tail-to-tail arrangement, such that their daunosamine moieties lie in the minor groove and are within 7 Å; of one another. The proximity of their reactive amine groups suggested a simple design strategy. By simply crosslinking two monomers through their amines using an appropriate linker, a new type of *bis*anthracycline might be synthesized. Monomers crosslinked through their amine groups would feature an optimal stereochemical fit within the DNA minor groove, with no hindrance to intercalation, and with added favorable minor groove interactions from the linker. The design was realized using a *p*-xylenyl linker to covalently link two daunorubicin molecules to form *bis*daunorubicin, WP631 (Figure 6).[33]

*Bis*intercalators ought to possess enormously enhanced binding affinity relative to their monointercalating substituents. The binding constant for a *bis*intercalator should equal the *square* of the monomer binding constant,[39] $K_B = K_M \times K_M = K_M^2$. Since daunorubicin binds to DNA with $K_M \approx 10^6$ M^{-1}, bisdaunorubicin ought to bind to DNA with picomolar affinity, $K_B \approx 10^{12}$ M^{-1}. Such a tight binding poses practical problems for the experimental determination of binding constants using traditional spectrophotometric titrations.[40] For accurate determinations of binding isotherms, one must work at concentrations near the reciprocal of the binding constant. At picomolar concentrations, it becomes difficult to accurately monitor absorbance or fluorescence. Fortunately, the melting methods described here do not depend on a signal from the binding ligand, and can be used to measure binding affinity.

For WP631, ultratight binding was in fact observed, and measured by the thermal denaturation methods described in this chapter.[31,33] A binding constant of 3.1×10^{11} M^{-1} was determined, with a binding enthalpy of -30.2 kcal mol^{-1}. The binding constant is nearly the square of that measured for the monomer daunorubicin under the same conditions. WP631 was subsequently found to exert potent biological activity and proved to be an effective inhibitor of the DNA binding of the transcription factor SP1.[41–43]

While the binding affinity of WP631 to DNA was impressive indeed, the fact that it was one order of magnitude lower than expected suggested that improvements to the design could be made. The thermodynamic profile obtained for WP631 suggested that rotation around linker bonds might contribute to an unfavorable entropic cost for immobilization of the linker. An improved design was thus suggested. Immobilization of linker bonds and

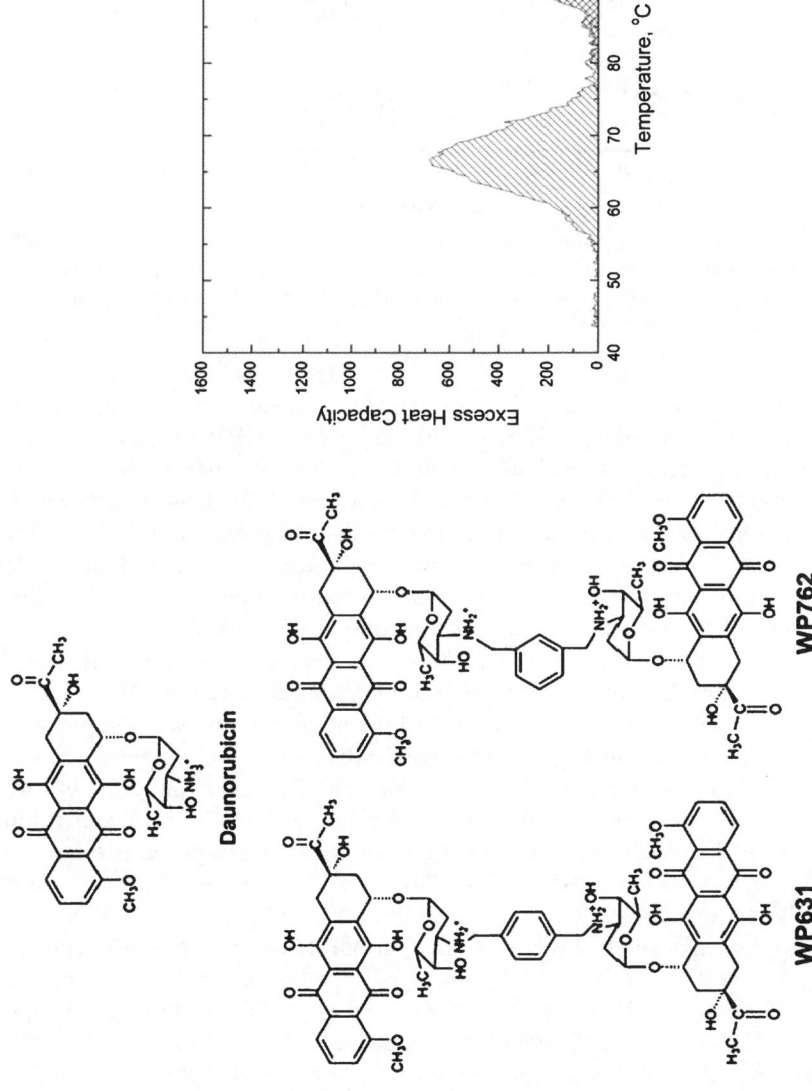

Figure 6 *Bisanthracycline compounds. The parent antibiotic daunorubicin (monomer) is shown along with WP631 and WP762, in which monomers are linked by a p-xylenyl and m-xylenyl linker, respectively. The results of a DSC experiment with herring sperm DNA and WP762 are shown on the right side*

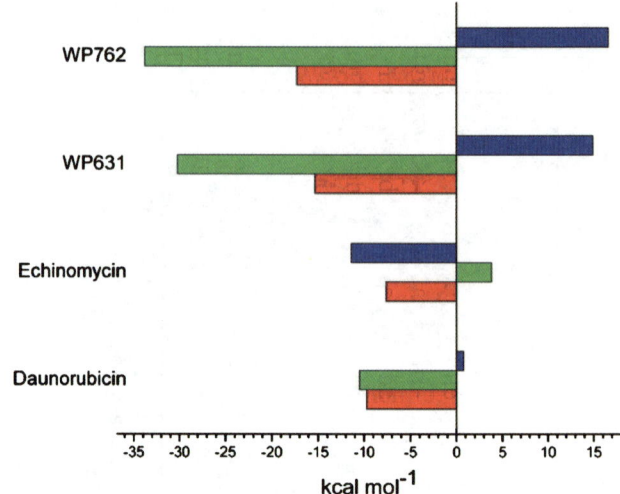

Figure 7   *Thermodynamic profiles for the binding of ligands to DNA as determined by thermal denaturation methods. Red fill represents binding free energy, green fill represents binding enthalpy, and blue fill represents the entropic contribution, -TΔS, all in kcal mol^{-1}*

optimization of the fit of the linker into the minor groove might improve the already ultratight binding. This proved to be correct. WP762 (Figure 6) was a member of the second generation of *bis*anthracyclines that featured an *m*-xylenyl linker.[44] The different linker slightly adjusted the length between the monomer units, and reduced rotation around linker bonds. Sample DSC data are shown in Figure 6 for WP762, and a complete binding and molecular modeling study of its binding to DNA has recently appeared.[44] The binding constant of WP762 was in fact increased in comparison to WP631, to a value of 7.3×10^{12} M^{-1}. Its binding enthalpy was increased to -33.8 kcal mol^{-1}. Anthracycline binding thermodynamics are compared in Figure 7.

These studies on the *bis*anthracyclines illustrate and emphasize the power of thermal denaturation methods to provide quantitative characterization of ultratight binding. Complete thermodynamic profiles may be obtained in these cases where optical methods cannot be used because of the low concentrations required to properly characterize the equilibrium. At the required low concentrations, the necessary absorbance or fluorescence signals are too weak to be accurately measured.

Another example of the advantage of thermal denaturation approaches for the characterization of binding thermodynamics is provided by the natural product echinomycin (Figure 8). Echinomycin was the first *bis*intercalator to be discovered.[34] Echinomycin is uncharged and hydrophobic. Studies of its DNA binding were hampered by its poor solubility, about 5 μM in aqueous solution. Out of necessity (and perhaps desperation), phase partition methods were devised to study echinomycin binding to DNA.[45] While the phase partition

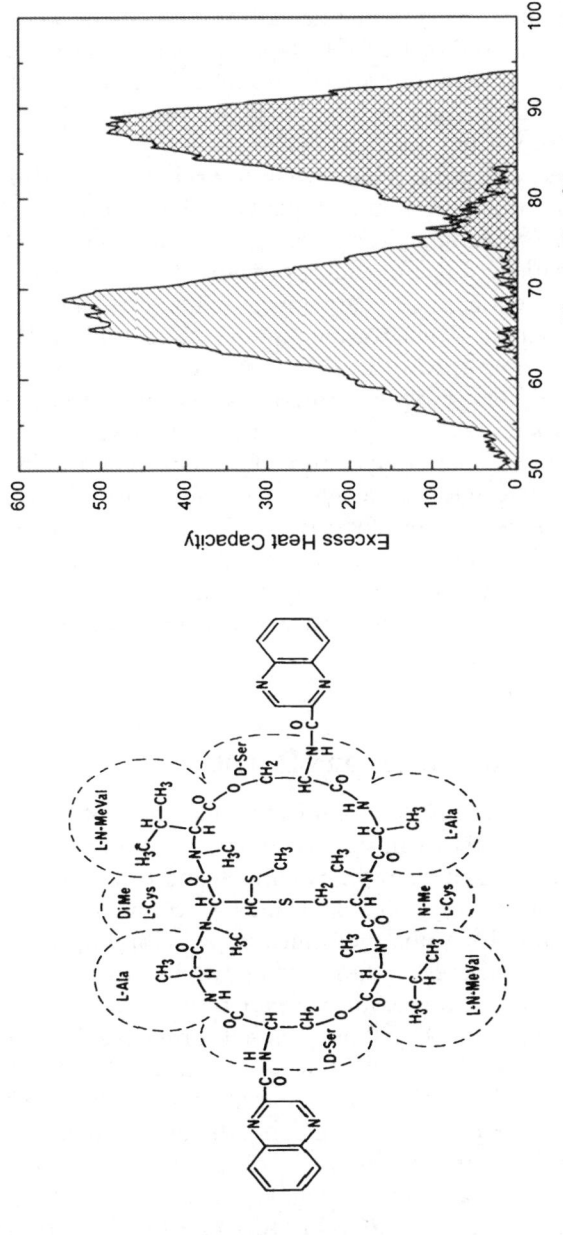

Figure 8 *Echinomycin (left) and its effect on the denaturation of herring sperm DNA (right)*

method can reliably determine binding constants, studies of the temperature dependence of binding constants is both difficult and fraught with error because of the simultaneous temperature dependence of the drug partition coefficient. Proper evaluation of the thermodynamic profile was thus difficult, and it was only recently that thermal denaturation methods were successfully applied to alleviate the difficulties.[46] In this case, since thermal denaturation studies are initiated with a preformed drug–DNA complex, the solubility problem is lessened. The requisite complex can be formed by the "solid-shake" protocol devised by Waring and co-workers, in which a solution containing DNA is equilibrated with solid echinomycin to bring the lattice to any desired degree of saturation. Figure 8 shows sample DSC data. Optical melting studies over a wide range of binding ratios were also done, and may be found in the original publication.[46] These studies provided the first direct calorimetric determination of the enthalpy of echinomycin binding to DNA, and yielded a complete thermodynamic profile for the binding interaction. It was found that $\Delta G^{\circ} = -7.6$ kcal mol^{-1}, $\Delta H = +3.8$ kcal mol^{-1}, and $\Delta S = +38.9$ cal mol^{-1} K^{-1} at 20°C. The binding reaction is clearly entropically driven, a hallmark of a process that is predominantly stabilized by hydrophobic interactions. The binding thermodynamics of echinomycin provide an interesting contrast to the *bis*anthracylines, as summarized in Figure 7. Echinomycin binding is entropically driven, with a positive binding enthalpy, whereas binding of the *bis*anthracyclines is enthalpically driven, and is opposed by entropy. The difference in binding thermodynamics points to clear differences in the molecular forces that drive the binding of the two types of *bis*intercalators.

7.2.5   Summary: Advantages and Pitfalls

These studies collectively illustrate significant advantages of the thermal denaturation methods. First, they can be used to study ultratight binding interactions, since no optical signal from the binding ligand is required. Second, they can be used for even sparingly soluble ligands, since the drug–DNA complex required for the methods can be preformed by equilibrating DNA with solid ligand. Finally, DSC measures binding enthalpy as directly as possible, yielding complete thermodynamic profiles for complex formation.

There are possible pitfalls. For simple, closed form equations to be used for analysis of T_m shift, the DNA lattice must be saturated. If the lattice is not saturated, complex multiphasic melting curves will result, which cannot be easily analyzed or interpreted. The magnitude of the increase in melting temperature, even at saturation, is not a simple, unambiguous measure of affinity. K and n values are tightly coupled, and many combinations may combine to produce similar T_m shifts. In order to estimate K, some value of n must be known or assumed. Finally, while DSC returns an estimate of binding enthalpy at the melting temperature, it is difficult to obtain values for the heat capacity change for binding, $\Delta C_p = \mathrm{d}\Delta H/\mathrm{d}T$. Neglect of ΔC_p may lead to systematic errors in the extrapolation of binding constants to lower temperatures.

Finally, it is necessary to caution that all the discussion presented here refers to polymeric DNA. The theories of McGhee and Crothers assumed a long nucleic acid lattice with homogeneous binding sites. The analysis described here is inappropriate for use with short oligonucleotides.

7.3 Thermal Denaturation: New Tricks

7.3.1 Melting Mixtures to Assess Sequence- and Structural-Selectivity

Nucleic acids are an important target for new therapeutic agents.[47–53] Two fundamental strategies exist for targeting nucleic acids. The first is to target specific sequences that are vital for the control of gene expression (transcription factor binding sites, for example) with small molecule inhibitors.[54–56] A second strategy is to target non-canonical structures that may regulate gene activity. Daunorubicin and its synthetic enantiomer, for example, can act as mutual allosteric effectors to switch DNA between right- and left-handed forms.[57] G-quadruplexes are of intense current interest as structurally unique targets, and have been implicated as important structural elements in both the biology of telomeres and in the regulation of gene expression[51,58,59](see also Chapters 9 and 10).

The evolution of new tools for the study of sequence- and structural-selective ligand binding is important for efficient drug discovery. Chemical and enzymatic footprinting methods revolutionized studies of sequence-selective recognition of DNA by small molecules, enabling identification of ligand binding sites to base pair resolution.[60–62] A competition dialysis method was recently devised that provides a rapid means of quantitatively evaluating ligand sequence- and structural-selectivity by using an array of nucleic acid samples.[63–66] Thermal denaturation methods provide a thermodynamically sound approach for quantifying drug–nucleic acid interactions as described here, but have been hampered by the relatively slow rate at which data can be acquired. While thermal denaturation methods are powerful, rigorous, and readily automated, they are hampered by comparatively low-throughput. Samples are generally run serially, and a typical melting curve takes several hours to accumulate. Comparison of ligand binding to many sequences or structures is thus time consuming. In an attempt to overcome such problems, a rapid and frugal thermal denaturation assay using molecular beacons was devised to study ligand interactions with duplex, triplex, and quadruplex nucleic acids.[67]

Another simple new thermal denaturation assay that greatly facilitates comparison of ligand binding to different nucleic acid sequences or structures was recently developed in our laboratory.[68] The assay is based on the simple expediency of melting mixtures of polynucleotides with different sequences or structures whose melting temperatures are well resolved. Addition of ligand to such mixtures at appropriate molar ratios results in a shift of the melting temperature of the nucleic acid containing the preferred structure or sequence, providing a clear indication of ligand selectivity. Two applications of the assay will be illustrated here, first with a mixture to test sequence selectivity and second with a mixture to test structural selectivity. These initial applications utilize assays that contain four–five sequences or structures. The concept is robust, however, and can potentially be expanded to include large numbers of samples of particular interest.

A few additional comments concerning the strategy and design of the assay are needed. As shown above, the melting of nucleic acids in the presence of ligands is complex. The apparent T_m shift has a simple quantitative meaning only under conditions where the nucleic acid lattice is fully saturated with ligand. At ligand concentrations where the lattice is not fully saturated, melting curves become multiphasic due to ligand redistribution, and the shift in T_m has no simple interpretation or meaning. By design, the new assay utilizes low molar ratios of ligand, where the lattice is far from saturation. This is opposite from the typical experimental design usually used in melting experiments,[69] but is an essential condition for visualizing sequence or structural selectivity. Sequence- or structural-selective binding is most clearly manifested in the limit as the binding ratio approaches zero.[70,71] The intent is to provide an assay that provides a rapid, qualitative demonstration of selective binding, one that clearly shows a T_m shift for the melting of the preferred nucleic acid. The assay is, by design, optimized for demonstrating selectivity in binding, and is not intended to be used for quantitative analysis of T_m shifts. Once the assay identifies interesting types of selective binding, rigorous biophysical studies can follow to characterize the thermodynamics of binding quantitatively. The situation is analogous to early footprinting methodologies that focused on the qualitative identification of preferred binding sequences rather than on quantitative analysis of binding affinities.

7.3.1.1 *Assay for Sequence Selectivity*

An assay for base and sequence selectivity was devised by preparing an equimolar mixture of [Poly (dA-dT)]$_2$, Poly (dA)·poly (dT), Poly (dA-dC)·poly (dG-dT), Poly (dC)·poly (dG), and [Poly(dG-dC)]$_2$. The T_m values of these duplexes are well separated. Collectively, the polynucleotide mixture contains 8 of the 10 unique dinucleotide steps (AA = TT, AT, TA, AC, CA, GG = CC, GC, CG), and 8 of the 32 unique triplet sequences (AAA, GGG, ATA, TAT, GCG, CGC, ACA, CAC). Figure 9 shows the results of addition of several compounds with known sequence preferences to the mixture. Low molar ratios of compound ($\sim$0.01 drug bp^{-1}) were added to the mixture because such conditions are optimal for the selection of a particular binding site from the mixture.

Netropsin (Figure 9A) shows a clear preference for binding to poly (dA)·poly (dT) and to a lesser extent, [poly (dA-dT)]$_2$, as is evident from the shifts in peaks 2 and 1 (Figure 9A). This result is fully consistent with the known

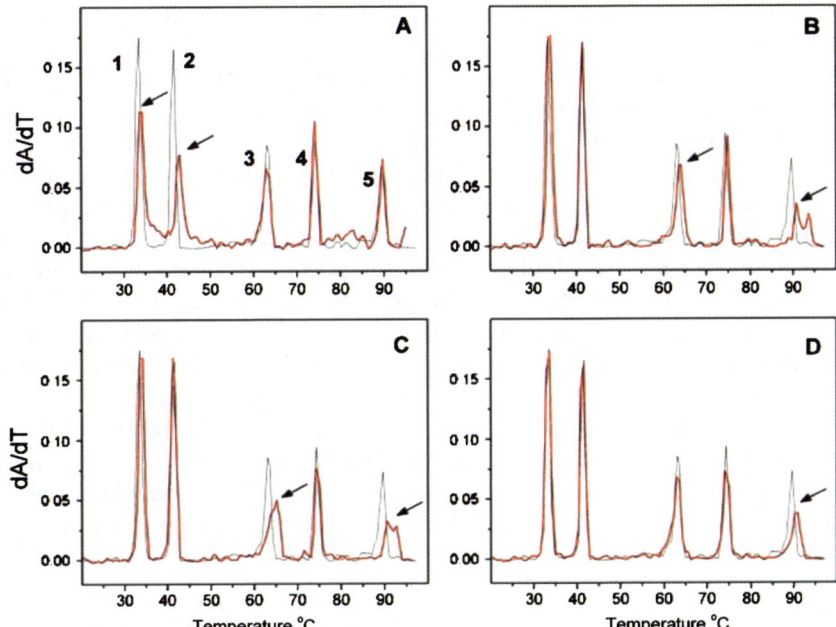

Figure 9 *Sequence selectivity revealed by melting studies of polynucleotide mixtures. Each panel shows the melting of a mixture of [poly (dAdT)]$_2$ (peak 1), poly (dA)·poly (dT) (peak 2), poly (dAdC)·poly (dGdT) (peak 3), poly (dG)·poly (dC) (peak 4), and [poly (dGdC)]$_2$ (peak 5) as the solid black line. The concentration of each polynucleotide is 40 μM (bp); total polynucleotide concentration is 200 μM (bp). The red line in each panel shows the effect of additions of low molar ratios of each of the test compounds. The arrows indicate the peaks that are altered by addition of the compound. (A). Netropsin at 2 μM. (B). Ethidium at 2 μM. (C). Daunorubicin at 2μM. (D) Actinomycin d at 1.5 μM*

preference of this groove binder for AT-rich sequences revealed in the very first footprinting studies.[60,62] Competition dialysis also revealed the strong preference of netropsin for [Poly (dA-dT)]$_2$ and Poly (dA)·poly (dT), along with a general preference for AT-rich natural DNA samples.[65]

Ethidium is widely, but erroneously, thought to lack sequence selectivity. It can, in fact, discriminate between sequences as revealed in low-temperature footprinting experiments.[72] From an analysis of the changes in patterns of digestion by DNAse I, Fox and Waring deduced that ethidium binds best to regions of mixed nucleotide sequence, especially those containing alternating purines and pyrimidines. Exactly that preference is revealed in Figure 9B. T_m shifts are evident for only poly (dA-dC)·poly (dG-dT) and [Poly(dG-dC)]$_2$.

The anticancer agent daunorubicin binds preferentially (but not exclusively) to triplet sequences in which an AT base pair is flanked by adjacent GC base pairs.[73–75] Daunorubicin shows a general preference for GC-rich DNA with an alternating purine-pyrimidine sequence.[70] Such preferences are observed in

Figure 9C, where daunomycin elevates the T_m of poly (dA-dC) · poly (dG-dT) and [Poly(dG-dC)]$_2$.

Actinomycin binds selectively to 5'GpC steps in duplex DNA.[60,62,76] That sequence preference is clearly observed in Figure 9D, where actinomycin only alters the T_m of [Poly(dG-dC)]$_2$.

Collectively, the data shown in Figure 9 confirm the utility of the melting of mixtures assay. The results obtained are fully consistent with the known sequence preferences of the standard compounds used for validation of the assay. There are significant advantages of the thermal denaturation assay over both footprinting and competition dialysis methods. Both footprinting and dialysis are time consuming, with at least 24 h needed in each case to execute the experiment and process the data. Competition dialysis, in its simplest form, requires that ligands to possess convenient absorbance or fluorescence signals for concentration determinations. The thermal denaturation assay described here requires no signal from the ligand, and can be completed in a few hours with real-time data display. As a standard spectrophotometric assay, the method is clearly amenable to automation and multiplexing.

7.3.1.2 Assay for Structural Selectivity

A mixture was made to study ligand selectivity for four different structures, DNA, RNA, and two types of DNA:RNA hybrids. In this mixture, DNA is represented by poly (dA) · poly (dT), RNA by poly (rA) · poly (rU), hybrid I by poly (dA) · poly (rU), and hybrid II by poly (rA) · poly (dT). DNA:RNA hybrids are structures of profound biological importance, and are an emerging target for drug design efforts.[77-79] Results are shown in Figure 10.

The classical groove-binder netropsin (Figure 10A) showed a clear preference for DNA as represented by poly (dA) · poly (dT), as was expected. Netropsin is known to convert non-B-DNA structures back to a standard B-form that contains the preferred minor groove geometry to which it binds most avidly.[80,81] Netropsin does not bind to A-form RNA.[82]

Ethidium was previously shown by competition dialysis to bind preferentially to an RNA:DNA hybrid structure.[78] Figure 10B shows that same preference in the melting assay. Ethidium shifts the T_m of poly (rA) · poly (dT), but does not appreciably alter the melting temperatures of any other polynucleotides in the mixture at the molar ratio added.

With the structural melting assay validated by netropsin and ethidium, an unknown compound was studied to illustrate the utility of the method. A derivative of the natural product β lapachone[83] clearly shows a strong preference for the DNA:RNA hybrid poly (dA) · poly (rU) (Figure 10C). That is a wholly novel type of nucleic acid structural recognition that is under more detailed biophysical study in our laboratory. The tantalizing result is shown here only to illustrate the utility of the method.

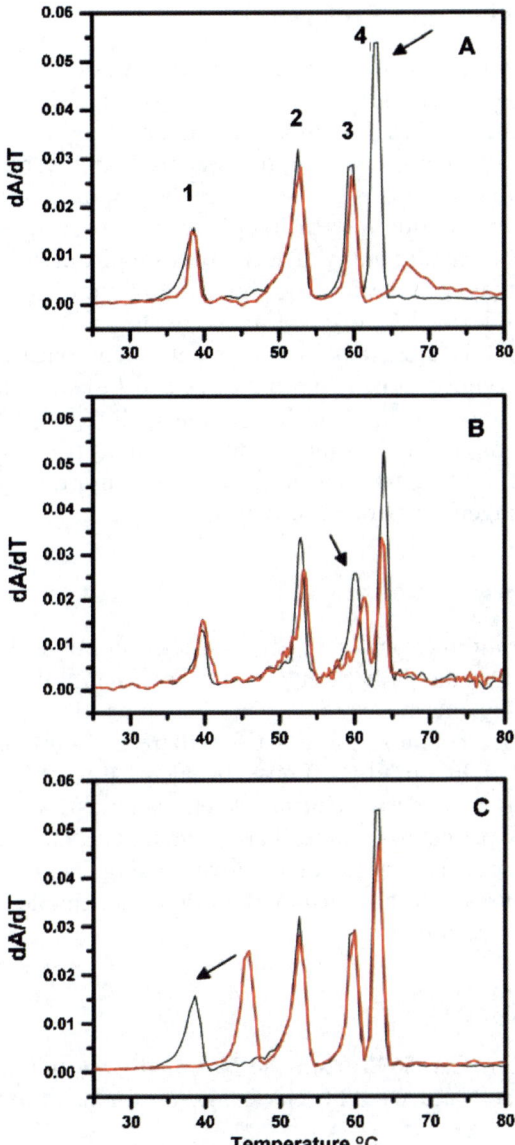

Figure 10 *Structural selectivity revealed by melting studies of polynucleotide mixtures. Each panel shows the melting of a mixture of DNA (poly (dA)ñpoly (dT); peak 4), RNA (poly(rA)·poly (rU); peak 2), a DNA:RNA hybrid (poly (dA)·poly (rU); peak 1), and an RNA:DNA hybrid (poly(rA)·poly (dT); peak 3) as the solid black line. The concentration of each polynucleotide structure was 10 μM (bp); total polynucleotide concentration is 40 μM (bp). The red line in each panel shows the effect of addition of ligand. (A). Netropsin at 1.6 μM. (B) Ethidium at 1.5 μM. (C) A semi-synthetic derivative of the natural product β-lapachone*

7.3.2   Advantages and Prospects

New types of thermal denaturation assays have been described for the study of sequence- and structural-selective nucleic acid binding. Mixtures of polynucleotides or oligonucleotides of different sequences or structures with well-resolved melting temperatures can be prepared and subjected to thermal denaturation. Addition of ligands at low molar ratios results in elevation of the T_m of the polynucleotide with the most preferred sequence or structure. The advantages of the assay are many. The assay is simple, direct, inexpensive, and rapid. As a spectrophotometric procedure, the method is amenable to automation and multiplexing. Proof-of-principle of the method has been provided here with simple mixtures that are nonetheless complicated enough to be interesting. The composition of the mixtures can be designed in accord with the particular interests of the investigator and with regard to the nucleic acid being targeted. Simple four- to five-component mixtures have been described here, but there is nothing to prevent the design of more complex mixtures to provide more stringent tests of selectivity.

7.4   Summary

Thermal denaturation methods utilizing spectrophotometric or calorimetric detection offer fundamentally sound means for quantitatively characterizing drug–DNA binding thermodynamics. The methods are particularly valuable for characterizing the binding interactions of recalcitrant ligands with poor solubility or ultratight binding affinity. In such cases, thermal denaturation may offer the only means for quantitative characterization. Thermal denaturation of mixtures provides an extrathermodynamic, qualitative approach for identifying sequence- and structural-selective binding. That application is in its infancy, but promises to be a robust method that is amenable to automation for high-throughput screening.

Acknowledgments

Supported by grant CA35635 from the National Cancer Institute. We thank Dr. Nichola Garbett and Dr. Patricia Raggazon for their comments.

References

1. J.D. Watson and F.H. Crick, *Nature*, 1953, **171**, 737.
2. R. Thomas, *Biochim. Biophys. Acta*, 1954, **14**, 231.
3. P. Doty, *J. Cell Physiol.*, 1957, **49**, 27.
4. P. Doty and S.A. Rice, *Biochim. Biophys. Acta*, 1955, **16**, 446.
5. J. Marmur and P. Doty, *Nature*, 1959, **183**, 1427.
6. P. Doty, *Harvey Lect.*, 1961, **55**, 103.
7. C.A. Thomas and P. Doty, *J. Am. Chem. Soc.*, 1956, **78**, 1854.
8. P. Doty, *Proc. Natl. Acad. Sci. USA*, 1956, **42**, 791.

9. S.A. Rice and P. Doty, *J. Am. Chem. Soc.*, 1957, **79**, 3937.
10. P. Ehrlich and P. Doty, *J. Am. Chem. Soc.*, 1958, **80**, 4251.
11. J. Marmur and P. Doty, *J. Mol. Biol.*, 1962, **5**, 109.
12. L. Peller, *J. Physic. Chem.*, 1959, **63**, 1194.
13. J.A. Schellman, *J. Physic. Chem.*, 1958, **62**, 1485.
14. J.M. Sturtevant and P. Geiduschek, *J. Am. Chem. Soc.*, 1958, **80**, 2911.
15. D.M. Crothers, *Biopolymers*, 1971, **10**, 2147.
16. Y.S. Lazurkin, M.D. Frank-Kamenetskii and E.N. Trifonov, *Biopolymers*, 1970, **9**, 1253.
17. J.D. McGhee, *Biopolymers*, 1976, **15**, 1345.
18. C.H. Spink and S.E. Wellman, *Methods Enzymol.*, 2001, **340**, 193.
19. D. Lando, *J. Biomol. Struct. Dyn.*, 1994, **12**, 343.
20. A.S. Benight, *Biopolymers*, 2003, **69**, 406.
21. G.E. Plum, in: *Current Protocols in Nucleic Acid Chemistry*, S.L. Beaucage D.E. Bergstrom, G.D. Glick, R.A. Jones (eds), Wiley, New York, 2000, pp. 7.3.1–7.3.17.
22. J.D. Puglisi and I. Tinoco Jr., *Methods Enzymol*, 1989, **180**, 304.
23. W.D. Wilson, F. Tanious, M. Fernades-Saiz and C.T. Rigl, in: *Drug–DNA Interaction Protocols*, K.R. Fox (ed), **vol 90**, Humana Press, Totowa, NJ, 1997, pp. 219–240.
24. C. Bailly and J.B. Chaires, *Bioconjug Chem*, 1998, **9**, 513.
25. C. Zimmer and U. Wahnert, *Prog. Biophys. Mol. Biol.*, 1986, **47**, 31.
26. S. Glasstone, *Textbook of Physical Chemistry*, D. Van Nostrand Company, Inc., Princeton, NJ, 1946.
27. I. Haq, T.C. Jenkins, B.Z. Chowdhry, J. Ren and J.B. Chaires, *Methods Enzymol.*, 2000, **323**, 373.
28. K.J. Breslauer, D.P. Remeta, W.Y. Chou, R. Ferrante, J. Curry, D. Zaunczkowski, J. G. Snyder and L.A. Marky, *Proc. Natl. Acad. Sci. USA*, 1987, **84**, 8922.
29. L.A. Marky and K.J. Breslauer, *Proc. Natl. Acad. Sci. USA*, 1987, **84**, 4359.
30. L.A. Marky, J. Curry and K.J. Breslauer, *Prog. Clin. Biol. Res.*, 1985, **172B**, 155.
31. F. Leng, W. Priebe and J.B. Chaires, *Biochemistry*, 1998, **37**, 1743.
32. G.G. Hu, X. Shui, F. Leng, W. Priebe, J.B. Chaires and L.D. Williams, *Biochemistry*, 1997, **36**, 5940.
33. J.B. Chaires, F. Leng, T. Przewloka, I. Fokt, Y.H. Ling, R. Perez-Soler and W. Priebe, *J. Med. Chem.*, 1997, **40**, 261.
34. M.J. Waring and L.P. Wakelin, *Nature*, 1974, **252**, 653.
35. F. Arcamone, *Doxorubicin Anticancer Antibiotics*, Academic Press, New York, 1981.
36. J.W. Lown (ed), *Anthracycline and Anthracenedione-Based Anticancer Agents*, Elsevier, Amsterdam, 1988.
37. W. Priebe (Ed), *Anthracycline Antibiotics: New Analogues, Methods of Delivery and Mechanisms of Action*, American Chemical Society, Washington, DC, 1995.

38. C.A. Frederick, L.D. Williams, G. Ughetto, G.A. van der Marel, J.H. van Boom, A. Rich and A.H. Wang, *Biochemistry*, 1990, **29**, 2538.
39. D.M. Crothers and H. Metzger, *Immunochemistry*, 1972, **9**, 341.
40. F.G. Loontiens, P. Regenfuss, A. Zechel, L. Dumortier and R.M. Clegg, *Biochemistry*, 1990, **29**, 9029.
41. S. Mansilla, W. Priebe and J. Portugal, *Biochemistry*, 2004, **43**, 7584.
42. J. Portugal, B. Martin, A. Vaquero, N. Ferrer, S. Villamarin and W. Priebe, *Curr. Med. Chem.*, 2001, **8**, 1.
43. B. Martin, A. Vaquero, W. Priebe and J. Portugal, *Nucleic Acids Res.*, 1999, **27**, 3402.
44. J. Portugal, D.J. Cashman, J.O. Trent, N. Ferrer-Miralles, T. Przewloka, I. Fokt, W. Priebe and J.B. Chaires, *J. Med. Chem.*, 2005, **48**, 8209.
45. M.J. Waring, L.P. Wakelin and J.S. Lee, *Biochim. Biophys. Acta*, 1975, **407**, 200.
46. F. Leng, J.B. Chaires and M.J. Waring, *Nucleic Acids Res.*, 2003, **31**, 6191.
47. L.H. Hurley, *J. Med. Chem.*, 1989, **32**, 2027.
48. L.H. Hurley, *Biochem. Soc. Trans.*, 2001, **29**, 692.
49. L.H. Hurley, *Nat. Rev. Cancer*, 2002, **2**, 188.
50. L.H. Hurley and F.L. Boyd, *Trends Pharmacol. Sci.*, 1988, **9**, 402.
51. L.H. Hurley, R.T. Wheelhouse, D. Sun, S.M. Kerwin, M. Salazar, O.Y. Fedoroff, F.X. Han, H. Han, E. Izbicka and D.D. Von Hoff, *Pharmacol. Ther.*, 2000, **85**, 141.
52. J.L. Mergny and C. Helene, *Nature Medicine*, 1998, **4**, 1366.
53. S. Neidle and M.A. Read, *Biopolymers*, 2000, **56**, 195.
54. J.E. Darnell Jr., *Nat. Rev. Cancer*, 2002, **2**, 740.
55. P.B. Dervan, *Science*, 1986, **232**, 464.
56. P.B. Dervan, *Bioorg. Med. Chem.*, 2001, **9**, 2215.
57. X. Qu, J.O. Trent, I. Fokt, W. Priebe and J.B. Chaires, *Proc. Natl. Acad. Sci. USA*, 2000, **97**, 12032.
58. J.L. Mergny and C. Helene, *Nat. Med.*, 1998, **4**, 1366.
59. S. Neidle and G. Parkinson, *Nat. Rev. Drug Discov.*, 2002, **1**, 383.
60. M.J. Lane, J.C. Dabrowiak and J.N. Vournakis, *Proc. Natl. Acad. Sci. USA*, 1983, **80**, 3260.
61. C.M. Low, H.R. Drew and M.J. Waring, *Nucleic Acids Res.*, 1984, **12**, 4865.
62. M.W. Van Dyke, R.P. Hertzberg and P.B. Dervan, *Proc. Natl. Acad. Sci. USA*, 1982, **79**, 5470.
63. J.B. Chaires, in: *Current Protocols in Nucleic Acid Chemistry*, S.L. Beaucage *et al.* (eds), **vol. 1**, Wiley, Inc., New York, 2002, pp. 8.3.1–8.3.8.
64. J.B. Chaires, *Curr. Med. Chem. Anti-Cancer Agents*, 2005, **5**, 339.
65. J. Ren and J.B. Chaires, *Biochemistry*, 1999, **38**, 16067.
66. J. Ren and J.B. Chaires, *Methods Enzymol.*, 2001, **340**, 99.
67. R.A. Darby, M. Sollogoub, C. McKeen, L. Brown, A. Risitano, N. Brown, C. Barton, T. Brown and K.R. Fox, *Nucleic Acids Res.*, 2002, **30**, e39.
68. X. Shi and J.B. Chaires, *Nucleic Acids Res.*, 2006, **34**, e14.

69. W.D. Wilson, F.A. Tanious, M. Fernandez-Saiz and C.T. Rigl, *Methods Mol. Biol.*, 1997, **90**, 219.
70. J.B. Chaires, in: *Advances in DNA Sequence Specific Agents*, L.H. Hurley (ed), **vol. 1**, JAI Press Inc., Greenwich, CT, 1992, pp. 3–23.
71. W. Muller and D.M. Crothers, *Eur. J. Biochem.*, 1975, **54**, 267.
72. K.R. Fox and M.J. Waring, *Nucleic Acids Res.*, 1987, **15**, 491.
73. L. Cai, L. Chen, S. Raghavan, R. Ratliff, R. Moyzis and A. Rich, *Nucleic Acids Res.*, 1998, **26**, 4696.
74. J.B. Chaires, *Biophys Chem*, 1990, **35**, 191.
75. J.B. Chaires, K.R. Fox, J.E. Herrera, M. Britt and M.J. Waring, *Biochemistry*, 1987, **26**, 8227.
76. K.R. Fox and M.J. Waring, *Nucleic Acids Res*, 1984, **12**, 9271.
77. R. Francis, C. West and S.H. Friedman, *Bioorganic Chemistry*, 2001, **29**, 107.
78. J. Ren, X. Qu, N. Dattagupta and J.B. Chaires, *J. Am. Chem. Soc.*, 2001, **123**, 6742.
79. C. West, R. Francis and S.H. Friedman, *Bioorg. Med. Chem. Lett.*, 2001, **11**, 2727.
80. H. Fritzsche and A. Rupprecht, *J. Biomol. Struct. Dyn.*, 1990, **7**, 1135.
81. C. Zimmer, C. Marck and W. Guschlbauer, *FEBS Lett.*, 1983, **154**, 156.
82. C. Bailly, P. Colson, C. Houssier and F. Hamy, *Nucleic Acids Res.*, 1996, **24**, 1460.
83. K.C. De Moura, K. Salomao, R.F. Menna-Barreto, F.S. Emery, C. Pinto Mdo, A.V. Pinto and S.L. de Castro, *Eur. J. Med. Chem.*, 2004, **39**, 639.

CHAPTER 8

Computer Simulations of Drug–DNA interactions: A Personal Journey

FEDERICO GAGO

Department of Pharmacology, University of Alcalá, E-22871 Madrid, Spain

8.1 Introduction

Deoxyribonucleic acid (DNA), as the cell's repository of genetic information, has been one of the prototypical cancer targets for decades.[1–3] At its lowest level of organization, the three-dimensional (3D) structure of this biological macromolecule looks deceptively simple and monotonous: hydrogen-bonded base pairs are stacked along the axis of a right-handed double helix with the sugar–phosphate backbone of each strand on the outside, winding up in antiparallel orientation.[4] As a consequence of this arrangement, the ability of a given stretch of DNA to act as a molecular recognition target appears to be limited, at first sight, to functional group discrimination along the major and minor grooves that lie between the phosphodiester linkages of both strands.[5] While this is true for many sequence-specific DNA-binding proteins[6,7] and low-molecular weight ligands (including natural and synthetic information-reading lexitropsins[8] as well as many clinically used antitumour drugs[9]), it is also perceived nowadays that additional potential for specific recognition is provided by sequence-dependent DNA microheterogeneity.[10,11]

The use of models helped Watson and Crick to suggest a structure for the DNA molecule and to hint at a "possible copying mechanism for the genetic material".[12] In the past few years, the way molecular structure is perceived by researchers, educators, and students alike has changed dramatically, thanks to spectacular advances in structural biology[13,14] and to the availability of high-quality molecular visualization programs from the public domain (*e.g.* RasMol,[15] Swiss-PdbViewer – a.k.a. DeepView – ,[16] ViewerLite,[17] VMD,[18] PyMol,[19] BALLView,[20] *etc.*), together with extremely effective plug-ins (*e.g.* Chime[21]) that can be installed *gratis* on one's favourite Internet navigator and allow an easy and highly informative display of both shape and properties of

152

macromolecules. As shown for DNA, the mere visualization of a single molecular structure can help to answer some of the simplest questions regarding function, but more complex problems may require the simulation of a process, such as molecular motion or the binding of a ligand. In fact, Sir David Phillips' comment in 1981 that "Brass models of DNA and a variety of proteins dominated the scene and much of the thinking"[22] summarizes the progressive realization that biological molecules are not rigid structures, but dynamic systems whose internal motions play important functional roles. The long-range goal of computer-based molecular modelling, as an '*in silico*' approach to biochemistry and pharmacology,[23] is to describe biomolecular interactions in terms of the general laws of chemistry and physics that have been so successfully applied to small molecules. For most practical purposes, the tool of choice to deal with large biological macromolecules is classical mechanics in conjunction with molecular dynamics (MD) simulation methods.[24]

By generating hundreds or thousands of time-linked low-energy system configurations of biological macromolecules, MD simulations attempt to capture accurately "the jigglings and wigglings", as Feynman *et al.*[25] put it, of the atoms that make up all living things. MD methods have benefited enormously from the recent convergence of advances in force–field parameters,[26] improved treatment of long-range electrostatic interactions,[27] and greatly boosted computer power (speed usually increases by a factor of 2 or so every 18 months, in accordance with Moore's law). Thus, it is now feasible for many research groups to obtain multi-nanosecond trajectories of solvated macromolecular systems in a reasonable time. The ensemble of successively created "snapshots", each containing a set of Cartesian coordinates, allows the creation of "molecular movies" and various analyses of molecular motions in atomic detail. The information derived from these simulations can be used to (i) extract structural and energetic information that is usually beyond current experimental possibilities,[28] (ii) provide independent accounts of experimentally observed behaviour,[29] (iii) help in the interpretation of biochemical or pharmacological results,[30] and (iv) open new avenues for research by posing relevant questions that can guide the design of new experiments.

In the DNA field, current equilibration and sampling times usually range from a few nanoseconds[31–33] to tens[34–36] or hundreds[37] of nanoseconds. Although these times are still not sufficiently long[38,39] to monitor large-scale structural changes (in the absence of distance or angular restraints) or to achieve convergence of ion distributions,[34,36] relaxation times for helicoidal parameters have been shown to be in the order of 0.5 ns.[36] Thus, trajectories lasting a few nanoseconds can provide adequate dynamic descriptions of DNA and DNA–ligand complexes, because the characteristic structural parameters (*e.g.* roll, twist, and tilt) settle much more rapidly than ion occupancies.[39]

The initial feeling that theoretical calculations were a waste of time (not least because of the near impossibility of treating such complex systems as macromolecules in living cells with the level of detail that was thought to be necessary to add something of importance to our knowledge[40]) has given way, little by little, to a renewed wave of confidence in simulation methodologies.

Ongoing improvements in molecular mechanics force fields,[41,42] the almost systematic incorporation of solvent molecules and counterions into the models,[43,44] and a reliable treatment of long-range electrostatics by applying particle mesh Ewald methods to cubic and octahedral solvent "boxes"[45] are making it possible to simulate the dynamics of relatively large systems (including such highly charged ones as nucleic acids[44,46,47]) over considerably long times, sometimes approaching the microsecond time scale,[48,49] without the instabilities that plagued earlier simulations. Moreover, use of the computationally much faster Generalized Born (GB) model of implicit solvation[50] allows the exclusion of explicit solvent molecules[51] and extends the length of some of these MD simulations even further, as does the use of coarse-grained representations of biomolecules.[52] These methodologies are now well suited to study conformational changes that are coupled to function[53] and also, in conjunction with quantum mechanical methods,[54] the course of reactions in which covalent bonds are broken and formed such as those taking place in the active sites of enzymes.[55]

A number of tools also exist that can 'pump' some extra energy into the system to accelerate the crossing of local barriers ('targeted' or 'steered' MD) thus allowing the study of processes that would normally occur too slowly in a standard MD run based only on random thermal fluctuations.[56–57] The repertoire of macromolecules relevant to cancer that can be studied by these methods has also expanded enormously in the last 20 years, from just a handful of 'classical' targets (*e.g.* short DNA oligonucleotides, dihydrofolate reductase, or thymidylate synthase) to DNA–protein complexes,[59] RNA,[60] and whole protein families (*e.g.* growth factor receptors,[61] kinases,[62] and phosphatases[63]), both alone and in complex with ligands.

Researchers arguably have on their laptops today more computing power than was available on most mainframe computers when MD simulations of macromolecular systems started at the end of the 1970s.[64] Moreover, a realistic alternative to large-scale 'supercomputers' these days is to use massively grid-distributed computing[65] to harness the power of dozens or hundreds of relatively inexpensive personal computers (PCs) towards a common goal such as the folding of a protein[66] or the docking of potential ligands into a protein-binding site.[67]

Spectacular advances in the molecular biology of DNA-binding proteins and the concurrent crystallographic studies of binary and ternary complexes involving drugs, DNA, and associated proteins are progressively narrowing the gap between our perception of the structural aspects of drug–DNA complexes and our limited understanding of how these drugs really work in living cells. In the following sections, I will provide a succinct overview of some of these advances in relation to my own research, since I was a postdoctoral student in Graham Richards' laboratory to the present day.

My introduction to molecular modelling and computer simulations originated at the Physical Chemistry Laboratory of Oxford University after completing my doctorate in Madrid on the study of hydrophobic parameters derived from liquid chromatography measurements. Prof. Richards was well known in Spain for his book "Quantum Pharmacology" and I had the privilege of listening to one of his talks when our paths coincided in Barcelona on the occasion of a seminar organized by Miguel Martín at Universitat Autónoma. Incidentally, Corwin Hansch also attended that meeting and showed a fascinating video providing atomic details of the docking of an inhibitor in the active site of an enzyme, possibly alcohol dehydrogenase. I was so struck by the power of these computational methods and the intrinsic beauty of molecular graphics that I decided to try my luck adventuring as a novice in this emerging field. The paper I was given to start with on joining Graham's group in April 1987 was one in which the authors reported on the differential binding of some natural and synthetic compounds to DNA tracts rich in adenine and thymine in preference to regions rich in guanines and cytosines.[68] Could theoretical methods shed some light on this issue?

8.2 Minor Groove DNA Binders

Netropsin and distamycin (Figure 1) are well-characterized DNA minor groove-binding ligands that have been used extensively as models for the study of drug–DNA interactions in a variety of contexts.[68] Both antibiotics were among the set of small molecules whose discriminatory ability for different DNAs had been evaluated using an ethidium displacement assay that allowed an indirect estimate of drug–DNA association constants to be obtained.[68] For netropsin, additional spectroscopic and thermodynamic data[69] showed that the free energies for netropsin binding to both poly[d(AT)]·poly[d(AT)] and poly(dA)·poly(dT) were very similar and about 4 kcal mol^{-1} more favourable than the free energy of binding to poly[d(GC)]·poly[d(GC)].

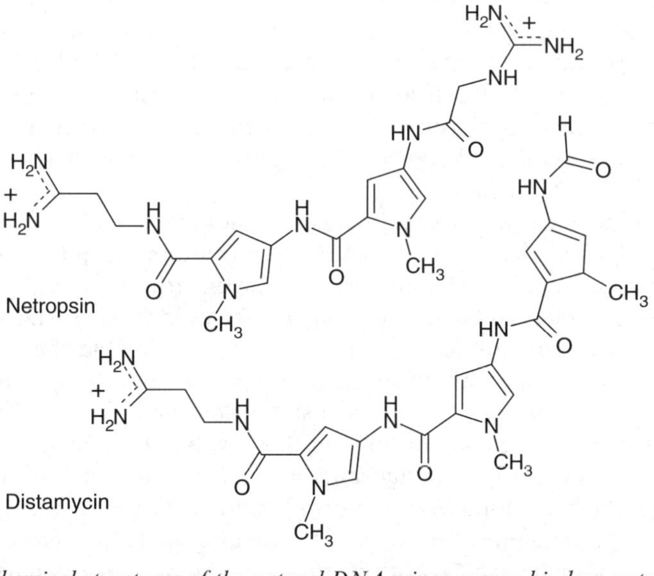

Figure 1 *Chemical structures of the natural DNA minor groove binders, netropsin (top), and distamycin (bottom)*

We were aware that an accurate computation of binding free energies is hampered by uncertainties regarding both the particular conformation of the interacting partners in the unbound state and the validity of the energy partitioning schemes, as well as limitations with respect to some energy contributions that are not easily amenable to calculation (*e.g.* hydrophobic and entropic effects). For this reason, we initially focussed our efforts on the binding of these and other compounds (including the dye Hoechst 33258) to dodecanucleotides containing alternating runs of AT and GC. By using a molecular mechanics force field and performing energy minimization and decomposition, we were able to show that the differences in interaction energies between the drugs and the two DNA sequences correlated satisfactorily with the selectivity ratio provided by experiment.[70] Furthermore, when we replaced the pyrrole rings in netropsin and distamycin with imidazole, we could also observe an increased preference for the alternating runs of Gs and Cs due to the new hydrogen bonds between the imidazole nitrogens of these "lexitropsins" and the amino groups of guanines, although no significant loss of affinity was detected for ATAT stretches. We therefore supported the view that, to have molecules with the potential to bind to predetermined sequences in a highly specific manner, not only "G,C-reading" but also "A,T-rejecting" elements had to be incorporated into the ligands. In this respect, a structural factor that we considered relevant to the specificity of the interaction, in agreement with other authors, was the different width of the minor groove permissible or attainable in the two types of complexes. The same year this work was published, the unanticipated discovery of the first 2:1 polyamide–DNA complex paved the way to design strategies that used imidazole in place of pyrrole rings to "read" the exocyclic amino group of guanine very effectively.[71] More recent results have suggested that the language of current design motifs for polyamide sequence recognition should be extended to include the use of "words" for recognizing two adjacent base pairs, rather than "letters" for binding to single base pairs.[72] All in all, this field of molecular recognition of DNA from the perspective of the minor groove is a beautiful example of maturation from serendipity to successful design at the interface of chemistry, biology, and (possibly) human medicine.[73]

The major problem with our estimates of binding energies for netropsin and the other drugs was that they did not really capture the essence of the binding process, that is, the difference in energy between drug molecules in solution, interacting with water and ions, and drug molecules in the drug–DNA complex. In fact, computer simulations (now and then) are not capable of calculating this difference directly, especially when large molecules are at play. The difficulty lies in the fact that the binding process should be simulated slowly enough to achieve thermodynamic equilibrium and this involves the generation of a huge number of representative configurations of the system along the reaction coordinate, which is impractical in most situations. However, the non-physical process of perturbing one of the reactants into another (which has been dubbed "computational alchemy"[74]), both in the free state and in the bound state, is more easily amenable to computation[75,76] and provides an accurate relative free energy change (ΔG) that can be related to the physically relevant one through a thermodynamic cycle (Figure 2).

Stimulated by impressive results produced by Graham and his postdoctoral student Chris Reynolds on the calculation of redox potentials,[77] and inspired by the illuminating paper on the thermodynamics of netropsin binding to several polynucleotides of defined sequence (which also contained data for poly[d(IC)] · poly[d(IC)]),[69] I realized that we could attempt to apply the same methodology to the problem of DNA–ligand binding. Why not "mutate" inosines (Figure 3) to guanosines (and the reverse) in both the free DNA and its complex with netropsin to assess the effect on the calculated binding energies of growing/removing the exocyclic amino group of guanine? Despite the enormity of the task ahead of us and the lack of precedent for work of this sort, I was encouraged to go ahead. An advantage over simulations of ligand binding to proteins, which generally need to address substantial molecular reorganization that can span long time scales, the conformational changes involved in the alchemical changes studied here were relatively minor, as they did not significantly affect the double-helical structure of the DNA. We managed to borrow computer time from every single machine we could lay hands on, and the work was completed in little more than one year.

The "mutation" of the inosine C2 hydrogen into an amino group in the ICIC sequence was made possible with the aid of two "dummy atoms" (D), which were characterized as having both the point charge and the non-bonded energy parameter set to zero. During the simulations of free and netropsin-bound DNA, the parameters for each HD_2 group were linearly increased in small increments so that at the end of the perturbation they attained the values corresponding to the standard NH_2 group. Upon doing this, we realized that the DNA minor groove became progressively less deep, more polar, and less narrow. In its complex with netropsin, these changes were accompanied by the

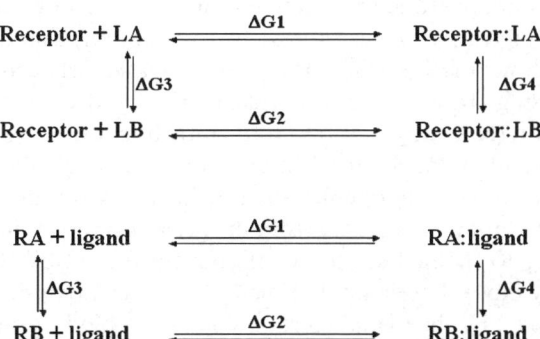

Figure 2 *Examples of typical thermodynamic cycles used in free energy perturbation simulations. **Top**: a ligand A is converted into a ligand B, both binding to a common receptor. **Bottom**: the same ligand binds with different affinities to receptors A and B. The free energy differences calculated experimentally ($\Delta\Delta G_{binding}=\Delta G2-\Delta G1$) must be the same as those calculated from the non-physical perturbations of A into B ($\Delta\Delta G_{binding}=\Delta G4-\Delta G3$)*

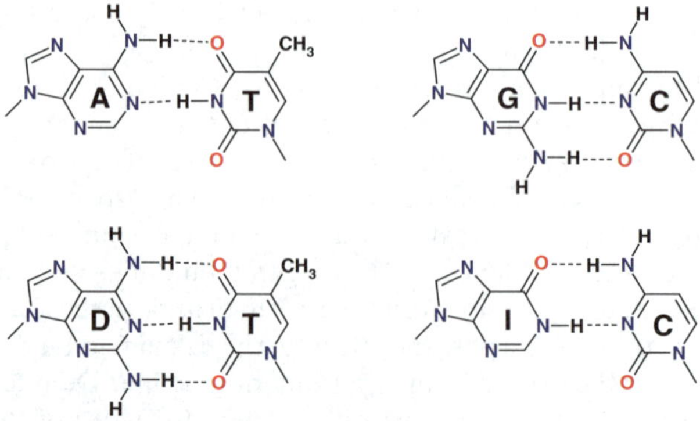

Figure 3 *Chemical structures of natural and modified DNA base pairs in the Watson–Crick arrangement. Hydrogen bonds are depicted as broken lines. D stands for 2,6-diaminopurine, whereas I stands for the hypoxanthine-containing nucleoside, inosine*

extrusion of part of the drug molecule into the solvent and a decrease in binding free energy that was in very good agreement with the experimental value.[78] The calculations, therefore, were able to reproduce the observed preference of netropsin for binding to ICIC over GCGC in a quantitative manner while at the same time providing a good rationale for the specificity of the association that pointed, once again, to minor groove width as an important determinant of complex stability.

8.3 Natural Bifunctional Intercalators and Hoogsteen Base Pairing

Having first studied ligands with a tendency to bind to AT-rich DNA better than to GC-containing sequences, on my return to Alcalá I became interested in intercalators, which mostly get sandwiched at CpG.[79] The process of DNA intercalation allows the insertion of a planar ligand between successive base pairs by increasing their vertical separation (rise) and changing their relative degree of rotation (twist angle) through the introduction of crankshaft motions in the sugar–phosphate backbone.[4] Interest in bifunctional intercalators stems not only from the possibility of enhancing their binding affinity over that of the corresponding monomers, but also from the greater opportunities for imposing selective binding to defined sequences that are afforded by the bracketing of two (or possibly more) base pairs between the intercalative sites. Thus, the size of the site occupied by the ligand increases, *i.e.* for a binding site covering 4 bp, the number of distinguishable sequences is 136 versus only 10 unique dinucleotide steps at which monointercalation can take place. The design problem here is to select a suitable linker with the ability to recognize functional groups in either the minor or the major groove. As is usually the case, Nature anticipated by providing us with some extremely elegant and very interesting examples of bifunctional intercalators in the family of quinoxaline antibiotics represented by echinomycin (also known as quinomycin A) and its biosynthetic precursor, triostin A. These agents are primarily produced by *Streptomyces echinatus* and *S. triostinicus*, respectively, and for them a binding site size of 4 bp was early demonstrated and then repeatedly confirmed.[80] These natural *bis*-intercalators were shown to have a definite preference for binding to a 5′-CpG-3′ core flanked by an A:T base pair on either side, as opposed to the natural crescent-shaped non-intercalating ligands mentioned above (*e.g.* netropsin and distamycin), which show a strong preference for binding to the minor groove of A,T-rich DNA regions. Rather strikingly, a sequence-dependent rearrangement in the base pairs adjacent to the CpG core was manifested in cases where the base located on the 5′ side of the CpG-binding step was a purine: this base rotates 180° about the glycosidic bond to adopt a *syn* orientation relative to the sugar, but remains hydrogen-bonded to the opposing thymine making use of a so-called Hoogsteen scheme (Figure 4), as first reported by an eponymous German investigator when he solved the X-ray crystal structure of the 1:1 complex of 1-methylthymine with 9-methyladenine.[81] Intriguingly, this base-pairing rearrangement was detected in d(ACGT)$_2$, d(GCGC)$_2$ (at low pH), d(CGTACG)$_2$, d(ACGTACGT)$_2$, d(ACGTATACGT)$_2$, and d(GCGTACGC)$_2$ (again at acidic pH), but not in d(AAACGTTT)$_2$, d(TCGA)$_2$, d(CCGG)$_2$, and d(TCGATCGA)$_2$.

Curious about this incompletely understood phenomenon, I proposed that my first, highly motivated Ph.D. student José Gallego should undertake a theoretical investigation of the factors that might be determining these conformational preferences in the base pairs flanking both sides of the CpG-binding step. Because Kollman and co-workers had previously focused on the relative stability of the d(CGTACG)$_2$:(triostin A)$_2$ complex with the central AT base pairs in either Hoogsteen pairing (as found in a crystal structure reported in 1984[82]) or Watson–Crick pairing using molecular mechanics,[83] we chose to study the complexes between echinomycin and the DNA tetramers d(ACGT)$_2$ and d(TCGA)$_2$, for which conformational information was available from nuclear magnetic resonance (NMR) spectroscopy.[84] We tried to address two main issues: (i) why the terminal base pairs in the d(ACGT)$_2$ complex (hereafter termed A$_H$) appeared Hoogsteen-paired, while those in the d(TCGA)$_2$ complex (T$_W$) did not and (ii) why the former complex was more stable than the latter.

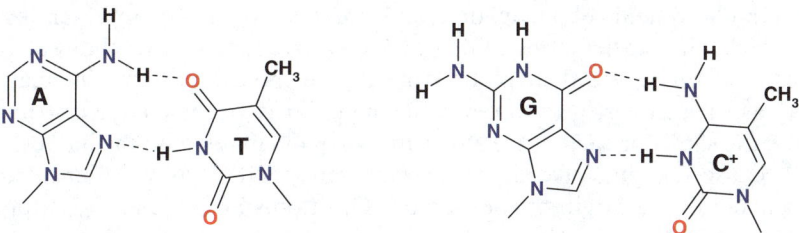

Figure 4 *Adenine:thymine (left) and guanine:cytosine base pairs (right) in the Hoogsteen configuration*

To this end we analysed the behaviour of the d(ACGT)$_2$:echinomycin and d(TCGA)$_2$:echinomycin complexes in which both terminal AT base pairs were in either Hoogsteen (A$_H$ and T$_H$) or Watson–Crick conformation (A$_W$ and T$_W$) by means of four independent MD simulations in aqueous solution each lasting 40 ps (a considerable time in those days for our modest computational resources!).[85]

None of the complexes underwent any large conformational changes with respect to the corresponding initial structure during the simulations in water and the DNA bases remained paired, except for some fraying effects that were more noticeable in the A$_W$ complex. The conformations for the experimentally found complexes, A$_H$ and T$_W$, were in qualitative agreement with the nuclear Overhauser effects (NOE) detected in the NMR spectra. We considered and discussed the three major types of interaction that echinomycin establishes with the DNA molecules separately: (i) van der Waals and electrostatic interactions between the depsipeptide part of the drug and the minor groove of the DNA molecules; (ii) hydrogen bonds between the alanine residues of echinomycin and the guanine bases of DNA; and (iii) stacking interactions between the quinoxaline chromophores and the adjacent base pairs. In agreement with NMR and X-ray results, the A$_H$ model was found to be stabilized by two strong hydrogen bonds between the amino groups of alanines and N3 of guanines, and two more between the carbonyl groups of alanines and the 2-amino groups of guanines, one of which was only slightly weaker due to the asymmetrical binding of echinomycin. In the other three models, one of the hydrogen bonds between an alanine residue and the DNA was lost. Although these results argued in favour of the greatest stabilization taking place in the A$_H$ complex, they provided no clues about the origin of the Watson–Crick to Hoogsteen rearrangement.

We then focused our attention on the quinoxaline residues that, according to our calculations, were contributing almost 50% of the total interaction energy between echinomycin and the DNA tetramers. We performed a detailed analysis of these interactions in the four models, because it was generally accepted that the flanking sequence specificity of this antibiotic rested largely on the aromatic stacking interactions of its quinoxaline chromophores[80,86] and also because the dependence of DNA stacking interactions on the *composition* and *orientation* of the nucleic acid bases had been shown to be largely dominated by the electrostatic term.[87] We considered the entire quinoxaline-2-carboxamide system as a coplanar and conjugated moiety representing the molecular entity participating in the stacking interactions rather than solely the quinoxaline rings. We then plotted the dipole moment of this intercalating moiety as the simplest way to represent the polarity of the charge distribution and did the same for each of the adjacent base pairs making up the intercalation site. Because the sum of charges must be zero for the dipole moment to be independent of the origin of coordinates, C1$'$ atoms of DNA and C$_\alpha$ atoms of serines were included in the calculations as buffers so that each of these systems attained electrical neutrality.

We made three significant observations relevant to the problem in hand: (i) in the Hoogsteen arrangement, the dipole moment of an AT base pair increases to more than 5 D from the value of about 2 D found in the Watson–Crick conformation; (ii) the quinoxaline-2-carboxamide chromophores of echinomycin have a high dipole moment of 4–5 D, with the negative pole located on the pyrazine-2-carboxamide region and the positive pole on the benzene ring region; and (iii) in the experimentally found A_H complex, the high dipole moment of each AT base pair in the Hoogsteen conformation and that of the stacked echinomycin chromophore lie in opposite directions, which gives rise to a favourable dipolar interaction. By contrast, the dipole moment of the AT base pair reverses its orientation in the imaginary Hoogsteen complex T_H, so that both dipole vectors point in the same direction and the ensuing dipolar interaction is unfavourable.

Taken together, our interpretation of these results[85] was that a Hoogsteen base-pair scheme should be preferred for binding of echinomycin to the $d(ACGT)_2$ tetramer, whereas the classical Watson–Crick hydrogen-bonding arrangement would be more stable for the TCGA sequence. It was encouraging to note that this was precisely what had been detected experimentally.

When GC rather than AT pairs flank the echinomycin CpG-binding step, Hoogsteen base pairing is likewise observed (at acidic pH), but only when the purine base (guanine) is on the 5' side of the CpG sites, *i.e.* it is detected in $d(GCGC)_2$ and $d(GCGTACGC)_2$[88,89], but not in $d(CCGG)_2$ regardless of pH.[90] The need for a low pH is justified, because the Hoogsteen hydrogen-bonding scheme in GC pairs requires the protonation of the cytosine base (Figure 4).

In an attempt to rationalize these observations we extended our previous work by modelling and studying by means of MD simulations in aqueous solution the complexes of echinomycin with $d(GCGC)_2$ and $d(CCGG)_2$ in which both terminal G:C base pairs adopt either a Hoogsteen arrangement (named G_H and C_H, respectively) or a Watson–Crick conformation (G_W and C_W). In this case, we found that the conformations of the experimentally detected G_H and C_W complexes were not significantly altered during the course of the 70 ps simulation, as assessed by the low root-mean-square deviation (rmsd) from the initial structure, whereas comparatively larger rmsd values were observed for C_H and G_W.[91] Additional calculations also predicted a large increase in the interaction energy between echinomycin and the $d(GCGC)_2$ duplex upon protonation of the terminal cytosines with subsequent Hoogsteen pair formation. In agreement with previous work,[83,85] we found that about 50 per cent of the total interaction energy in all the complexes appeared to be contributed by the van der Waals interactions involving the quinoxaline ring systems alone. This term, however, decreased when the terminal guanine bases in $d(CCGG)_2$ adopted a Hoogsteen conformation. On the other hand, in the G_H complex a marked gain in electrostatic stacking interaction energy was apparent between the chromophores of echinomycin and the terminal Hoogsteen-paired bases. We construed that the combination of these two factors could account for the experimental observation that at low pH, Hoogsteen base pair formation is only observed for the echinomycin–d(GCGC)$_2$ complex.[84]

All in all, our findings confirmed early theoretical evidence that for stacked DNA bases the electrostatic contribution plays a decisive role in stabilizing the stacking interaction[92] and also that this term is very sensitive to the relative orientation of the planar systems involved.[93] In our view, the conformational changes brought about by the binding of echinomycin and triostin A to short oligonucleotides provide an excellent test case for probing the nature and dependence of stacking interactions in DNA–drug complexes using theoretical methods. But could this methodology be extended to the issue of binding selectivity?

8.4 *Bis*-Intercalation of Echinomycin and Related Bifunctional Agents in Relation to Binding Sequence Preferences

The origin of the selectivity of triostin A and echinomycin for binding to CpG steps was widely accepted to arise mostly from the hydrogen bonds between the depsipeptide part of these antibiotics and the exocyclic amino group of guanine in the minor groove.[80,82] Traditional work on structure–affinity relationships (SAR) for the quinoxaline family of antibiotics dealt with the effects that introduction of new substituents or removal of existing ones had on the binding properties of a given drug, as assessed mainly by DNA footprinting experiments. The best-known examples are probably those provided by des-N-tetramethyl triostin A (TANDEM) and [N-MeCys^3,N-MeCys^7] TANDEM (CysMeTANDEM), triostin A analogues lacking either all of the N-methyl groups of cysteines and valines or just those of the latter' that bind better to a central TpA core. This change could also be rationalized on the basis of the lack of suitably placed hydrogen-bonding functionalities in the modified depsipeptide that could interact with the 2-amino group of guanine. An elegant complementary SAR approach was later adopted in which the nature of the bases making up the DNA target was changed (Figure 3) rather than the structure of the ligand, and this was accomplished in a series of unequivocal experiments where inosine (I) took the place of guanosine in the DNA molecule (which is tantamount to removing the exocyclic 2-amino group selectively from the minor groove) and where 2,6-diaminopurine (2-aminoadenine) was used to pair with thymine in place of adenine (which gives rise to DNA containing an amino group at every step). As expected, echinomycin did not bind to CpI steps, lending further credence to the crucial role played by this group in the interaction of this ligand with the DNA minor groove.[94] Introduction of the extra amino group in the minor groove of DT regions, on the contrary, led to a redistribution of binding sites relative to normal DNA, and echinomycin was shown to bind to any pyrimidine–purine combination other than the usual CpG step.[95] Strikingly enough, not only was the selectivity drastically changed but also the affinity of echinomycin for these new binding sites in the modified

DNA turned out to be at least one order of magnitude greater than that for normal DNA.

Because our previous studies had highlighted an unfavourable electrostatic interaction between the quinoxaline-2-carboxamide system of echinomycin and the base pairs making up the central CpG step,[85,91] we reasoned that not only hydrogen bonding but also stacking interactions might be playing a role in determining the binding preferences of echinomycin. To test this hypothesis, we generated 13 molecular models of 1:1 complexes of echinomycin with standard and modified DNA hexamers and evaluated the different contributions to their relative stability.[96] The whole set of sequences contained every combination of binding sites experimentally probed by echinomycin. The aim was to get information both on the stacking interactions between the quinoxaline-2-carboxamide system and any DNA base pair as well as all the possible hydrogen-bonding arrangements between the depsipeptide and the DNA atoms in the minor groove.

The energy analysis revealed that the interactions involving the alanine residues and the quinoxaline-2-carboxamide chromophores of echinomycin could by themselves account for the observed binding selectivities. Furthermore, by plotting these energy components versus the total interaction energy a good linear correlation was detected and the different hexamers appeared clustered in two distinct subsets. One subset encompassed a family of good binding sites that presented both a central dinucleotide step endowed with full hydrogen-bonding capabilities in the minor groove and an arrangement of base pairs that gave rise to an overall favourable stacking interaction with the chromophores of the antibiotic. The sequences in the other subset shared poorer hydrogen-bonding possibilities and overall weaker interactions with the quinoxaline-2-carboxamide moieties. The most favourable binding sites were indeed those in which the adenines in the DNA had been replaced with 2,6-diaminopurine (D), and most notably the sequence DTDT, in good agreement with the experimental observations.[95]

It was particularly illustrative to view the footprinting results in the light of calculations that decomposed the binding enthalpy due to stacking interactions with both the sandwiched base pairs and those flanking the *bis*-intercalation site into van der Waals and electrostatic components. By first examining those complexes with a common central CpG step, it was possible to see that the minor differences between them were, as expected, restricted to interactions with the flanking bases. Nonetheless, the electrostatic component appeared to be more discriminating than the corresponding van der Waals term, as most clearly illustrated by CCGG (apparently a weaker binding site for echinomycin than ACGT) for which this electrostatic component was found to be slightly repulsive. The fact that the electrostatic interaction between the antibiotic chromophores and the central CpG step was calculated to be unfavourable strongly suggested to us that this negative effect had to be outweighed by the very favourable electrostatic and hydrogen-bonding interactions established with the minor groove. Furthermore, it was indicative that the different charge distribution of the base pairs in a TpD step (which also presents two 2-amino groups in the minor groove) could give rise to an attractive electrostatic stacking interaction with the quinoxaline-2-carboxamide system. This turned out to be the case, and the improved calculated binding energy was in nice accord with the larger experimental association constant. The results accounting for the enhanced affinity of echinomycin for DTDT over ACGT were reinforced by the interaction energies obtained for "mixed" sequences, for which values half way between those in each of these two complexes were found. The fact that in the modified DNA employed in the experiments the originally protected CpG sites were adjacent to these D-containing high-affinity sites could explain why binding to the canonical CpG steps was precluded.

Directed biosynthesis has produced a number of analogues of echinomycin and triostin containing substituted chromophores. Prominent among these is the *bis*-quinoline derivative of echinomycin designated 2QN,[97] for which an interesting correlation between spectroscopic and thermodynamic properties has been found[98] that made researchers think that the intercalated ring systems could largely control the sequence selectivity. The footprinting results, however, showed no great differences from echinomycin, including findings that the overriding determinant of preferred binding sites was the 2-amino group exposed in the minor groove of the DNA double helix and also that sequences surrounding the TpD steps were bound with much higher affinity than were standard CpG steps. In line with previous calculations, we detected greater stabilization in a 2QN-d(GDTDTC)$_2$-modelled complex containing a central TpD step than in the equivalent 2QN-d(GACGTC)$_2$ complex containing a CpG step.[97] We calculated and displayed the molecular electrostatic potential (MEP) to aid in visualizing the origin of this difference. The MEPs of both 2QN and echinomycin revealed a distinctive pattern that was considered relevant to DNA binding, namely, a positive region surrounding the intercalating rings and a negative region (which emanates from the sulfur atoms and most of the depsipeptide carbonyl groups) on the opposite side of the molecule (Figure 5). The resulting electrostatic asymmetry was deemed to assist the productive approximation and correct orientation of this type of ligand with respect to the DNA molecule prior to intercalation, possibly compensating for the lack of a net positive charge. On the other hand, comparison of the MEPs calculated for the two DNA hexanucleotides in the complexed state using a subtractive method revealed substantial differences not only in the major groove, as expected from the reversal in the positions of the O and NH$_2$ groups located in it (Figure 3), but also in the spaces between the base pairs that furnish the intercalation cavities and are occupied by the chromophores of the drug upon binding (Figure 5). We suggested that these differences could modulate the DNA-binding specificity of echinomycin and related analogues acting in concert with the well-established hydrogen-bonding interactions involving the exocyclic amino groups in the minor groove.

The preference of TANDEM and CysMeTANDEM for TpA steps rather than the CpG steps recognized by triostin A, echinomycin, and 2QN in standard DNA was early rationalized on the basis of lost hydrogen-bonding

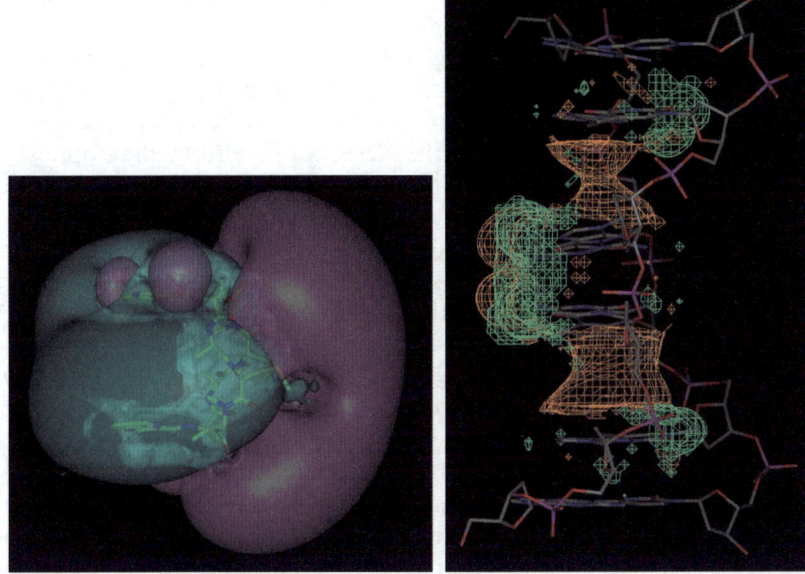

Figure 5 *Left: Molecular electrostatic potential (MEP) calculated for echinomycin and displayed as solid contours coloured in pink and cyan for negative and positive regions, respectively. Right: MEP difference map for d(GACGTC)₂ and d(GDTDTC)₂ in an aqueous medium. The MEP calculated for the latter sequence was subtracted from that calculated for the former to highlight those regions, where the electrostatic potential differs between the two DNA molecules. The mesh contours in orange and green delineate those regions where this difference is ≥ 0.5 or ≤ −0.5, respectively (see text for details)*

possibilities with the 2-amino group of guanines due to formation of two intramolecular hydrogen bonds between the CO of alanines and the NH of valines within the demethylated analogues. The G → I replacement in the DNA molecule, however, has no influence on the interaction of TANDEM with TpA steps (*i.e.* TANDEM does not bind to CpI steps),[99] and the binding of CysMeTANDEM to CpI has been shown to be much weaker than binding to the high-affinity TpA sites,[100] despite the fact that the minor grooves of CpI and TpA steps afford the same hydrogen-bonding possibilities (Figure 3). Furthermore, the affinity of CysMeTANDEM[80,101] and TANDEM[80] for TpA-containing sequences is clearly affected by the flanking bases, with the tetra-nucleotides ATAT and TTAA being the best and poorest binding sites, respectively.

Since these observations also supported the view that the binding specificity of this type of *bis*-intercalators depends on some additional factors beyond hydrogen-bonding interactions between the depsipeptide and the minor groove, we tried to understand their origin using theoretical methods. The problem, as explained in Section 1, is that ligand–DNA interaction energies calculated using molecular mechanics (either on a single structure that is taken to represent the ensemble average of each complex or on multiple snapshots periodically extracted along an MD trajectory for averaging purposes) are too crude to approximate the binding free energies that truly relate to the association constants determined experimentally. Although this relatively simple approach had shown its merits in detecting trends[96] and can be complemented with continuum methods that consider the desolvation effects that oppose the favourable electrostatic interactions,[102] it neglects entropic contributions and does not usually take into account the changes in internal energy undergone by either the drug or the DNA molecules upon binding. For the ligands, these changes are generally assumed to be small and of similar magnitude for all the complexes considered, but the situation can be rather different for the DNA molecules due to the sequence-dependent microheterogeneity mentioned in the introduction. For example, it is conceivable that intercalation at certain sequences might be favoured simply because of the inherent tendency of some dinucleotide steps to underwind and roll.[120]

To further test the hypothesis that stacking interactions can play a decisive role in modulating the preferential binding of these *bis*-intercalating ligands, the computational tool of choice is the free energy perturbation or the thermody-namic integration method.[75,76] However, these studies were not undertaken in my laboratory until recently when a talented Ph.D. student, Esther Marco (later assisted by an enthusiastic new recruit, Ana Negri), was able to complete several thermodynamic cycles in an attempt to understand the origins of both the remarkably increased affinity of 2QN for DTDT relative to GCGC sites and the notable loss of binding affinity between CysMeTANDEM and ICIC compared to ATAT. These studies encompassed the perturbation of whole base pairs (from normal DNA to inosine- or 2,6-diaminopurine-containing DNA), and the calculations were performed in such a way that the contributions of the base pairs on either side of the intercalated chromophores could be separately assessed.[103] The calculated differences in binding free energy for the two DNA *bis*-intercalating agents, 2QN, and CysMeTANDEM, were found to be in very good qualitative agreement with the experimental data. Thus, binding of 2QN to d(GCGGCGCCGC)$_2$ was disfavoured over binding to d(GCGDTDTCGC)$_2$ by 1.8 kcal mol^{-1}, in accordance with the $\sim$100-fold increase in affinity for 2QN binding to a DTDT site compared to a GCGC site in DNA,[97] whereas binding of CysMeTANDEM to d(CTCICICCAG)$_2$ ap-peared to be even more markedly disfavoured (by 4.4 kcal mol^{-1}) over binding to d(CTCATATCAG)$_2$, consonant with the fact that this ligand binds with very high affinity to TpA steps but does not bind significantly to CpI steps.[100] Furthermore, our calculations clearly pointed to the electrostatic contribution to the stacking interactions involving the distal bases as the critical determinant of binding selectivity for the two *bis*-intercalating ligands. This finding rein-forces the view that the recognition site for these *bis*-intercalators definitely extends beyond the central dinucleotide step to cover 4 bp.

In conclusion, although the hydrogen bonds between the depsipeptide of the *bis*-intercalating drug and the functional groups present in the DNA minor groove need to be correct (because there is a penalty when they are missing or wrong), the selectivity patterns can be adequately explained only when the stacking interactions emanating from the base pairs that make up the intercalating site on both sides of the chromophore are taken into account.

8.5 Binding Preferences of Synthetic Pyridocarbazole Bis-Intercalators

Ditercalinium and Flexi-Di are the best-studied representatives of a class of synthetic *bis*-intercalators that, contrary to the natural *bis*-intercalators just described, bind to DNA from the major groove. Both are dimers of a quaternized 10-methoxy-7*H*-pyrido[4,3-*c*]carbazolium chromophore[104] incorporating either a rigid or a flexible linker (Figure 6). It had been shown by NMR spectroscopy[105] and X-ray crystallography[106] that each of the heteroaromatic rings of ditercalinium was able to intercalate into one of the contiguous CpG steps of a d(CGCG)$_2$ oligonucleotide, whereas binding of the drug to d(GCGC)$_2$ showed one ring intercalated at the CpG step and the other stacked atop one of the external base pairs.[107] My attention was drawn to the fact that none of these linkers alone would appear able to discriminate between GpC and CpG steps, since N7 and O6 atoms of guanines in both of them lie diagonally across the major groove in equivalent positions thus shaping two regions of very negative electrostatic potential. Besides, the observation in the minor groove of TpA steps of linking chains similar to those of ditercalinium and Flexi-Di[108] cast doubt on the reported fundamental role of the spacers in dictating the preference of these drugs for the major groove. In the light of our previous findings, we suspected that stacking

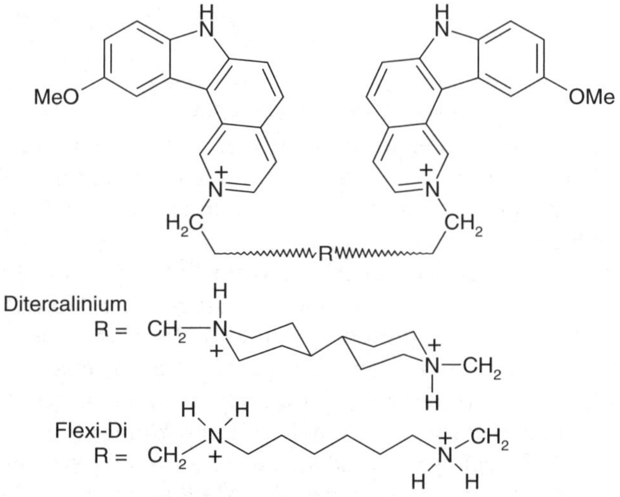

Figure 6 *Chemical structures of the synthetic bis-intercalators ditercalinium and Flexi-Di*

interactions involving the positively charged chromophores could have an important bearing on these experimental observations and could also possibly account for additional crystallographic evidence suggesting significant bending of the DNA helical axis towards the minor groove. In fact, the crystallographers initially suggested that this unusual deformation of the DNA complexed with ditercalinium ($\sim 15°$ kink) could arise from the limited flexibility of the drug's linker, but they were surprised to find that the degree of bending was even higher in the crystal structure of the same DNA sequence complexed with Flexi-Di.[109]

To investigate the origins of both the sequence-binding preferences and the conformational changes brought about in the DNA molecule upon complexation with these drugs, I gave this project to Beatriz de Pascual-Teresa, who had just joined my group after a postdoctoral stay at the University of California. Preliminary calculations on the X-ray crystallographic structures showed that the electrostatic energy component of the stacking interactions between the pyridocarbazolium rings and the DNA bases was favourable for the inner base pairs clamped by them, but was unfavourable for the base pairs making up the outer boundaries of the *bis*-intercalation site.[110] Moreover, the electrostatic interaction was found to be attractive for the internal guanines but repulsive for the internal cytosines, which nicely accounted for the experimentally observed poor van der Waals contacts between the latter and the drug chromophores.[106,109] We also established that as a result of drug binding the underwound central base pair step was stabilized relative to a GpC step in a standard DNA helix.

We then constructed computer models of the complexes of ditercalinium and Flexi-Di with the alternating hexanucleotide d(GCGCGC)$_2$, in which each drug embraced the central GpC-binding site. The complexes were fully solvated and neutralized by addition of sodium counterions, and MD simulations were run for 0.5 ns. To assess the effect of the positively charged spacers, two additional simulations were performed, in which the charges of the linkers were "switched off".

Our results reinforced the idea that the hydrogen-bonding potential of the linkers was effectively reduced in the aqueous medium due to competing interactions with water, their location in the major groove being essentially the result of the marked electrostatically directed orientational preferences displayed by the quaternized pyridocarbazolium chromophores with regard to the highly polarized G:C base pairs.[87]

Taken together, the theoretical results from both avenues of work on *bis*-intercalators provide quantitative support for qualitative descriptions that can be formulated based on interpretation of the MEPs of the stacked systems.[111] Thus, the graphical representation of these quantum mechanically calculated MEPs on a plane close to the recognition surface can be of help to understand electrostatic complementarity and repulsive interactions. For a G:C pair, the most negative MEP region is found in the surroundings of the N7 and O6 atoms of guanine, whereas the most positive MEP region is located around the cytosine ring (Figure 7).

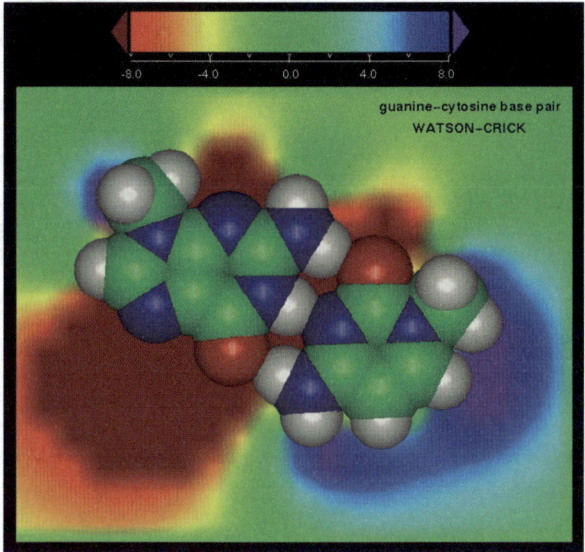

Figure 7   *Molecular electrostatic potential of a G:C base pair as calculated on a plane 1.7 Å below the atom centres, which approximates the recognition surface, and colour coded such that regions of 8 kcal mol^{-1} appear in blue and regions of –8 kcal mol^{-1} are coloured in red, with intermediate values ramping smoothly. More positive and more negative regions are coloured in violet and brown, respectively. Atoms are represented as CPK spheres: C, green; O, red; N, blue; and H, white*

8.6   Sequence Selectivity of Actinomycin D

The clinical importance of actinomycin D (Figure 8) as an antitumour agent and its marked and unusual preference for binding to 5′-GpC-3′ (GpC) steps, as well as other secondary sites such as GpG (CpC) and GpT (ApC) makes this antibiotic an attractive model for the computational study of ligand–DNA interactions. Binding of actinomycin D to DNA involves intercalation of the 2-aminophenoxazin-3-one planar chromophore between two base pairs and fitting of the two cyclic pentadepsipeptides into the minor groove, each extending for two base pairs on either sides of the intercalation site.

The solution of several crystal structures[112–115] strengthened the belief that guanine–threonine hydrogen bonds are the factors responsible for the specificity of actinomycin binding to GpC steps, a hypothesis that was supported by theoretical studies.[116] Nevertheless, this specificity is retained in both actinomine, an actinomycin analogue in which the cyclic pentapeptides have been replaced by *N,N*-diethylethylen–diamine side chains, and 2-aminophenoxazin-3-one, neither of which can form hydrogen bonds with the 2-amino groups of guanine. These observations lend significance to early reports that pointed to electronic interactions between the phenoxazone chromophore and the relatively polarized G:C base pair as the source of specificity for guanine over the

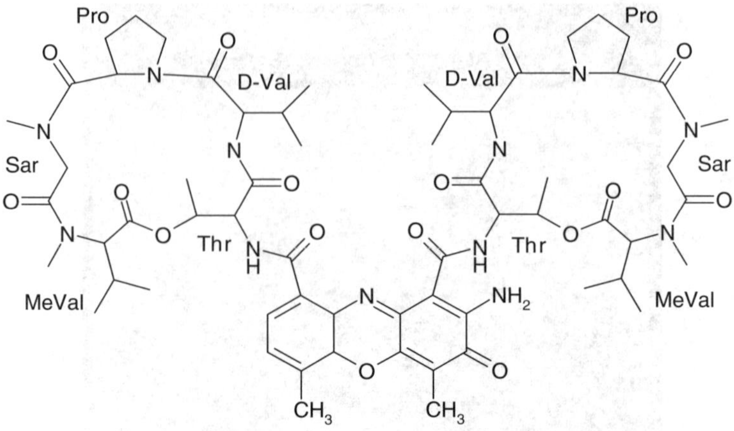

Figure 8 *Chemical structure and amino acid composition of actinomycin D*

other common bases.[117] Furthermore, when 2,6-diaminopurine (D) was used in place of adenine in a DNA fragment, footprinting experiments showed that the sequence specificity of actinomycin was drastically altered, and the drug was found to bind to any 5′-purine-pyrimidine-3′ step different from the standard GpC site.[95] Because the minor grooves of GpC and DpT steps present similar hydrogen-bonding capabilities, as explained above (Figure 3), this interaction could not adequately explain the observed change in binding preferences.

To gain a more detailed understanding of the interactions that govern the sequence-specific binding of actinomycin to DNA, we built molecular models of 14 double-helical DNA hexamers that contained a variety of binding sites in the central region, namely, the canonical GpC step, secondary sites such as GpG and GpT, non-preferred sequences such as CpG, ApT, ApA, and TpA, and the optimal site DpT. Besides, GpC and DpT steps were embedded in different contexts so as to assess the influence of flanking base pairs.[118] The whole set of complexes thus contemplated all possible hydrogen-bonding arrangements between the depsipeptides and the DNA minor groove, as well as all possible combinations of stacking interactions. After energy refinement, intermolecular interaction energies were calculated and partitioned into different van der Waals and electrostatic components. The resulting data matrix was subjected to principal component (PC) analysis as a means to highlight the SAR for this important antitumour antibiotic.

The total interaction energies were found to reflect well the relative affinities of actinomycin D for the different sites: the canonical GpC step was energetically favoured over any other central step in the family of natural sequences, followed by the secondary sites GpG and GpT, although the most favourable binding energies in the whole set were those belonging to sequences in which the adenines had been replaced with D, in accord with the experimental evidence. Regarding the flanking sequences, pyrimidines were preferred over purines on the 5′ side of the GpC intercalation site, and GGCC appeared as the least favoured sequence. Perhaps more interesting was the finding that the first two PC readily explained over 90% of the variance contained in the original data matrix reproducing remarkably well the experimentally observed sequence preferences of actinomycin D. The different clusters of sites were easily discriminated along the axis of the first PC, which was made up by the electrostatic interactions of the peptides with the central base pairs, but also by the van der Waals and electrostatic interactions between the phenoxazone chromophore and the same base pairs. Thus, actinomycin D was shown to discriminate among the different sequences by an interplay of hydrogen bonding and stacking interactions. The importance of the stacking term increases for distinguishing amongst G:C-containing sites, such as GpC, GpG, and CpG, and is crucial for favouring DpT over GpC central steps. These results indicate that the presence of the 2-amino groups not only alters the hydrogen-bonding potential or the geometry of the minor groove, as discussed above, but also imparts distinct stacking properties to the base pairs, both on steric and electrostatic grounds.[111] The simple and intuitive model of like charges repelling one another and unlike charges attracting one another is also valid for molecules and molecular fragments, and the energy minimum found in each of the intercalated ligand–DNA complexes discussed so far can be thought of as the balance point of minimized repulsive and maximized attractive stacking interactions between electric quadrupole moments.[119] Invited by Prof. Pier de Benedetti in 1998, I reviewed stacking interactions and intercalative DNA binding in a companion to *Methods in Enzymology*.[120]

8.7 Binding of the Potent Antitumour Agent Trabectedin to DNA

The amino group of guanine was again the centre of attention when I became involved in studying the mechanisms of action of several interesting natural products of marine origin that are currently being developed by the pharmaceutical and biotechnological company PharmaMar, based in Madrid (Spain) and Cambridge (USA). The compound most advanced into the clinic is trabectedin (Figure 9), a potent antitumour ecteinascidin that was originally isolated from the sea squirt *Ecteinascidia turbinata*[121,122] and presently enjoys orphan drug status for the treatment of soft tissue sarcomas in both Europe and the USA. Ecteinascidins, which are structurally related to microbially derived safracins and saframycins, undergo nucleophilic attack by the exocyclic 2-amino group of guanine in the minor groove of double-stranded DNA when this base is found in a suitable triplet context.[123] Sequence selectivity has been shown to operate predominantly through a set of well-defined hydrogen-bonding rules such that the preferred target triplets are 5'-RGC and 5'-YGG, where R and Y stand for purine and pyrimidine, respectively, and the underlined base is the guanine to which the drug becomes covalently bonded.

Unrestrained MD simulations in aqueous solution were first used by my student Raquel García-Nieto to explore the stability and behaviour of the precovalent complexes between trabectedin and two DNA nonamers containing in

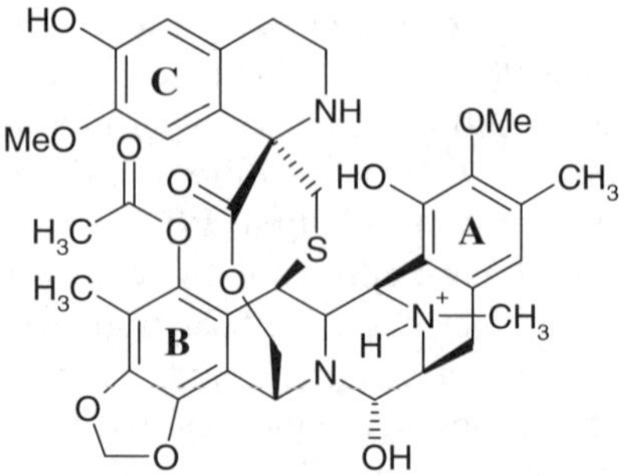

Figure 9 *Chemical structure of trabectedin with its three main tetrahydroisoquinoline-fused subunits labelled A, B, and C*

the central region either 5′-AGC or 5′-CGA.[30] The AGC-containing complex was very stable for the whole length of the 1.4 ns simulation and led to a suitable geometry for adduct formation. On the contrary, the complex containing the disfavoured CGA triplet (which contains a single mismatch relative to CGG) did not result in a stable association as the drug slid one step "upstream" along the minor groove. These results lent support to the proposed role of a well-defined hydrogen-bonding network in the stabilization of these complexes.[123]

We then simulated the covalent complexes between trabectedin and DNA nonamers containing the 5′-AGC and 5′-CGG target sites. In both cases, the bonded drug displayed the predicted binding mode, giving rise to a widening of the minor groove, reducing the twist angle at ApG and CpG steps, and introducing a significant positive roll (relative rotation of the base pairs about the long axis of the base step) at the base pair step involved in covalent bond formation that caused smooth bending of the helix towards the major groove, in very nice qualitative and quantitative agreement with independent results from gel electrophoresis experiments.[124] We found the so-called "bending dials",[125] which are polar plots that display both the magnitude of the axis deflection angle (θ) and its orientation relative to the major groove (φ), the most convenient pictorial representation (Figure 10) of the local helicoidal parameters, roll and tilt, as calculated by the program CURVES developed in Prof. Lavery's group.[126] Dots on the Northern hemisphere of the dial reflect bending that compresses the major groove, whereas those on the left-hand side of the vertical line indicate negative tilt compressing the sugar–phosphate backbone of the complementary strand.

Moreover, by comparing the dynamic behaviour of these duplexes in the absence and in the presence of bonded drug, we assessed the distinct intrinsic bendabilities of AGC and CGG tracts, which were found to be in consonance with their intrinsic twisting and bending propensities as deduced from crystal structures.[10] Thus, the reported significant bias of CpG steps towards bending into the major groove was already clearly apparent in the simulation of the free DNA, whereas the described undertwisting of ApG steps was much stronger in

5' base

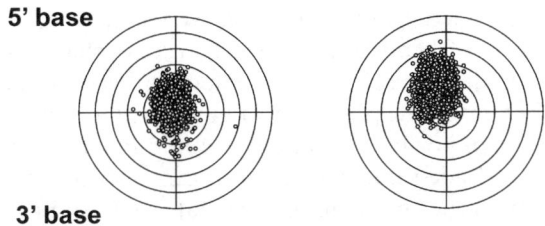

3' base

Figure 10 *Example of typical "bending dials" showing the introduction of positive roll and negative tilt in a trabectedin-bonded DNA step (right) relative to the same step in a free DNA duplex (left). The radial and angular coordinates are, respectively, θ and φ (running clockwise from the top), as defined in the text, and each concentric ring indicates a 10° deflection of the helical axis. Individual points are for structures separated by 1 ps throughout the course of a 1.5 ns MD trajectory so that each ensemble of 1500 points constitutes a probability density*

the presence of the drug, which also induced bending by increasing a previously non-existent roll (Figure 10).

Quite unexpectedly, we later discovered that the peculiar distortion that trabectedin inflicts on the DNA molecule is strikingly similar to that observed in the complexes of DNA with the zinc fingers of typical transcription factors such as EGR-1 (also known as Zif268).[127] In our view, this raises the interesting possibility that trabectedin might be preferentially targeting the minor groove of those DNA regions in which a protein that contains one or more zinc fingers in its DNA-binding domain is already bound in the major groove. The rationale is that the protein-induced DNA structural rearrangement might facilitate the docking of trabectedin over binding to a naked DNA site. This hypothesis led us to model the DNA-binding domain of the ubiquitous transcription factor Sp1, which consists of a tandem array of three Cys_2-His_2 zinc finger modules,[59] as well as to the study during 2 ns of MD simulation of a covalent complex between three trabectedin molecules and an oligonucleotide containing three adjacent target sites.[128] The representative sequence chosen was TGGCGGCGG, which is found in the 5'-upstream promoter region of the human p73 gene encoding a protein with structural and functional homologies with the p53 tumour suppressor protein.

The results suggested that tandem binding of several trabectedin molecules to suitable adjacent DNA sites is sterically and energetically feasible, which made us think about the as-yet untested possibility of cooperative binding. This complex displayed a very stable behaviour, in agreement with the fact that both TGG and CGG triplets represent equally optimal binding sites for the drug. The expected intermolecular hydrogen-bonding scheme between each trabectedin molecule and the Pyr-G-G sites was maintained for the whole length of the simulation. The DNA minor groove was notably wider than in the parallel simulation of the free oligonucleotide, and the ensuing compression of the major groove was comparable to that found in previously studied complexes: increases in positive roll were observed at successive steps and were interrupted only at GpC steps, while negative tilt values were calculated for the CpG and TpG steps, indicative of bending into the sugar–phosphate backbone of their complementary strand. Both roll and tilt components contributed to bending of the double helix but, given the proximity of binding sites around one turn of the helix, the alternation of positive roll at Pyr/Pur and Pur/Pur steps with zero roll at GpC steps on the 3′ side led to a continuous gentle writhe that did not change the overall direction of the helix. This means that the local curvature induced by just one molecule, as reported earlier,[30,123] is cancelled by binding of the others, resulting in a DNA conformation that is best described as intermediate between A- and B-form DNA, such as that found in DNA–RNA hybrids and in the complexes of DNA with zinc fingers. To assess the possible dependence of the observed conformation of the free oligonucleotide on the starting geometry, this was built in both A and B forms,[129] and each was simulated under identical conditions. Interestingly, the A form evolved towards a conformation remarkably similar to that obtained when the simulation started from B-DNA, which was intermediate between A- and B-type DNA as well.[130]

Triplets recognized by the zinc fingers of EGR1 and the related Sp1-like transcription factors include GCG, GAG, GGG, and TGG, whereas favoured sites for trabectedin are CGG, TGG, AGC, and GGC.[123] Although different combinations of these triplets afford binding sites for both proteins and drugs, the majority of optimal trabectedin sites would be located at the junctions between the triplets that are recognized by successive zinc fingers rather than at the triplets themselves (TGG would be the most obvious exception). Remarkably, direct superposition of the trabectedin-bonded TGG, CGG, and CGG triplets onto the corresponding DNA consensus triplets recognized by each EGR1 zinc finger (*i.e.*, GCG, GGG, and GCG) showed a good match (rmsd=1.1 Å using the C1′ atoms) for the three triplets. Moreover, equally good structural similarity (rmsd=0.9 Å) could be obtained when one complex was shifted 1 bp upstream or downstream for the superposition, the best match ultimately depending on the particular sequences that can be simultaneously recognized by the drug and the different zinc fingers of a particular transcription factor or DNA-binding protein. In any case, our findings lend further support to the putative existence of a protein–DNA–trabectedin ternary complex[123] and open new avenues for research.

In the last case presented, binding of trabectedin took place in a head-to-tail fashion on only one of the DNA strands. However, when we realized that the activity parameters of trabectedin against the 60 human tumour cell lines in the Anticancer Drug Screen of the National Cancer Institute (NCI) were highly correlated with those of chromomycin A_3 using the COMPARE algorithm,[131] and knowing that this latter anticancer antibiotic binds as a dimer to adjacent TGG sequences on different strands, we decided to simulate the structure of the complex formed between the self-complementary dodecanucleotide d(GTATGGCCATAC)$_2$ and two trabectedin molecules, each covalently bonded to a different strand in tail-to-tail fashion.[132]

For comparison and completeness, the same sequence was also simulated under identical conditions in the absence of any drug. This allowed us to assess the conformational preferences of (i) the TGG triplet, which is ubiquitously over-represented in human viral and eukaryotic sequences; (ii) its complementary sequence CCA, which is an important component of the so-called "CCAAT box" that is present in a large number of gene promoters[133]; and (iii) the juxtaposition of the two (*i.e.* TGGCCA), which has been dubbed by some X-ray crystallographers as a "natural bending element".[134]

In line with previous observations,[128] bonding of trabectedin resulted in increased DNA stabilization as assessed by smaller rmsd fluctuations in the simulation of the complex relative to the simulation of the free oligonucleotide. In the 2:1 complex, each subunit A of trabectedin protrudes perpendicularly off the helix in front of the guanine to which it is bonded, the B subunits stack over the sugar rings of the complementary cytosines in a manner reminiscent of typical non-intercalative minor groove binders, and the C subunits expose one flat side to the solvent, while the other side makes extensive contacts with the sugar–phosphate backbone of the two nucleosides downstream from each covalently bonded guanine. Additionally, in this tail-to-tail arrangement both B subunits establish favourable van der Waals contacts at the junction of both DNA triplets.

The greatest unwinding as a consequence of trabectedin bonding takes place at the TpG (=CpA) steps, which also show the largest values of roll angle, in good accord with the known inverse relationship between this parameter and helical twist.[10] By contrast, the GpC step displays both negative roll and the largest helical twist within the TGGCCA sequence. Overall, the conformational features of the complex were reminiscent of those previously found in the 3:1 head-to-tail complex described above and correspond, once more, to an intermediate form between A- and B-DNA.

Remarkably, this 2:1 trabectedin complex and the recently solved complex between an Mg^{2+}-chelated chromomycin A_3 dimer and a duplex octanucleotide d(TTGGCCAA)$_2$[135] revealed several features in common, such as (i) an unwound DNA with a considerably widened minor groove in the virtually superimposable target TGGCCA region (rmsd of only 2.1 Å over phosphate backbone atoms); (ii) a similar pattern of roll values consisting of increased positive roll at TpG steps and small positive roll at GpG steps, together with negative roll at the central GpC step which is more marked in the trabectedin complex; and (iii) negative slide at both GpG steps, in good agreement with the finding of glycosyl torsional angles, χ, closer to those of A-DNA.

Our simulation of the free dodecamer showed two distinct bends produced by rolling at each TpG step, but no bending at the junction between TGG and CCA, in agreement with results from a two-dimensional NMR spectroscopy study reporting that GGC in the related self-complementary decamer d(CAT-GGCCATG)$_2$ forms a tight stack with parallel bases, and that high positive roll is present at both TpG steps.[136] This result, however, is at odds with earlier X-ray crystallographic observations showing that stacked B-DNA double helices of general sequence C-C-A-x-x-x-x-T-G-G exhibit the same 23° bend across the -T-G-G C-C-A- non-bonded junction in the crystal lattice[134] that is encountered in the middle of the C-A-T-G-G-C-C-A-T-G decamer.[137] Since high roll at GGC is not seen in GpC steps from other sequences, the most likely explanation is that this structural effect, together with the unusually high positive slide adopted by the CpA(=TpG) steps and the non-standard B_{II} backbone conformation in these crystal forms, is due to strong intermolecular interactions in the crystal lattice.

Finally, both trabectedin and chromomycin A_3 have been shown to exert at least part of their cytotoxicity by interfering with cell replication and transcription. In light of the structural information obtained, it can be proposed that these effects are largely due to the stalling of replication and transcription forks through stabilization of the helical structure of duplex DNA and prevention of strand separation.

8.8 Lamellarins as Topoisomerase I Poisons

A long-standing goal of molecular modelling studies for many years was to understand the way camptothecin (CPT) and other related compounds, such as the clinically useful antineoplastic agent topotecan (Figure 11), specifically block the religation step that follows single-strand nicking of DNA by eukaryotic topoisomerase I (Top1), an enzyme that mediates relaxation of supercoiled DNA in cells. However, the precise atomic details of the interaction with CPT-like Top1 inhibitors were fully understood only when topotecan was cocrystallized with the "covalent binary complex,"[138] despite the existence of a variety of proposed models[139] and the availability of X-ray crystal structures of Top1 interacting both covalently and non-covalently with DNA[140] (see also chapter 2 by Marchand and Pommier elsewhere in this volume). These notable advances in structural biology fuelled my interest in this anticancer target and paved the way for theoretical studies dealing with a large protein–DNA–ligand ensemble that could be treated in the presence of explicit water molecules,

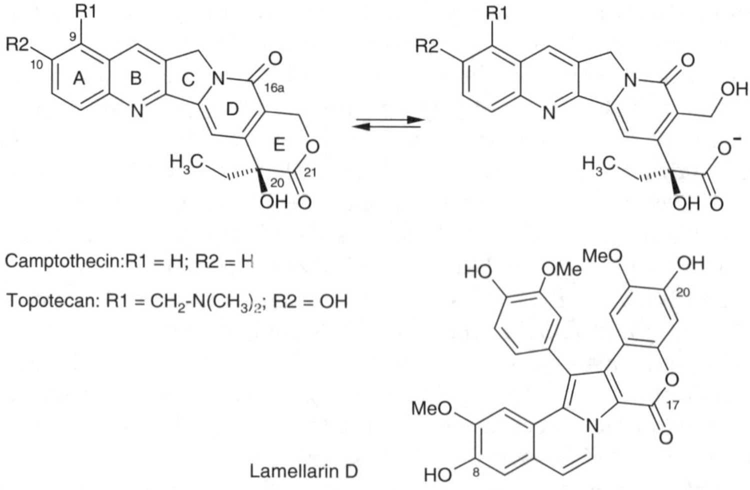

Camptothecin: R1 = H; R2 = H

Topotecan: R1 = CH$_2$-N(CH$_3$)$_2$; R2 = OH

Lamellarin D

Figure 11 *Chemical structures of camptothecin, topotecan, and lamellarin D*

thanks to concomitant improvements in the capacity and power of computers. Thus, when Christian Bailly proposed a collaboration to look at some novel agents that were also capable of converting Top1 into a cellular poison, I was keen to accept. The molecules belonged to the class of 6*H*-[1]benzopyrano-[4,3;4,5]pyrrolo[2,1-*a*]isoquinoline alkaloids known as lamellarins, which had been originally isolated from a prosobranch mollusc (*Lamellaria* sp.), but were later identified in several other marine organisms (see chapter 3 by Dias and Bailly elsewhere in this volume). One of the most cytotoxic compounds in this series is lamellarin D (LMD) (Figure 11), some derivatives of which have already been selected by PharmaMar for preclinical development.[141]

We used the crystal structure of human Top1 covalently linked to a double-stranded DNA molecule with bound topotecan, at 2.10 Å resolution (PDB code: 1k4t),[138] to model the drug-free covalent complexes between Top1 and two different DNA 15-mers containing either a CpG or a TpG intercalation site. These were known sites of cleavage induced by LMD and/or CPT in Top1-mediated DNA cleavage assays.[142] Our work aimed to validate a previously proposed mode of LMD binding to the binary complex[143] using automated docking techniques and MD simulations, and to delineate the structure–activity relationships of lamellarins. For comparison purposes, the same MD protocol was applied to a similar complex containing CPT for which no such information was available at the time we started our studies.

LMD and CPT were each automatically docked into the intercalation site within the DNA–Top1 complex using the Lamarckian genetic algorithm implemented in AutoDock 3.0.[144] In the orientation scored best by the program, the ring skeleton of CPT was practically superimposable (rmsd of 0.3 Å) on that of topotecan, as found in the X-ray crystal structures, and the 20-hydroxyl was found to be engaged in a hydrogen-bonding interaction with the side chain carboxylate of Asp533 (*cf.* page 38). The orientation chosen for LMD as representative of the mode of binding of this drug to the DNA–Top1 complex presented (i) the exocyclic phenyl ring in the major groove; (ii) the C8 and C20 hydroxyl groups at hydrogen-bonding distances from the side chains of Asn722 and Glu356, respectively; and (iii) the keto group (O17) facing the guanidinium group of Arg364. Thus, C10 and C16a of CPT appeared to be positionally equivalent to C20 and C8 in LMD despite the fact that the concave and convex sides of both drugs are found in different grooves of the DNA molecule. This result strongly suggests that docking and modelling approaches based simply on molecular similarity can be very misleading, especially in the absence of 3D structural information about the receptor-binding site.

Interestingly, when the X-ray crystal structures of the human Top1–DNA complex bound with the lactone form of CPT and representative members of the indenoisoquinoline and indolocarbazole classes of Top1 poisons were published,[145] the superposition of these ternary complexes revealed (i) the common presence of a hydrogen-bond acceptor on the minor groove side of the drug molecules (in the case of CPT, between the pyridine N1 atom and the Nε of Arg364); (ii) the presence of substituents on the major groove side; (iii) important contacts with amino acid residues Asn352 and Glu356; and also (iv) that the base-stacking interactions between drug molecules and the cleaved strand side of the DNA duplex are not spatially conserved.

The feasibility of our proposed binding orientations for CPT and LMD was ascertained by carrying out MD simulations of the complexes in aqueous solution, which showed notably stable behaviour during the whole 1.5 ns trajectories, with the fully intercalated drugs giving rise to stacking interactions with the DNA bases and to a number of other stabilizing interactions with both the protein and the surrounding solvent. The hydrogen bonds involving the hydroxyls at C8 and C20 of LMD were often broken and reformed as they exchanged with similar interactions involving neighbouring water molecules.

The binding energy analysis of the Top1–DNA–CPT complex and the two Top1–DNA–LMD complexes (with the drug intercalated at TpG or CpG) revealed a predominance of van der Waals interactions involving the nucleic bases making up the intercalation site and electrostatic interactions with some crucial protein residues (Arg488, Lys532, Asp533, Arg364, Glu356, and Asn722). In the three cases studied, the most favourable and quantitatively more important interaction was with G_{+1}, although in the CG–LMD complex the magnitude of this van der Waals term was similar to that between LMD and the guanine complementary to C_{-1}. Remarkably, the interaction with T_{-1} was more favourable for CPT (by ~ 3 kcal mol^{-1}) than it was for LMD. On the other hand, the lack of detectable Top1-induced C$^{\downarrow}$G sites in the presence of CPT could be explained by the steric clash that would arise between the exocyclic 4-amino group of C_{-1} and the sp^3 carbon in ring C. By contrast, both TpG and CpG steps provided equally good binding sites for LMD, in good accord with the experimental findings.

Since our simulations consistently showed the 20-hydroxyl of LMD to be engaged in a direct or water-mediated hydrogen bond with the carboxylate of Glu356, we decided to test whether the loss of this interaction might account for the deleterious effect that removing the 20-hydroxyl group of LMD has both on Top1 inhibition and cytotoxicity.[142] To this end, we made use of a thermodynamic cycle (Figure 2) that allowed us to calculate the difference in binding free energies for these two molecules by using statistical mechanical information generated from a set of MD simulations during which this hydroxyl was converted into a hydrogen when the drug was free in solution and also when it was part of each of the two ternary complexes studied. The calculated free energy difference was 2.6 and 2.9 kcal mol^{-1} for the CG and TG Top1–DNA complexes, respectively, favouring LMD over the analogue without the 20-hydroxyl, which nicely accounted for the decreased potency detected in the experimental assays. This finding lends credence to the proposed binding mode, which is additionally supported by the observation that incorporation of a hydroxyl at the C10 position of CPT (positionally equivalent to the 20-OH of LMD), as in topotecan (Figure 11) and other simpler analogues, results in greatly improved Top1 inhibitory activity.

8.9 Concluding Remarks

Molecular modelling and computer simulation techniques have come of age and have proved their value in a variety of research fields, including the study of DNA–ligand and DNA–protein interactions. This chapter seeks to provide a summary of my research activity in this area, which has naturally evolved in consonance with the increased availability of structural data on biological macromolecules, advances in force fields including improved treatment of long-range electrostatic interactions, and amazing developments in computer technology, most significantly processing speeds, multi-processor programming, and data storage capacity. I consider myself fortunate to have been led to study the DNA double helix, which despite its overt simplicity continues to yield surprises and awaits full understanding of its encoded message in all free-living organisms. It was also a happy coincidence that my chosen postdoctoral field has continuously benefited from a vast array of programs, interdisciplinary knowledge, and expertise contributed by many talented scientists from all over the world.

For a number of years it has also been a privilege to assist my colleagues at the Organic and Pharmaceutical Chemistry Department in their attempts to synthesize new DNA-interacting compounds with potential anticancer activity.[146–149] This collaboration has had a major impact on my career, because it has highlighted the tremendous challenges involved in the developments of new drugs. In parallel, I have also marvelled at the great chemical diversity and tremendous potential of marine natural products, which can yield promising drugs with unique mechanisms of action. While it is true that we have learnt many lessons from past studies, it is not less true that the discoveries of today are posing even more challenging questions for tomorrow. I can only trust that we will continue producing new generations of enthusiastic and gifted students capable of answering them, and passing to those students the baton that they in turn will pass on to future scientists.

Acknowledgements

For the personal work described in this article, I was assisted by a gifted and dedicated group of co-workers and collaborators who have enriched this entire research effort. I thank all of them for their enthusiasm and helpful discussions over the years. I am also grateful to the Spanish Comisión Interministerial de Ciencia y Tecnología, the National Foundation for Cancer Research, and PharmaMar Laboratories for financial support.

References

1. L.H. Hurley and F.L. Boyd, DNA as a target for drug action, *Trends Pharmacol. Sci.*, 1988, **9**, 402–407.
2. L.H. Hurley, DNA and associated targets for drug design, *J. Med. Chem.*, 1989, **32**, 2027–2033.

3. L.H. Hurley, DNA and its associated processes as targets for cancer therapy, *Nat. Rev. Cancer*, 2002, **2**, 188–200.
4. W. Saenger,*Principles of Nucleic Acid Structure*, Springer-Verlag, New York, 1984.
5. L.H. Hurley, Secondary DNA structures as molecular targets for cancer therapeutics, *Biochem. Soc. Trans.*, 2001, **29**, 692–696.
6. N.M. Luscombe, R.A. Laskowski and J.M. Thornton, Amino acid–base interactions: a three-dimensional analysis of protein–DNA interactions at an atomic level, *Nucl. Acids Res.*, 2001, **29**, 2860–2874.
7. M. Suzuki, S.E. Brenner, M. Gerstein and N. Yagi, DNA recognition code of transcription factors, *Protein Eng.*, 1995, **8**, 319–328.
8. D.S. Goodsell, Sequence recognition of DNA by lexitropsins, *Curr. Med. Chem.*, 2001, **8**, 509–516.
9. X.L. Yang and A.H. Wang, Structural studies of atom-specific anticancer drugs acting on DNA, *Pharmacol. Ther.*, 1999, **83**, 181–215.
10. A.A. Gorin, V.B. Zhurkin and W.K. Olson, B-DNA twisting correlates with base-pair morphology, *J. Mol. Biol.*, 1995, **247**, 34–48.
11. M.A. Young, G. Ravishanker, D.L. Beveridge and H.M. Berman, *Biophys. J.*, 1995, **68**, 2454–2468.
12. J.D. Watson and F.H.C. Crick, Molecular structure of nucleic acids, *Nature*, 1953, **171**, 737–738.
13. H.M. Berman, J. Westbrook, Z. Feng, G. Gilliland, T.N. Bhat, H. Weissig, I.N. Shindyalov and P.E. Bourne, The Protein Data Bank, *Nucl. Acids Res.*, 2000, **28**, 235–242.
14. H.M. Berman, W.K. Olson, D.L. Beveridge, J. Westbrook, A. Gelbin, T. Demeny, S.-H. Hsieh, A.R. Srinivasan and B. Schneider, The nucleic acid database a comprehensive relational database of three-dimensional structures of nucleic acids, *Biophys. J.*, 1992, **63**, 751–759.
15. http://www.umass.edu/microbio/rasmol/.
16. http://www.expasy.org/spdbv/.
17. http://www.accelrys.com/viewer/.
18. http://www.ks.uiuc.edu/Research/vmd/.
19. W.L. DeLano, *The PyMOL Molecular Graphics System* (http://pymol.sourceforge.net/). DeLano Scientific LLC, San Carlos, CA, 2004.
20. http://www.ballview.org/.
21. http://www.mdlchime.com/chime/.
22. D.C. Phillips, in *Biomolecular Stereodynamics*, *II*, R.H. Sarma (ed), Adenine Press, Guilderland, NY, 1981, 497.
23. F. Gago, Modelling and simulation: a computational perspective in anticancer drug discovery, *Curr. Med. Chem. Anti-Cancer Agents*, 2004, **4**, 401–403.
24. M. Karplus, Molecular dynamics simulations of biomolecules, *Acc. Chem. Res.*, 2002, **35**, 321–323.
25. R.P. Feynman, R.B. Leighton and M. Sands, *The Feynman Lectures in Physics*, Addison-Wesley, Reading, MA, 1963, vol I, 36.

26. T.E. Cheatham, P. Cieplak and P.A. Kollman, A modified version of the Cornell *et al.* force field with improved sugar pucker phases and helical repeat, *J. Biomol. Struct. Dyn.*, 1999, **16**, 845–862.
27. T.E. Cheatham III, J.L. Miller, T. Fox, T.A. Darden and P.A. Kollman, Molecular dynamics simulation on solvated biomolecular systems: the particle mesh Ewald method leads to stable trajectories of DNA, RNA and proteins, *J. Am. Chem. Soc.*, 1995, **117**, 4193–4194.
28. R. Kazlauskas, Modeling – a tool for experimentalists, *Science*, 2001, **293**, 2277–2279.
29. D. Strahs and T. Schlick, A-tract bending: insights into experimental structures by computational models, *J. Mol. Biol.*, 2000, **301**, 643–663.
30. R. García-Nieto, I. Manzanares, C. Cuevas and F. Gago, Bending of DNA upon binding of ecteinascidin 743 and phthalascidin 650 studied by unrestrained molecular dynamics simulations, *J. Am. Chem. Soc.*, 2000, **122**, 7172–7182.
31. D. Barsky, N. Foloppe, S. Ahmadia, D.M. Wilson and A.D. MacKerell, New insights into the structure of a basic DNA from molecular dynamics simulations, *Nucl. Acids Res.*, 2000, **28**, 2613–2626.
32. E. Giudice, P. Varnai and R. Lavery, Base pair opening within B-DNA: free energy pathways for GC and AT pairs from umbrella sampling simulations, *Nucl. Acids Res.*, 2003, **31**, 1434–1443.
33. N. Spackova, E. Cubero, J. Sponer and M. Orozco, Theoretical study of the guanine → 6-thioguanine substitution in duplexes triplexes and tetraplexes, *J. Am. Chem. Soc.*, 2004, **126**, 14642–14650.
34. P. Varnai and K. Zakrzewska, DNA and its counterions a molecular dynamics study, *Nucl. Acids Res.*, 2004, **32**, 4269–4280.
35. D.L. Beveridge, G. Barreiro, K.S. Byun, D.A. Case, T.E. Cheatham III, S.B. Dixit, E. Giudice, F. Lankas, R. Lavery, J.H. Maddocks, R. Osman, E. Seibert, H. Sklenar, G. Stoll, K.M. Thayer, P. Varnai and M.A. Young, Molecular dynamics simulations of the 136 unique tetranucleotide sequences of DNA oligonucleotides. I. Research design and results on d(CpG) steps, *Biophys. J.*, 2004, **87**, 3799–3813.
36. S.Y. Ponomarev, K.M. Thayer and D.L. Beveridge, Ion motions in molecular dynamics simulations on DNA, *Proc. Natl. Acad. Sci. USA*, 2004, **101**, 14771–14775.
37. N.K. Banavali and B. Roux, Free energy landscape of A-DNA to B-DNA conversion in aqueous solution, *J. Am. Chem. Soc.*, 2005, **127**, 6866–6876.
38. T. Schlick, D.A. Beard, J. Huang, D.A. Strahs and X. Qian, Computational challenges in simulating large DNA over long times, *Comput. Sci. Eng.*, 2000, **2**, 38–51.
39. T.E. Cheatham III, Simulation and modeling of nucleic acid structure dynamics, and interactions, *Curr. Opin. Struct. Biol.*, 2004, **14**, 360–367.
40. M. Karplus, Molecular dynamics simulations of biomolecules, *Acc. Chem. Res.*, 2002, **35**, 321–322.

41. T.E. Cheatham III, P. Cieplak and P.A. Kollman, A modified version of the Cornell *et al.* force field with improved sugar pucker phases and helical repeat, *J. Biomol. Struct. Dyn.*, 1999, **16**, 845–862.

42. A. D. MacKerell Jr. and N.K. Banavali, All-atom empirical force field for nucleic acids: II. Application to molecular simulations of DNA and RNA in solution, *J. Comput. Chem.*, 2000, **21**, 105–120.

43. M. Feig and B.M. Pettitt, Sodium and chlorine ions as part of the DNA solvation shell, *Biophys. J.*, 1999, **77**, 1769–1781.

44. T.E. Cheatham and P.A. Kollman, Molecular dynamics simulation of nucleic acids, *Annu. Rev. Phys. Chem.*, 2000, **51**, 435–471.

45. C. Sagui and T.A. Darden, Molecular dynamics simulations of biomolecules: long-range electrostatic effects, *Annu. Rev. Biophys. Biomol. Struct.*, 1999, **28**, 155–179.

46. E. Giudice and R. Lavery, Simulations of nucleic acids and their complexes, *Acc. Chem. Res.*, 2002, **35**, 350–357.

47. T.E. Cheatham, Simulation and modeling of nucleic acid structure, dynamics and interactions, *Curr. Opin. Struct. Biol.*, 2004, **14**, 360–367.

48. Y. Duan and P.A. Kollman, Pathways to a protein folding intermediate observed in a 1-microsecond simulation in aqueous solution, *Science*, 1998, **282**, 740–744.

49. R. Elber, A. Ghosh and A. Cardenas, Long time dynamics of complex systems, *Acc. Chem. Res.*, 2002, **35**, 396–403.

50. D. Bashford and D. Case, Generalized Born models of macromolecular solvation effects, *Annu. Rev. Phys. Chem.*, 2000, **51**, 129–152.

51. W.L. Jorgensen, J. Chandrasekhar and J.D. Madura, Comparison of simple potential functions for simulating liquid water, *J. Chem. Phys.*, 1983, **79**, 926–935.

52. V. Tozzini, Coarse-grained models for proteins, *Curr. Opin. Struct. Biol.*, 2005, **15**, 144–150.

53. M.A. Young, S. Gonfloni, G. Superti-Furga, B. Roux and J. Kurigan, Dynamic coupling between SH2 and SH3 domains of c-Src and Hck underlies their inactivation by C-terminal tyrosine phosphorylation, *Cell*, 2001, **105**, 115–126.

54. D.G. Truhlar, J. Gao, C. Alhambra, M. García-Viloca, J. Corchado, M.L. Sánchez and J. Villa, The incorporation of quantum effects in enzyme kinetics modeling, *Acc. Chem. Res.*, 2002, **35**, 341–349.

55. J. Mendieta, S. Martín-Santamaría, E.-M. Priego, J. Balzarini, M.-J. Camarasa, M.-J. Pérez-Pérez and F. Gago, Role of His-85 in the catalytic mechanism of thymidine phosphorylase as assessed by targeted molecular dynamics simulations and quantum mechanical calculations, *Biochemistry*, 2004, **43**, 405–414.

56. B. Isralewitz, M. Gao and K. Schulten, Steered molecular dynamics and mechanical functions of proteins, *Curr. Opin. Struct. Biol.*, 2001, **11**, 224–230.

57. J. Ma and M. Karplus, Molecular switch in signal transduction: reaction paths of the conformational changes in ras p21, *Proc. Natl. Acad. Sci. USA*, 1997, **94**, 11905–11910.

58. F. Rodríguez-Barrios, J. Balzarini and F. Gago, The molecular basis of resilience to the effect of the Lys103Asn mutation in non-nucleoside HIV-1 reverse transcriptase inhibitors studied by targeted molecular dynamics simulations, *J. Am. Chem. Soc.*, 2005, **127**, 7570–7578.
59. E. Marco, R. García-Nieto and F. Gago, Assessment by molecular dynamics simulations of the structural determinants of DNA binding specificity for transcription factor Sp1, *J. Mol. Biol.*, 2003, **328**, 9–32.
60. G.J.R. Zaman, P.J. Michiels and C.A. van Boeckel, Targeting RNA: new opportunities to address drugless targets, *Drug Discov. Today*, 2003, **8**, 297–306.
61. T. Hunter, Signaling – 2000 and beyond, *Cell*, 2000, **100**, 113–127.
62. S.R. Hubbard, Protein tyrosine kinases: autoregulation and small-molecule inhibition, *Curr. Opin. Struct. Biol.*, 2002, **12**, 735–741.
63. S. Klumpp and J. Krieglstein, Serine/threonine protein phosphatases in apoptosis, *Curr. Opin. Pharmacol.*, 2002, **2**, 458–462.
64. J.A. McCammon, B.R. Gelin and M. Karplus, Dynamics of folded proteins, *Nature*, 1977, **267**, 585–590.
65. A. Chien, I. Foster and D. Goddette, Grid technologies empowering drug discovery, *Drug Discov. Today*, 2002, **15**, S176–S180.
66. Folding@home project: http://www.stanford.edu/group/pandegroup/folding/.
67. E.K. Davies and W.G. Richards, The potential of Internet computing for drug discovery, *Drug Discov. Today*, 2002, **7**, S99–S103.
68. B.C. Baguley, Nonintercalative DNA-binding antitumour compounds, *Mol. Cell. Biochem.*, 1982, **43**, 167–181.
69. L.A. Marky and K.J. Breslauer, Origins of netropsin binding affinity and specificity: correlations of thermodynamic and structural data, *Proc. Natl. Acad. Sci. USA.*, 1987, **84**, 4359–4363.
70. F. Gago, C.A. Reynolds and W.G. Richards, The binding of non-intercalative drugs to alternating DNA sequences, *Mol. Pharmacol.*, 1989, **35**, 232–241.
71. P.B. Dervan and B.S. Edelson, Recognition of the DNA minor groove by pyrrole-imidazole polyamides, *Curr. Opin. Struct. Biol.*, 2003, **13**, 284–299.
72. K.L. Buchmueller, A.M. Staples, C.M. Howard, S.M. Horick, P.B. Uthe, N.M. Le, K.K. Cox, B. Nguyen, K.A. Pacheco, W.D. Wilson and M. Lee, Extending the language of DNA molecular recognition by polyamides: unexpected influence of imidazole and pyrrole arrangement on binding affinity and specificity, *J. Am. Chem. Soc.*, 2005, **127**, 742–750.
73. P.B. Dervan, Molecular recognition of DNA by small molecules, *Bioorg. Med. Chem.*, 2001, **9**, 2215–2235.
74. T.P. Straatsma and J.A. McCammon, Computational alchemy, *Ann. Rev. Phys. Chem.*, 1992, **43**, 407–435.
75. P. Kollman, Free energy calculations: applications to chemical and biochemical phenomena, *Chem. Rev.*, 1993, **93**, 2395–2417.

76. T. Rodinger and R. Pomès, Enhancing the accuracy, the efficiency and the scope of free energy simulations, *Curr. Opin. Struct. Biol.*, 2005, **15**, 164–170.

77. C.A. Reynolds, P.M. King and W.G. Richards, Computed redox potentials and the design of bioreductive agents, *Nature*, 1988, **334**, 80–82.

78. F. Gago and W.G. Richards, Netropsin binding to poly[d(IC)]·poly[-d(IC)] and poly[d(GC)]·poly[d(GC)]: a computer simulation, *Mol. Pharmacol.*, 1990, **37**, 341–346.

79. For a review, see B. Pullman and J. Jortner (eds), *Molecular Basis of Specificitiy in Nucleic Acid–Drug Interactions*, Kluwer Academic Publishers, Dordrecht, 1990.

80. M.J. Waring, The molecular basis of specific recognition between echinomycin and DNA, *Mol. Pharmacol.*, 1990, **37**, 225–245.

81. K. Hoogsteen, The structure of crystals containing a hydrogen-bonded complex of 1-methylthymine and 9-methyladenine, *Acta Cryst.*, 1959, **2**, 822–823.

82. A.H.-J. Wang, G. Ughetto, G.J. Quigley, T. Hakoshima, G.A. van der Marel, J.H. van Boom and A. Rich, The molecular structure of a DNA–triostin A complex, *Science*, 1984, **225**, 1115–1121.

83. U.C. Singh, N. Pattabiraman, R. Langridge and P.A. Kollman, Molecular mechanical studies of d(CGTACG)$_2$: complex of triostin A with the middle AT base pairs in either a Hoogsteen or Watson–Crick pairing, *Proc. Natl. Acad. Sci. USA*, 1986, **83**, 6402–6406.

84. X. Gao and D.J. Patel, NMR Studies of echinomycin bisintercalation complexes with d(A1-C2-G3-T4) and d(T1-C2-G3-A4) duplexes in aqueous solution: sequence-dependent formation of hoogsteen A1ñT4 and Watson–Crick T1ñA4 base pairs flanking the bisintercalation site, *Biochemistry*, 1988, **27**, 1744–1751.

85. J. Gallego, A.R. Ortiz and F. Gago, A molecular dynamics study of the bis-intercalation complexes of echinomycin with d(ACGT)$_2$ and d(TCGA)$_2$: rationale for sequence-specific Hoogsteen base-pairing, *J. Med. Chem.*, 1993, **36**, 1548–1561.

86. T.V. Alfredson and A.H. Maki, Phosphorescence and optically detected magnetic resonance studies of echinomycin–DNA complexes, *Biochemistry*, 1990, **29**, 9052–9064.

87. R. Rein, Studies of biomolecular interactions: principles of nucleic acid structure and function from the point of view of constituent interactions, in *Perspectives in Quantum Chemistry and Biochemistry*, vol II, B. Pullman (ed),Wiley, New York, 1978, 307–362.

88. A.H.-J. Wang, G. Ughetto, G.J. Quigley and A. Rich, Interactions of quinoxaline antibiotic and DNA: the molecular structure of a triostin A–d(GCGTACGC) complex, *J. Biomol. Struct. Dyn.*, 1986, **4**, 319–342.

89. J.A. Cuesta-Seijo and G.M. Sheldrick, Structures of complexes between echinomycin and duplex DNA, *Acta Crystallogr. D Biol. Crystallogr.*, 2005, **61**, 442–448.

90. X. Gao and D.J. Patel, Antitumour drug–DNA interactions: NMR studies of echinomycin and chromomycin complexes, *Quart. Rev. Biophys.*, 1989, **22**, 93–138.

91. J. Gallego, F.J. Luque, M. Orozco and F. Gago, Binding of echinomycin to d(GCGC)$_2$ and d(CCGG)$_2$: distinct stacking interactions dictate the sequence-dependent formation of Hoogsteen base pairs, *J. Biomol. Struct. Dyn.*, 1994, **12**, 111–129.

92. M. Aida and C. Nagata, An *ab initio* molecular orbital study on the stacking interaction between nucleic acid bases: dependence on the sequence and relation to the conformation, *Int. J. Quantum Chem.*, 1986, **29**, 1253–1261.

93. R.L. Ornstein, R. Rein, D.L. Breen and R.D. MacElroy, An optimized potential function for the calculation of nucleic acid interaction energies. I. Base stacking, *Biopolymers*, 1978, **17**, 2341–2360.

94. C. Marchand, C. Bailly, M. McLean, S.E. Moroney and M.J. Waring, The 2-amino group of guanine is absolutely required for specific binding of the anti-cancer antibiotic echinomycin to DNA, *Nucl. Acids Res.*, 1992, **20**, 5601–5606.

95. C. Bailly, C. Marchand and M.J. Waring, New binding sites for anti-tumor antibiotics created by relocating the purine 2-amino group in DNA, *J. Am. Chem. Soc.*, 1993, **115**, 3784–3785.

96. J. Gallego, F.J. Luque, M. Orozco, C. Burgos, J. Alvarez-Builla, M.M. Rodrigo and F. Gago, DNA sequence-specific reading by echinomycin: role of hydrogen bonding and stacking interactions, *J. Med. Chem.*, 1994, **37**, 1602–1609.

97. C. Bailly, S. Echepare, F. Gago and M.J. Waring, Recognition elements that determine affinity and sequence-specific binding to DNA of 2QN, a biosynthetic bis-quinoline analogue of echinomycin, *Anti-Cancer Drug Des.*, 1999, **14**, 291–303.

98. T.V. Alfredson, A.H. Maki and M.J. Waring, Optically detected triplet-state magnetic resonance studies of the DNA complexes of the bisquinoline analogue of echinomycin, *Biochemistry*, 1991, **30**, 9665–9675.

99. C. Bailly and M.J. Waring, DNA recognition by quinoxaline antibiotics: use of base-modified DNA molecules to investigate determinants of sequence-specific binding of triostin A and TANDEM, *Biochem. J.*, 1998, **330**, 81–87.

100. K.J. Addess and J. Feigon, Sequence specificity of quinoxaline antibiotics. 2. NMR studies of the binding of [N-MeCys^3N-MeCys^7]TANDEM and triostin A to DNA containing a CpI step, *Biochemistry*, 1994, **33**, 12397–12404.

101. M. Lavesa Sánchez and K.R. Fox, Preferred binding sites for [N-MeCys^3N-MeCys^7]TANDEM determined using a universal footprinting substrate, *Anal. Biochem.*, 2001, **293**, 246–250.

102. B. Honig Sánchez and A. Nicholls, Classical electrostatics in biology and chemistry, *Science*, 1995, **268**, 1144–1149.

103. E. Marco, A. Negri, F.J. Luque Sánchez and F. Gago, Role of stacking interactions in the binding sequence preferences of DNA bis-intercalators: insight from thermodynamic integration free energy simulations. *Nucl. Acids Res.*, 2005, **33**, 6214–6224.

104. D. Pelaprat, A. Delbarre, I. Le Guen Sánchez and B.P. Roques, DNA intercalating compounds as potential antitumor agents. 2. Preparation and properties of 7*H*-pyridocarbazole dimmers, *J. Med. Chem.*, 1980, **23**, 1336–1343.

105. A. Delbarre, M. Delepierre, C. Garbay, J. Igolen, J.-B. Le Pecq Sánchez and B.P. Roques, Geometry of the antitumor drug ditercalinium bisin-tercalated into d(CpGpCpG)$_2$ by ^{1}H NMR, *Proc. Natl. Acad. Sci. USA*, 1987, **84**, 2155–2159.

106. Q. Gao, L.D. Williams, M. Egli, D. Rabinovich, S.L. Chen, G.J. Quigley Sánchez and A. Rich, Drug-induced DNA repair: X-ray structure of a DNA–ditercalinium complex, *Proc. Natl. Acad. Sci. USA*, 1991, **88**, 2422–2426.

107. M. Delepierre, C. Milhe, A. Namane, T.H. Dinh Sánchez and B.P. Roques, ^{1}H- and ^{31}P-NMR studies of ditercalinium binding to a d(GCGC)$_2$ and d(CCTATAGG)$_2$ minihelices: a sequence specificity study, *Biopolymers*, 1991, **31**, 331–353.

108. H.P. Spielmann, D.E. Wemmer Sánchez and J.P. Jacobsen, Solution structure of a DNA complex with the fluorescent bis-intercalator TOTO determined by NMR spectroscopy, *Biochemistry*, 1995, **34**, 8542–8553.

109. M.E. Peek, L.A. Lipscomb, J.A. Bertrand, Q. Gao, B.P. Roques, C. Garbay-Jaureguiberry Sánchez and L.D. Williams, DNA distortion in bis-intercalated complexes, *Biochemistry*, 1994, **33**, 3794–3800.

110. B. de Pascual-Teresa, J. Gallego, A.R. Ortiz Sánchez and F. Gago, Molecular dynamics simulations of the bis-intercalated complexes of ditercalinium and flexi-di with the hexanucleotide d(GCGCGC)$_2$ theoretical analysis of the interaction and rationale for the sequence binding specificity, *J. Med. Chem.*, 1996, **39**, 4810–4824.

111. J. Gallego, B. de Pascual-Teresa, A.R. Ortiz, M.T. Pisabarro Sánchez and F. Gago, Molecular electrostatic potentials of DNA base pairs and drug chromophores in relation to DNA conformation and bis-intercalation by quinoxaline antibiotics and ditercalinium, in *QSAR and Molecular Modelling: Concepts, Computational Tools and Biological Applications*, F. Sanz, J. Giraldo Sánchez and F. Manaut (eds), J. R. Prous, Barcelona, 1995, 274–281.

112. H.M. Sobell, S.C. Jain, T.D. Sakore Sánchez and C.E. Nordman, Stereo-chemistry of actinomycin–DNA binding, *Nat. New Biol.*, 1971, **231**, 200–205.

113. F. Takusagawa, M. Dabrow, S. Neidle Sánchez and H.M. Berman, The structure of a pseudo intercalated complex between actinomycin and the DNA binding sequence d(GpC), *Nature*, 1982, **296**, 466–469.

114. S. Kamitori Sánchez and F. Takusagawa, Crystal structure of the 2:1 complex between d(GAAGCTTC) and the anticancer drug actinomycin D, *J. Mol. Biol.*, 1992, **225**, 445–456.

115. S. Kamitori, Sánchez and F. Takusagaw, Multiple binding modes of anticancer drug actinomycin D: X-ray molecular, modeling and spectroscopic studies of d(GAAGCTTC)$_2$–actinomycin D complexes and its host DNA, *J. Am. Chem. Soc.*, 1994, **116**, 4154–4165.

116. T.P. Lybrand, S.C. Brown, S. Creighton, R.H. Shafer Sánchez and P.A. Kollman, Computer modeling of actinomycin D interactions with double-helical DNA, *J. Mol. Biol.*, 1986, **191**, 495–507.

117. W. Müller Sánchez and D.M. Crothers, Studies of the binding of actinomycin and related compounds to DNA, *J. Mol. Biol.*, 1968, **35**, 251–290.

118. J. Gallego, A. R. Ortiz, B. de Pascual-Teresa Sánchez and F. Gago, Structure–affinity relationships for the binding of actinomycin D to DNA, *J. Comp.-Aided Mol. Design*, 1997, **11**, 114–128.

119. J.H. Williams, The molecular electric quadrupole moment and solid-state architecture, *Acc. Chem. Res.*, 1993, **26**, 593–598.

120. F. Gago, Stacking interactions and intercalative DNA binding, *Methods*, 1998, **14**, 277–292.

121. K.L. Rinehart, Antitumor compounds from tunicate, . *Med. Res. Rev.*, 2000, **20**, 1–27.

122. I. Manzanares, C. Cuevas, R. García-Nieto, E. Marco Sánchez and F. Gago, Advances in the chemistry and pharmacology of ecteinascidins: a promising new class of anticancer agents, *Curr. Med. Chem. Anti-Cancer Agents*, 2001, **1**, 257–276.

123. F. Gago Sánchez and L.H. Hurley,*Devising a structural basis for the potent cytotoxic effects of ecteinascidin 743*, in*Small Molecule DNA and RNA Binders: From Synthesis to Nucleic Acid Complexes*, M. Demeunynck, C. Bailly and W.D. Wilson (eds), Wiley-VCH, Weinheim, Germany, 2002, 643–675.

124. M. Zewail-Foote Sánchez and L.H. Hurley, Ecteinascidin 743 a minor groove alkylator that bends DNA toward the major groove, *J. Med. Chem.*, 1999, **42**, 2493–2497.

125. G. Ravishanker, S. Swaminathan, D.L. Beveridge, R. Lavery Sánchez and H. Sklenar, Conformational and helicoidal analysis of 30 ps of molecular dynamics on the d(CGCGAATTCGCG) double helix: "curves", dials and windows, *J. Biomol. Struct. Dyn.*, 1989, **6**, 669–699.

126. R. Lavery Sánchez and H. Sklenar, The definition of generalized helicoidal parameters and of axis curvature for irregular nucleic acids, *J. Biomol. Struct. Dyn.*, 1988, **6**, 63–91.

127. R. García-Nieto, I. Manzanares, C. Cuevas Sánchez and F. Gago, Increased DNA binding specificity for antitumor ET743 through protein–DNA interactions?, *J. Med. Chem.*, 2000, **43**, 4367–4369.

128. E. Marco, R. García-Nieto, J. Mendieta, I. Manzanares, C. Cuevas Sánchez and F. Gago, A 3·(ET743)–DNA complex that both resembles

an RNA–DNA hybrid and mimics zinc finger-induced DNA structural distortions, *J. Med. Chem.*, 2002, **45**, 871–880.

129. S. Arnott Sánchez and D.W. Hukins, Optimised parameters for A-DNA and B-DNA, *Biochem. Biophys. Res. Commun.*, 1972, **47**, 1504–1509.

130. H.L. Ng, M.L. Kopka Sánchez and R.E. Dickerson, The structure of a stable intermediate in the A ↔ B DNA helix transition, *Proc. Natl. Acad. Sci. USA*, 2000, **97**, 2035–2039.

131. K.D. Paull, R.H. Shoemaker, L. Hodes, A. Monks, D. A. Scudiero, L. Rubinstein, J. Plowman Sánchez and M.R. Boyd, Display and analysis of patterns of differential activity of drugs against human tumor cell lines: development of mean graph and COMPARE algorithm, *J. Natl. Cancer Inst.*, 1989, **81**, 1088–1092.

132. E. Marco Sánchez and F. Gago, DNA structural similarity in the 2:1 complexes of the antitumor drugs Yondelis™ (trabectedin) and chromomycin A_3 with an oligonucleotide sequence containing two adjacent TGG binding sites on opposing strands, *Mol. Pharmacol.*, 2005, **68**, 1–9.

133. R. Mantovani, A survey of 178 NF-Y binding CCAAT boxes, *Nucl. Acids Res.*, 1998, **26**, 1135–1143.

134. K. Grzeskowiak, D.S. Goodsell, M. Kaczor-Grzeskowiak, D. Cascio Sánchez and R.E. Dickerson, Crystallographic analysis of C-C-A-A-G-C-T-T-G-G and its implications for bending in B-DNA, *Biochemistry*, 1993, **32**, 8923–8931.

135. M.-H. Hou, H. Robinson, Y.-G. Gao Sánchez and A.H.-J. Wang, Crystal structure of the $[Mg^{2+}$-(chromomycin $A_3)_2]$–d(TTGGCCAA)$_2$ complex reveals GGCC binding specificity of the drug dimer chelated by a metal ion, *Nucl. Acids Res.*, 2004, **32**, 2214–2222.

136. U. Dornberger, J. Flemming Sánchez and H. Fritzsche, Structure determination and analysis of helix parameters in the DNA decamer d(CAT-GGCCATG)$_2$ comparison of results from NMR and crystallography, *J. Mol. Biol.*, 1998, **284**, 1453–1463.

137. D.S. Goodsell, M.L. Kopka, D. Cascio Sánchez and R.E. Dickerson, Crystal structure of CATGGCCATG and its implications for A-tract bending models, *Proc. Natl. Acad. Sci. USA*, 1993, **90**, 2930–2934.

138. Staker, K. Hjerrild, M.D. Feese, C.A. Behnke, A.B. Burgin Jr., Sánchez and L. Stewart, The mechanism of topoisomerase I poisoning by a camptothecin analog, *Proc. Natl. Acad. Sci. USA*, 2002, **99**, 15387–15392.

139. C.J. Thomas, N.J. Rahier Sánchez and S.M. Hecht, Camptothecin: current perspectives, *Bioorg. Med. Chem.*, 2004, **12**, 1585–1604.

140. M.R. Redinbo, L. Stewart, P. Kuhn, J.J. Champoux Sánchez and W.G.J. Hol, Crystal structures of human topoisomerase I in covalent and non-covalent complexes with DNA, *Science*, 1998, **279**, 1504–1513.

141. C. Tardy, M. Facompré, M. Laine, B. Baldeyrou, D. García-Grávalos, A. Francesch, C. Mateo, A. Pastor, J.A. Jiménez, I. Manzanares, C. Cuevas Sánchez and C. Bailly, Topoisomerase I-mediated DNA cleavage as a guide to the development of antitumor agents derived from the marine

alkaloid lamellarin D: triester derivatives incorporating amino acid residues, *Bioorg. Med. Chem.*, 2004, **12**, 1697–1712.

142. E. Marco, W. Laine, C. Tardy, A. Lansiaux, M. Iwao, F. Ishibashi, C. Bailly Sánchez and F. Gago, Molecular determinants of topoisomerase I poisoning by lamellarins: comparison with camptothecin and structure-activity relationships, *J. Med. Chem.*, 2005, **48**, 3796–3807.

143. M. Facompré, C. Tardy, C. Bal-Mayeu, P. Colson, C. Pérez, I. Manzanares, C. Cuevas Sánchez and C. Bailly, Lamellarin D: a novel potent inhibitor of topoisomerase I, *Cancer Res.*, 2003, **63**, 7392–7399.

144. G.M. Morris, D.S. Goodsell, R. Huey, W.E. Hart, S. Halliday, R. Belew Sánchez and A.J. Olson, *AutoDock: Automated Docking of Flexible Ligands to Receptors, Version 3.0*, The Scripps Research Institute, La Jolla, CA, 1999.

145. B.L. Staker, M.D. Feese, M. Cushman, Y. Pommier, D. Zembower, L. Stewart Sánchez and A.B. Burgin, Structures of three classes of anticancer agents bound to the human topoisomerase I-DNA covalent complex, *J. Med. Chem.*, 2005, **48**, 2336–2345.

146. A. Molina, J.J. Vaquero, J.L. García-Navío, J. Alvarez-Builla, B. de Pascual-Teresa, F. Gago, M.-M. Rodrigo Sánchez and M. Ballesteros, Synthesis and DNA binding properties of γ-carbolinium derivatives and benzologues, *J. Org. Chem.*, 1996, **61**, 5587–5599.

147. J. Pastor, J. Siro, J.L. García-Navío, J.J. Vaquero, J. Alvarez-Builla, F. Gago, B. de Pascual-Teresa, M. Pastor Sánchez and M.M. Rodrigo, Azino fused benzimidazolium salts as DNA intercalating agents, *J. Org. Chem.*, 1997, **62**, 5476–5483.

148. A. Molina, J.J. Vaquero, J.L. García-Navío, J. Alvarez-Builla, B. de Pascual-Teresa, F. Gago and M.M. Rodrigo, Novel DNA intercalators based on the pyridazino[1',6':1,2]pyrido[4,3-*b*]indol-5-inium system, *J. Org. Chem.*, 1999, **64**, 3907–3915.

149. V. Martínez, C. Burgos, J. Alvarez-Builla, G. Fernández, A. Domingo, R. García-Nieto, F. Gago, I. Manzanares, C. Cuevas and J.J. Vaquero, Benzo[*f*]azino[2,1-*a*]phthalazinium cations novel DNA intercalating chromophores with antiproliferative activity, *J. Med. Chem.*, 2004, **47**, 1136–1148.

CHAPTER 9

The Discovery of G-Quadruplex Telomere Targeting Drugs

STEPHEN NEIDLE

Cancer Research UK Biomolecular Structure Group, The School of Pharmacy, University of London, WC1N 1AX, London, UK

9.1 Introduction

Cancer research is often multi-disciplinary. This is especially the case in the search for new small-molecule therapeutic agents, given the need to bring together chemical and biological insights. Traditionally this has been achieved by chemists doing chemistry and biologists being focussed on cellular, molecular and pharmacological issues. More recently, chemistry has extended beyond its traditional boundaries, with the rise of "new" subjects such as chemical biology and chemical genetics. My own laboratory, which is fundamentally chemistry-based, has long been focussed on understanding the structural basis of nucleic acid–small molecule interactions in the context of anticancer drugs, and of exploiting this information for rational drug design. It was, at least initially, a relatively small step to extend from purely medicinal chemistry and molecular structural investigations, to establishing parallel cell and molecular biological studies in cancer pharmacology. The resulting integrated approach to cancer-drug discovery has been applied, within the academic setting of the Institute of Cancer Research, and more recently at the School of Pharmacy in London, to the development of G-quadruplex telomere targeting agents, taking them from original concept through to the identification of potential clinical candidate molecules, over a 10-year period. This work did not take place in isolation from that of many others, and I am very grateful for their manifold contributions; however, this chapter is just an account of the work that has taken place in my own laboratory, so I ask for the indulgence of readers in not expecting a comprehensive review of the field.

9.2 Anthraquinones and Intercalation into Duplex DNA

Cancer chemotherapy is fundamentally concerned with exploiting selectivity between normal and neoplastic cells, by means of either molecular or pharmacological differences between them. One such difference resides in the ends of eukaryotic chromosomes, at the specialised nucleoprotein complexes termed telomeres,[1] whose function is to protect the ends from recombination and degradation. Telomeres were first identified well before the determination of the double-helical structure of DNA, and differences in gross telomere morphology between normal and cancer cells, in particular leukaemic cells, have a long history of investigation and have been extensively reported in the literature.[2]

Our involvement in this area developed by a circuitous route. In the late 1970s we were studying the structural basis of intercalation into duplex DNA by X-ray crystallography, using self-complementary dinucleoside phosphates as model nucleic acid duplexes. We had determined the mode of binding of the 3,6-diaminoacridine compound proflavine in the context of several ribo- and deoxy-sequences.[3–5] We then showed that simple molecular modelling using van der Waals potentials could quantitatively reproduce the experimental orientation and position of the bound drug, *i.e.*, could determine low-energy drug binding modes.[6] A subsequent fruitful collaboration with J.R. Brown and colleagues[7] applied this methodology to a series of disubstituted anthraquinone derivatives. The hypothesis driving this work was that DNA binding would correlate with cytotoxicity and antitumour activity, as had been found for other classes of intercalating cytotoxic agents such as the acridine drug m-AMSA.[8] These anthraquinones were being developed as readily accessible synthetic mimetics of the potent anthracycline natural product antibiotics daunomycin and doxorubicin, following on from the synthesis by the Lederle group[9] of the potent disubstituted anthraquinone derivative mitoxantrone, now an established clinical agent. Our own compounds did not show cytotoxic activity *in vitro* comparable to mitoxantrone, at least in the context of the assays then available to us. We then embarked on a series of structure–activity studies, which culminated in the development[10,11] of several series of diamido substituted anthraquinone derivatives (Figure 1). Again, none of these appeared to show sufficient potency with *in vitro* cancer cell-based screens to justify further development, although one compound in the series did show modest antitumour activity *in vivo*.

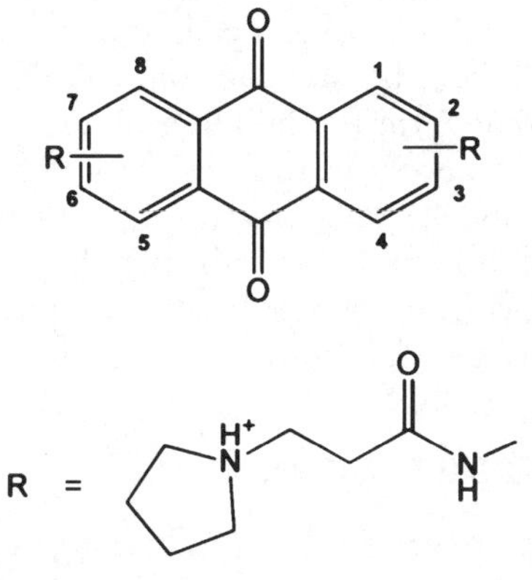

Figure 1 *Structure of disubstituted amidoanthraquinones with charged pyrrolidino side-chains*

9.3   Interactions with Higher-Order DNA

By the early 1990s cytotoxic agents were rapidly becoming unfashionable, and there were early stirrings of the emergence of molecular targeted agents in cancer. A large number of oncogenes had been identified, and various approaches to targeting the genes concerned were being developed, either antisense to mRNAs or antigene to DNA sequences. We were involved in aspects of one particular antigene approach in which a third oligonucleotide strand can be targeted to a particular duplex DNA sequence to form a triple helix. These structures can be stabilised by a variety of intercalating-type ligands, as shown by the pioneering work of C. Hélène and colleagues.[12] We subsequently found, in a fruitful collaboration with K.R. Fox[13] using both molecular modelling and gel-based measurements of stability, that the amidoanthraquinone compounds are able to stabilise parallel triplex DNA. Molecular modelling, even in the absence of experimental structural data, was able to predict that the stabilising effect depended on the precise pattern of substitution on the anthraquinone core, although the nature of the substituent was less critical, provided it terminates in a cationic group such as a pyrrolidino ring. It was surmised that the anthraquinone chromophore can become intercalated between successive base triplets, and that the cationic side-chains would occupy two grooves in the triple-stranded structure. Subsequent studies have demonstrated that both parallel and anti-parallel triplex DNA sequences can be stabilised by amido-anthraquinones.[14–16]

Just prior to this we had submitted a proposal in 1993 for funding a study to develop this concept further, to the then Cancer Research Campaign, citing the knowledge at that time about differences in telomere length between normal and leukaemic cells.[17] It had been long established that telomeric DNA comprises tandem repeats of guanine-rich non-coding DNA[1,2], and we hypothesised that the shortened telomeres in leukaemic cells could be more accessible to ligand binding – we could only surmise that this might lead to a differential cytotoxic effect. We examined by molecular modelling, the interaction of anthraquinone compounds with the four-stranded structure for poly G, and supposed that binding would occur in a manner analogous to their binding to triplex DNA, *i.e.*, by stacking onto the G-quartets of hydrogen-bonded guanine bases that form the building-blocks of poly G.[18] We hypothesised that this could provide a basis for selectivity, but perhaps unsurprisingly, the proposal was not funded. This was prior to the publication of the first crystal structure for a parallel DNA quadruplex by D.M.J. Lilley, B. Luisi *et al.*,[19] which showed at near-atomic resolution the structural arrangement that discrete G-rich sequences can fold into.

9.4   Telomerase and Cancer

By the mid-1990s, the telomere field was becoming very active, following the seminal discovery that the enzyme telomerase is expressed in tumour cells, but not in somatic cells.[20] This enzyme had been previously discovered and characterised by Blackburn and Greider in yeast[21] as being responsible for the synthesis of telomere repeats during replication. A key step was the establishment of an assay for telomerase activity that is able to cope with the low-copy numbers of the enzyme in cell extracts, and has high sensitivity.[22] This assay, the Telomerase Repeat Amplification Protocol (TRAP) is PCR-based and care has to be taken in order to eliminate false positives. A survey of telomerase levels in a range of human cancers showed that its expression was elevated in *ca.* 80–85% of them, a finding reinforced by a large number of studies using primary tumour material.[23] Subsequent surveys have confirmed the widespread prevalence of telomerase activity. The small percentage of cancers that are telomerase-negative (*e.g.*, those in the osteosarcoma category) maintain telomere lengths by the so-called Alternative Lengthening of Telomere (ALT) pathways, which involve recombination-type mechanisms.[24] ALT tumours tend to be less aggressive, and may be of lower clinical significance than telomerase-positive ones. Thus, telomerase emerged in the mid- to late-1990s as a major potential cancer therapy target, for almost all tumour types. This potential is still some way from being fully assessed in the clinic.

Subsequent studies by R.A. Weinberg *et al.*[25] have shown that telomerase expression is an essential element of the tumorigenesis pathway from a normal to a cancer cell. A number of studies have also examined inhibition of telomerase activity, using (i) a dominant-negative telomerase catalytic active-site mutant of the catalytic domain hTERT;[26]) (ii) oligonucleotides targeted against the RNA template in the 451 nt RNA domain (hTR) of the human telomerase complex;[27] or (iii) a small molecule (BIBR 1532) inhibitor of hTERT activity.[28] These studies have demonstrated the proof of principle that inhibition of telomerase results in time-dependent telomere shortening, followed by inhibition of cell growth and eventually senescence, although only after an extended time lag related to the time required to shorten telomeres of several kb in length, with shortening of *ca.* 50–100 nt occurring during each round of replication.

9.5 First-Generation G-Quadruplex Ligands

Prior to these studies, and soon after the initial discoveries of telomerase expression in tumours, Laurence Hurley (then at the University of Texas, Austin, now at the University of Arizona, Tucson) and I started discussing the possibility of screening compounds against telomerase. His group was in the process of establishing a direct, non-PCR telomerase assay, and we suggested that our library of anthraquinone compounds would be appropriate for study in this assay, not least because of our hypothesis that they would also bind to quadruplex DNA – even though we had not yet examined this experimentally. These ideas were inspired by an earlier study by T.R. Cech and colleagues,[29] which had shown that telomerase can be inhibited by folding a telomere primer into a higher-order structure by means of adjustments of potassium ion levels (quadruplex DNAs are not formed in the absence of potassium or other alkali metal ions), and also by a spectroscopic study by N.R. Kallenbach *et al.*,[30] showing that the dye ethidium can bind to DNA-containing guanine repeats.

Our collaborative project focussed on a series of 2,6-substituted amidoanthraquinones (Figure 1), and found that these compounds could inhibit telomerase, with one being an especially effective inhibitor.[31] The results of a telomerase assay are most often visualised on a non-denaturing gel, which shows the products formed by the enzyme elongating a primer-DNA sequence as a series of bands, each of which corresponds to the primer with successive telomeric hexanucleotide repeats. In the absence of any inhibitor, this ladder of bands extends indefinitely, but an inhibitor produces a truncated pattern. Our particular amidoanthraquinone compound (Figure 2) produced 50% inhibition of telomerase activity at a concentration of 23 μM, using a direct enzymatic assay, and not the TRAP assay procedure. It was also shown that the compound could bind to a quadruplex DNA sequence *in vitro*, using UV titration and chemical-shift NMR experiments. These results can be rationalised in terms of a model in which the 3′-end of telomeric DNA (which is single-stranded for the terminal 100–200 nucleotides – see Figure 3) is made inaccessible to the telomerase eleven-nucleotide RNA template by virtue of the induction of folding of the single-stranded DNA into a quadruplex structure by the amidoanthraquinone ligand (Figure 4).

We subsequently determined the structure–activity relationships for several regioisomeric series of substituted amidoanthraquinones, and were able to highlight the features contributing to maximal activity.[32,33] In parallel, molecular modelling studies were initiated,[34] that used the existing NMR structure of the sodium form of an intramolecular quadruplex formed by four repeats of the human telomeric DNA sequence, *i.e.*, d[AGGG(TTAGGG)$_3$] that had been determined by D.J. Patel and Y. Wang.[35] A fruitful collaboration with Lloyd

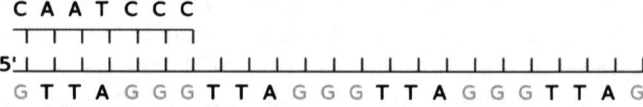

Figure 2   *Structure of 2,6-disubstituted amidoanthraquinone used in the first telomerase G-quadruplex ligand studies.[31] This compound has an $^{tel}EC_{50}$ value for telomerase inhibition of 23 μM*

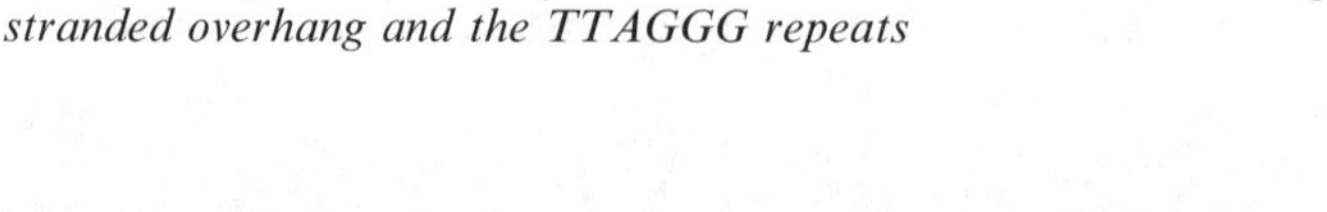

```
C  A  A  T  C  C  C
┬  ┬  ┬  ┬  ┬  ┬  ┬
5'└──┴──┴──┴──┴──┴──┴──┴──┴──┴──┴──┴──┴──┴──┴──┴──┴──┴──┴──┴──┴──┴──┴──┘
   G  T  T  A  G  G  G  T  T  A  G  G  G  T  T  A  G  G  G  T  T  A  G
```

Figure 3   *Schematic of the 3′-end of vertebrate telomeric DNA, showing the single-stranded overhang and the TTAGGG repeats*

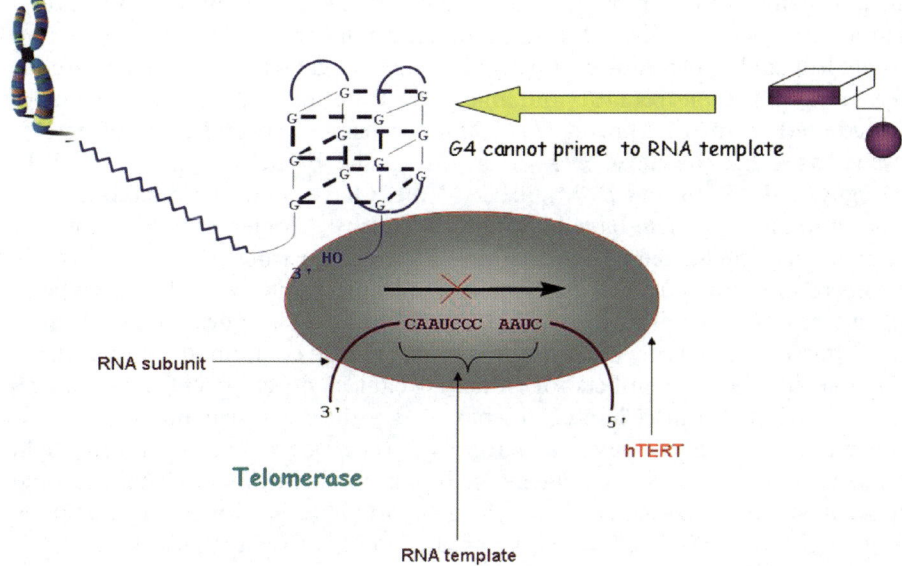

G4 cannot prime to RNA template

RNA subunit

Telomerase

RNA template

hTERT

Figure 4 *Schematic of the process envisaged for inhibition of telomere elongation by telomerase, showing the hTR template and the folding of the 3'-end of the DNA into a quadruplex on ligand binding*

Kelland was established at this time, when both of us were at the Institute of Cancer Research. His group set up telomerase activity assays using a modification of the TRAP methodology, and also started cell-based short-term growth inhibition assays. The most active amidoanthraquinone has telomerase inhibitory activity, expressed as $^{tel}EC_{50}$ values, in the low-micromolar range and cell-growth inhibition (IC_{50} values) at rather higher levels. It was apparent that significantly more quadruplex selectivity was required, and it was not obvious how this could be achieved with amidoanthraquinones. Moreover these compounds, although often straightforward to synthesise, are often challenging to purify because intermediates are exceptionally insoluble in most solvents. There are also difficulties in generating new intellectual property in the anthraquinone area.

At this time (1997) our work was subject to (yet another!) funding review by the then Cancer Research Campaign. It was pure bad luck that the review took place just a few days after the appearance of a paper that reported that telomerase-null mice did not suffer any ill-effects from the absence of telomerase, indicating that the presence of telomerase was not essential for cell viability. Unsurprisingly, the review panel was less than fully enthusiastic about telomerase as a therapeutic target. However we persisted in our view that it was important to develop selective ligands to more fully evaluate the hypothesis, and in due course some support for our work was forthcoming.

Our molecular modelling studies at that time highlighted the need to retain the core-structural feature of a planar aromatic polycyclic group, and several such systems were explored, such as substituted fluorenones[36] and several tetracyclic systems.[37,38] At the same time, a number of other groups started to publish analogous studies with a wide range of ligands, all of which share the features of a planar aromatic core together with cationic side-chains. Examples include tetra-*N*-methyl-porphyrins[39] and perylenes.[40] Several types of hetero-cyclic-based ligands have been described by the French group led by J.-L. Mergny, J.-F. Riou and P. Mailliet;[41,42] these have been of especial interest, since a number of them have high potency against telomerase, and a range of telomere-associated cellular effects have been demonstrated for them, notably telomere shortening as well as effective quadruplex binding. A distinct type of ligand has been devised by M.F.G. Stevens and colleagues[43] comprising a quaternised pentacyclic acridinium system; a lead compound in this series, RHPS4, has been identified for potential clinical development. The natural product telomestatin (Chapter 10) has a very different structure from these other agents, with no positive charges or basic side chains, and with eight oxazole rings in a cyclic arrangement. It is a potent telomerase inhibitor and quadruplex-binding agent.[44,45] We also reported studies on 3,6-disubstituted acridines and acridones,[46] which showed a profile of quadruplex binding and cellular behaviour that is similar to the amidoanthraquinones.

At this point, molecular modelling was employed in an attempt to rationally generate quadruplex-selective ligands that exploit the unique structural features of quadruplexes. The NMR structure of the human intramolecular quadruplex was used as a template with disubstituted acridines as starting-points. This suggested that substitution at the acridine 9-position, taken together with 3- and 6-side-chains, would lead to a molecule with selectivity for this structure compared to duplex DNA.[47] This hypothesis was based on the concept that all three substituents of a trisubstituted acridine would each be able to interact in a groove of a quadruplex structure (Figure 4) – the structural selectivity of the ligand then arises from the presence of four such grooves compared to two in duplex DNA.

Synthesis of a series of acridine compounds with an aniline group at the 9-position resulted in the BRACO-19 series (Figure 5). BRACO-19 itself, which has been used as the paradigm for the series, shows the predicted quadruplex DNA selectivity, with a 30-fold difference in association constants compared to duplex DNA, as measured using surface plasmon resonance methods by W.D. Wilson and colleagues in Atlanta.[47] It also has > 10-fold greater potency against telomerase than the most potent disubstituted acridine compounds. Conventional short-term growth inhibition IC_{50} values measured in a number of cancer cell lines are in the range 2–15 μM, providing a window between the minimum concentration required to achieve telomerase potency and short-term cell kill. It is presumed that telomerase inhibitors effect cell growth arrest and eventual cell kill by telomere-associated mechanisms rather than by generalised cell killing, so BRACO-19 has the appropriate properties for probing these. A series of structure–activity relationships has subsequently been explored for a wide range of BRACO-19 analogues, with in particular, variations in side-chain length and size.[48,49]

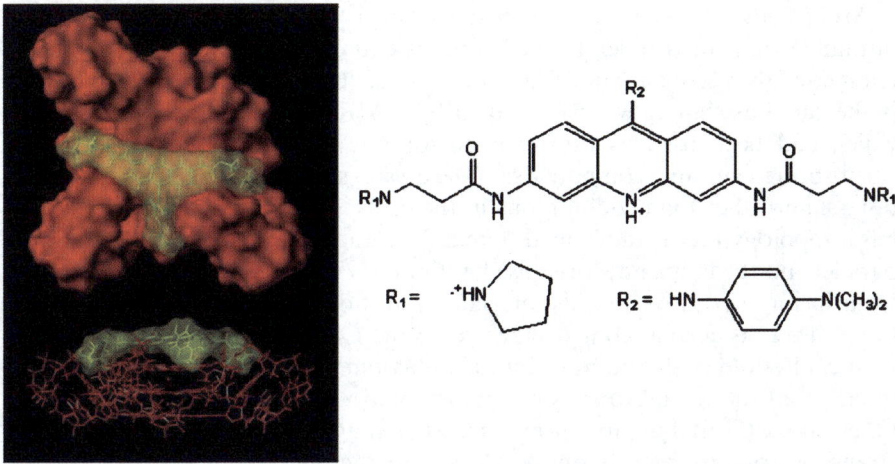

Figure 5 *(Left-hand side) two views of the computer-modelled structure of a tricyclic acridine bound to the parallel quadruplex structure. (Right-hand side) structure of BRACO-19*

9.6 Molecular Models for Quadruplex-Trisubstituted Acridine Complexes

The structure–activity studies that have been undertaken[48,49] to systematically probe the role of each component of the BRACO-19 moiety, have had the aim of optimising their quadruplex-binding and telomerase inhibitory features. The acridine ring itself, which is normally protonated at physiological pH, plays a pivotal role as the platform that interacts with the G-quartet feature in quadruplex structures by stabilising π–π stacking interactions. No experimental molecular structure is available for trisubstituted acridines bound to quadruplexes; instead molecular modelling studies have been extensively used to guide structure–activity relationships. The crystal structure of a complex between the dimeric quadruplex from *Oxytricha nova* and a disubstituted acridine is now available,[50] which has confirmed several of the essential structural features in the molecular model used for the earlier design of the trisubstituted acridines. Thus the platform acridine group is seen to be stacked on to the terminal G-quartet, and the two cationic side chains are each held in a quadruplex groove. This crystal structure, together with NMR studies in various laboratories (*e.g.*, with the perylene compound PIPER,[40] and the pentacyclic acridinium compound RHPS4[51]), has also demonstrated that the planar ligand groups do not intercalate between G-quartets, but stack externally – this is to be expected in view of the energy penalty required for mutual G-quartet unstacking. The *Oxytricha* quadruplex is a relevant model for the sodium form[35] of the human intramolecular quadruplex, with diagonal loops that connect nucleotide strands also forming the top of an acridine binding site.

Around this time we had also determined[52] the crystal structure of an intramolecular quadruplex formed from essentially four repeats of the human telomeric DNA sequence d(TTAGGG). This sequence, d[AGGG(TTAGGG)₃], is the same as that previously studied by NMR in sodium conditions,[35] and which had been used as the template for the structure-based design of the trisubstituted acridine compounds.[47] The crystal structure, which has associated potassium rather than sodium ions in the central ion channel of the structure, has a topology that is radically different from the sodium form with all strands parallel and the loops now organised at the sides of the G-quartets. The sodium form has anti-parallel strands and lateral and diagonal arrangements for the loops. There is current controversy[53] as to the nature of the species in solution with a physiological concentration of potassium ions, with some studies suggesting that the crystal form is present in solution, possibly in equilibrium with other species.[54] It has also been suggested that the features of the parallel-stranded structure may be most relevant to the densely packed environment within a telomere.[53] This in turn supports the concept that the parallel quadruplex structure is an appropriate model for the 3′ overhang of human telomeric DNA. It has been used by us in several molecular modelling studies with trisubstituted acridines in order to rationalise their quadruplex-binding and telomerase inhibition data, and possibly to direct the design of new ligands.[48,49]

The terminal G-quartets in the parallel structure of the 22-mer quadruplex are readily accessible for ligand binding. A recent molecular modelling study and quantitative energetic analysis by us[55] has systematically explored all possible positions for BRACO-19 binding, and has concluded that as a consequence there are potential alternative structures (see for example Figure 6) to the one previously modelled by us (Figure 7) using the anti-parallel sodium quadruplex structure.[54] In the newer model, the 9-anilino substituent is positioned above the central ion channel of the quadruplex rather than being in a groove. It is stabilised by direct interactions with the O6 substituent of one guanine, and enables the charged nitrogen atoms of each 3,6-side-chain to form hydrogen bonds with a phosphate group in a quadruplex groove. We note that

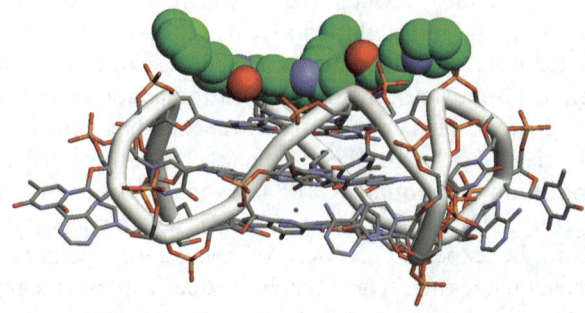

Figure 6   *Computer-modelled structure of BRACO-19, in green space-filled representation, bound to the parallel NMR structure[52] of the 22-mer intramolecular quadruplex formed from the sequence d[AGGG(TTAGGG)₃]*

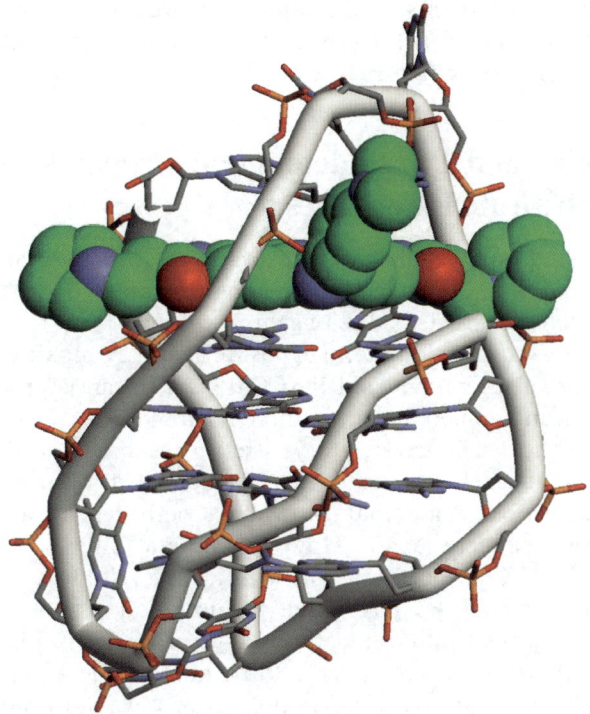

Figure 7 *Computer-modelled structure of BRACO-19, in green space-filled representation, bound to the anti-parallel crystal structure[35] of the 22-mer intramolecular quadruplex formed from the sequence d[AGGG(TTAGGG)₃]*

this arrangement cannot be readily formed with the anti-parallel sodium-form quadruplex since the diagonal loop sterically excludes the 9-anilino substituent from being positioned above the G-quartet. It is also 6.6 kcal mol^{-1} more stable than the three-groove arrangement when modelled into the parallel structure, probably as a result of the interactions detailed above.

These models have been found to be useful in rationalising the experimental trends found in several studies[48,49] with a range of modified substituents at the 3, 6- and 9-positions of the acridine ring.

(i) Systematically extending the length of the 3,6-substituents by –(CH₂)– groups results in decreased quadruplex affinity and telomerase inhibition. The model shows that extensions of the side-chains result in the terminal cationic groups no longer being able to hydrogen bond to phosphate groups, and that non-bonded interactions with the walls of the grooves are progressively lost.

(ii) These charged groups contribute significantly to the overall binding energy.

(iii) There is a critical size for the end groups, with binding and activity decreasing rapidly with increased ring size beyond six-membered rings.

(iv) Extended substituents at the 9-position are well tolerated. This is in accord with the model, since they are readily able to fit in the third-quadruplex groove (see Figure 6).

9.7 Cellular and Pharmacological Properties of Trisubstituted Acridines

A number of studies have shown that, in accord with the classic model for telomerase inhibition, G-quadruplex ligands are able to produce telomere shortening in cells exposed to these agents (see, *e.g.*, refs. 41,45 and 56–59). The effect is observed after long-term exposure at sub-cytotoxic doses, typically after a minimum of several weeks. BRACO-19, for instance, produces a 0.4 kb shortening in mean telomere length in the UXF1138L vulval-carcinoma line,[56] which has very short telomeres at the outset (a mean length of 2.7 kb). This would suggest that senescence would only be activated following an extended period of ligand exposure, and that this would strictly depend, as per the classic model, on mean telomere length. However, we and others have consistently observed[41,43,49,56,59,60] the occurrence of significant levels of cellular senescence, just a few days (typically <7 days) after the initiation of sub-cytotoxic ligand exposure. This has been found in a wide range of cancer cell lines, indicating early cell-growth arrest. There appears to be no dependence of this on mean telomere length. In accord with these observations, we have shown[60] that the expression of two senescence-associated genes, initially p21 and then p16[INK4a], is elevated upon exposure to BRACO-19. The p16[INK4a] pathway acts to inhibit the progression of the cell cycle *via* the inhibition of cyclins and cyclin-dependent kinases. Increased p16[INK4a] expression leads directly to the inhibition of cdk4 and cdk6 and the prevention of retinoblastoma protein phosphorylation. The p16[INK4a] pathway can act independently of p53 status. The activation of senescence then leads to rapid apoptosis.

We have suggested that two events can lead to this occurring in cancer cells.

(i) In general, within any population of cells there exists a sub-population with exceptionally short telomeres. This is commonly observed in Southern blots of telomere hybridisation, which show that the distribution of telomere lengths is broad, typically extending over several kb. It has been pointed out that such a sub-population is acutely sensitive to telomerase inhibition, which can then affect the population as a whole and drive it into senescence without altering telomere length overall.[61,62]

(ii) Physically uncapping the single-stranded 3′-end[63] of telomeric DNA from telomerase is known to lead to rapid senescence *via* activation of DNA damage response pathways.[64] Induction of G-quadruplex formation by ligands[65] will compete with hPOT1 binding to the single-strand overhang and displace telomerase end-capping, exposing the end.[66] Metaphase spreads of cells exposed to BRACO-19 show clear evidence of end-to-end chromosomal fusions,[60] supporting the hypothesis that

BRACO-19 acts at chromosome ends, exposing them, so that they are captured in metaphase as fusions. All this suggests that the G-quadruplex-ligand complex at the 3′-end of the telomere itself is the critical signal to the cell leading to senescence and cell death.

G-quadruplex ligands such as BRACO-19 then act not so much as pure telomerase inhibitors, but as telomere targeting agents with selectivity for cancer cells. This would make them rather more useful as anticancer agents, with a predicted profile of antitumour activity that would resemble conventional agents, and would be able to selectively kill target cells without an extended time lag following administration. An initial study with BRACO-19[67] and the A431 tumour xenograft model *in vivo* did not show single-agent activity, although activity was found in a combination study with paclitaxel. A subsequent xenograft study[56] with the vulval-carcinoma line UXF1138L and BRACO-19 alone showed highly significant activity with mean growth inhibition of 96%. This response was paralleled by loss of hTERT expression in treated tumours, measured by immunostaining, consistent with displacement of telomerase from telomeres and subsequent cytoplasmic degradation.

9.8 Conclusions

The targeting of cellular DNA with small molecules is inherently non-discriminatory and leads to generalised cellular cytotoxicity. Only when there is a therapeutic window, possibly due to adventitious pharmacological differences between normal and tumour cells, can such molecules be useful in the clinic.

The targeting of specific sequences, *e.g.*, of oncogenes, is described elsewhere in this volume. This approach is always going to be challenging since it relies on the ability of hydrogen bonds to drive selectivity, which has to overcome the inherent double-helical sameness of almost all the genome. Quadruplexes uniquely offer the opportunity for DNA structural specificity,[68] governed by the sequence-specificity of quadruplex-forming sequences. In the case of telomeric DNA, specificity for cancer cells is mediated by the expression of the telomerase catalytic subunit in the overwhelming majority of cancer cells.

We have shown that induction of intramolecular G-quadruplex structures in cancer cells by small molecules such as BRACO-19 does occur and can act as a selective cell-killing signal in cancer cells, both *in vitro* and *in vivo*, in accordance with this concept. However none of these molecules reported to date are specific for quadruplexes, and so all have at least some measure of duplex affinity, and hence some accompanying cytotoxicity. The observations of on-target anticancer activity for BRACO-19 *in vivo* show that this factor does not hinder the possibility of a therapeutic window. BRACO-19 and related compounds are currently being considered as candidates for clinical trial. One can envisage that a second generation of compounds with enhanced quadruplex selectivity is inherently possible, and is currently being developed in this and other laboratories, in our case with the continuing help of structure-based design principles.

Acknowledgements

The studies from my laboratory that I have outlined here would not have been possible without the dedication, skill and inspiration of many people over the past 10 years. In particular, the contributions of Tony Reszka, Gary Parkinson, Martin Read, Shozeb Haider, John Harrison, and more recently Chris Incles, Christoph Schultes, Mekala Gunaratnam and Cristina Martins, have all been crucial. We have been fortunate with a number of excellent collaborators, notably Lloyd Kelland and his group, initially when we were all at the Institute of Cancer Research, and more recently with his laboratory at Antisoma Ltd. David Wilson, Angelika Burger and Shankar Balasubramanian have all been important collaborators at various stages over the past 10 years. None of this would have happened without the initial collaboration with Laurence Hurley in 1996/1997. Since then we have gone our separate ways, but we both continue to uphold the quadruplex concept.

Cancer Research UK and its predecessor the Cancer Research Campaign, have generously funded the bulk of the work in my laboratory during this period. I am grateful to them for their willingness to (most of the time) take risks and to have faith in new, untried concepts. Other important funders have been the EU, the Association for International Cancer Research and most recently Antisoma Ltd.

References

1. (*a*) J. Meyne, R.L. Ratliff and R.K. Moyzis, *Proc. Natl. Acad. Sci. USA*, 1989, **86**, 7049–7053; (*b*) E.H. Blackburn, *Cell*, 2001, **106**, 661–673; (*c*) T. Cech, *Cell*, 2004, **116**, 273–279.
2. J.H. Ohyashiki, G. Sashida, T. Tauchi and K. Ohyashiki, *Oncogene*, 2002, **21**, 680–687.
3. S. Neidle, A. Achari, G.L. Taylor, H.M. Berman, H.L. Carrell, J.P. Glusker and W.C. Stallings, *Nature*, 1977, **269**, 304–307.
4. H.S. Shieh, H.M. Berman, M. Dabrow and S. Neidle, *Nucleic Acids Res.*, 1980, **8**, 85–96.
5. S. Neidle, H.M. Berman and H.S. Shieh, *Nature*, 1980, **228**, 129–133.
6. S.A. Islam and S. Neidle, *Acta Crystallogr.*, 1984, **B40**, 424–429.
7. S.A. Islam, S. Neidle, M. Gandecha, M. Partridge, H. Patterson and R. Brown, *J. Med. Chem.*, 1985, **28**, 857–864.
8. B.F. Cain, G.J. Atwell and W.A. Denny, *J. Med. Chem.*, 1975, **18**, 1110–1117.
9. K.C. Murdock, R.G. Child, P.F. Fabio, R.B. Angier, R.E. Wallace, F.E. Durr and R.V. Citarella, *J. Med. Chem.*, 1979, **22**, 1024–1030.
10. S. Neidle and D.A. Collier, *J. Med. Chem.*, 1988, **3**, 847–857.
11. M. Agbandje, T.C. Jenkins, R. McKenna, A.P. Reszka and S. Neidle, *J. Med. Chem.*, 1992, **35**, 1418–1429.
12. J.S. Sun and C. Hélène, *Nucleosides Nucleotides Nucleic Acids*, 2003, **22**, 489–505.

13. K.R. Fox, P. Polucci, T.C. Jenkins and S. Neidle, *Proc. Natl. Acad. Sci. USA*, 1995, **92**, 7887–7891.
14. M.D. Keppler, M.A. Read, P.J. Perry, J.O. Trent, T.C. Jenkins, A.P. Reszka, S. Neidle and K.R. Fox, *Eur. J. Biochem.*, 1999, **263**, 817–825.
15. M.D. Keppler, S. Neidle and K.R. Fox, *Nucleic Acids Res.*, 2001, **29**, 1935–1942.
16. M.D. Keppler, P.L. James, S. Neidle, T. Brown and K.R. Fox, *Eur. J. Biochem.*, 2003, **270**, 4982–4992.
17. D.J. Adamson, D.J. King and N.E. Haites, *Cancer Genet. Cytogenet.*, 1992, **61**, 204–206.
18. M. Gellert, E.R. Lipsett and D.R. Davies, *Proc. Natl. Acad. Sci. USA*, 1962, **48**, 2013–2018.
19. G. Laughlan, A.I. Murchie, D.G. Norman, M.H. Moore, P.C. Moody, D.M.J. Lilley and B. Luisi, *Science*, 1994, **265**, 520–524.
20. (*a*) N.W. Kim, M.A. Piatyszek, K.R. Prowse, C.B. Harley, M.D. West, P.L.C. Ho, G.M. Coviello, W.E. Wright, R.L. Weinrich and J.W. Shay, *Science*, 1994, **266**, 2011–2015; (*b*) C.M. Counter, H.W. Hirte, S. Bacchetti and C.B. Harley, *Proc. Natl. Acad. Sci. USA*, 1994, **91**, 2900–2904; (*c*) J.W. Shay and S. Bacchetti, *Eur. J. Cancer*, 1997, **33**, 787–791.
21. C.W. Greider and E.H. Blackburn, *Cell*, 1987, **51**, 887–898.
22. N.W. Kim and F. Wu, *Nucleic Acids Res.*, 1997, **25**, 2595–2597.
23. (*a*) J.W. Shay and S. Bacchetti, *Eur. J. Cancer*, 1997, **33**, 787–791; (*b*) L.R. Kelland, *Eur. J. Cancer*, 2005, **41**, 971–979.
24. A. Muntoni and R.R. Reddel, *Hum. Mol. Genet.*, 2005, **14**, R191–R196.
25. (*a*) W.C. Hahn, C.M. Counter, A.S. Lundberg, R.L. Beijersbergen, M.W. Brooks and R.A. Weinberg, *Nature*, 1999, **400**, 464–468; (*b*) S.D. Kendall, C.M. Linardic, S.J. Adam and C.M. Counter, *Cancer Res.*, 2005, **65**, 9824–9828.
26. W.C. Hahn, S.A. Stewart, M.W. Brooks, S.G. York, E. Eaton, A. Kurachi, R.L. Beijersbergen, J.H.M. Knoll, M. Meyerson and R.A. Weinberg, *Nat. Med.*, 1999, **10**, 1164–1170.
27. (*a*) B.S. Herbert, A.E. Pitts, S.I. Baker, S.E. Hamilton, W.E. Wright, J.W. Shay and D.R. Corey, *Proc. Natl. Acad. Sci. USA*, 1999, **96**, 14276–14281; (*b*) Z. Chen, K.S. Koeneman and D.R. Corey, *Cancer Res.*, 2003, **63**, 5917–5925.
28. (*a*) K. Damm, M. Pantic and A. Schnapp, *EMBO J.*, 2001, **20**, 6958–6968; (*b*) D.K. Barma, A. Elayadi, J.R. Falck and D.R. Corey, *Bioorg. Med. Chem. Lett.*, 2003, **13**, 1333–1336.
29. A.M. Zahler, J.R. Williamson, T.R. Cech and D.M. Prescott, *Nature*, 1991, **350**, 718–720.
30. Q. Guo, M. Lu, L.A. Marky and N.R. Kallenbach, *Biochemistry*, 1992, **31**, 2451–2455.
31. D. Sun, B. Thompson, B.E. Cathers, M. Salzar, S.M. Kerwin, J.O. Trent, T.C. Jenkins, S. Neidle and L.H. Hurley, *J. Med. Chem.*, 1997, **40**, 2113–2116.

32. P.J. Perry, S.M. Gowan, A.P. Reszka, P. Polucci, T.C. Jenkins, L.R. Kelland and S. Neidle, *J. Med. Chem.*, 1998, **41**, 3253–3260.
33. P.J. Perry, A.P. Reszka, A.A. Wood, M.A. Read, S.M. Gowan, H.S. Dosanjh, T.C. Jenkins, L.R. Kelland and S. Neidle, *J. Med. Chem.*, 1998, **41**, 4873–4884.
34. M.A. Read, A.A. Wood, J.R. Harrison, S.M. Gowan, L.R. Kelland, H.S. Dosanjh and S. Neidle, *J. Med. Chem.*, 1999, **42**, 4538–4546.
35. Y. Wang and D.J. Patel, *Structure*, 1993, **1**, 263–282.
36. P.J. Perry, M.A. Read, R.T. Davies, S.M. Gowan, A.P. Reszka, A.A. Wood, L.R. Kelland and S. Neidle, *J. Med. Chem.*, 1999, **42**, 2679–2684.
37. P.J. Perry, S.M. Gowan, M.A. Read, L.R. Kelland and S. Neidle, *Anti-Cancer Drug Design*, 1999, **14**, 373–382.
38. V. Caprio, B. Guyen, Y. Opoleu-Boahen, J. Mann, S.M. Gowan, L.R. Kelland, M.A. Read and S. Neidle, *Bioorg. Med. Chem. Lett.*, 2000, **10**, 2063–2066.
39. (*a*) D.F. Shi, R.T. Wheelhouse, D. Sun and L.H. Hurley, *J. Med. Chem.*, 2001, **44**, 4509–4523; (*b*) M.Y. Kim, M. Gleason-Guzman, E. Izbicka, D. Nishioka and L.H. Hurley, *Cancer Res.*, 2003, **63**, 3247–3256.
40. (*a*) O.Y. Fedoroff, M. Salazar, H. Han, V.V. Chemeris, S.M. Kerwin and L.H. Hurley, *Biochemistry*, 1998, **37**, 12367–12374; (*b*) L. Rossetti, M. Franceschin, S. Schirripa, A. Bianco, G. Ortaggi and M. Savino, *Bioorg. Med. Chem. Lett.*, 2005, **15**, 413–420.
41. (*a*) J.-L. Mergny, L. Lacroix, M.-P. Teulade-Fichou, C. Hounsou, L. Guittal, M. Hoarau, P.B. Arimondo, J.-P. Vigneron, J.-M. Lehn, J.-F. Riou, T. Garestier and C. Hélène, *Proc. Natl. Acad. Sci. USA*, 2001, **98**, 3062–3067; (*b*) J.-F. Riou, L. Guittat, P. Mailliet, A. Laoui, E. Renou, O. Petitgenet, F. Mégnin-Chanet, C. Hélène and J.-L. Mergny, *Proc. Natl. Acad. Sci. USA*, 2002, **99**, 2672–2677.
42. (*a*) P. Alberti, P. Schmitt, C.H. Nguyen, C. Rivalle, M. Hoarau, D.S. Grierson and J.L. Mergny, *Bioorg. Med. Chem. Lett.*, 2002, **12**, 1071–1074; (*b*) G. Pennarun, C. Granotier, L.R. Gauthier, D. Gomez, F. Hoffschir, E. Mandine, J.-F. Riou, J.-L. Mergny, P. Mailliet and F.D. Boussin, *Onco-gene*, 2005, **24**, 2917–2928.
43. (*a*) R.A. Heald, C. Modi, J.C. Cookson, I. Hutchinson, C.A. Laughton, S.M. Gowan, L.R. Kelland and M.F.G. Stevens, *J. Med. Chem.*, 2002, **45**, 590–597; (*b*) R.A. Heald and M.F.G. Stevens, *Org. Biomol. Chem.*, 2003, **1**, 3377–3389; (*c*) C. Leonetti, S. Amodei, C. D'Angelo, A. Rizzo, B. Benassi, A. Antonelli, R. Elli, M.F. Stevens, M. D'Incalci, G. Zupi and A. Biroccio, *Mol. Pharmacol.*, 2004, **66**, 1138–1146; (*d*) J.C. Cookson, F. Dai, V. Smith, R.A. Heald, C.A. Laughton, M.F. Stevens and A.M. Burger, *Mol. Pharmacol.*, 2005, **68**, 1551–1558.
44. (*a*) K. Shin-Ya, K. Wierzba, K. Matsuo, T. Ohtani, Y. Yamada, K. Furihata, Y. Hayakawa and H. Seto, *J. Am. Chem. Soc.*, 2001, **123**, 1262–1263; (*b*) M.-Y. Kim, H. Vankayalapati, K. Shin-ya, K. Wierzba and L.H. Hurley, *J. Am. Chem. Soc.*, 2002, **124**, 2098–2099.

45. (*a*) H. Tahara, K. Shin-ya, H. Seimiya, H. Yamada, T. Tsuruo and T. Ide, *Oncogene*, 2006, **25**, 1955–1966; (*b*) N. Binz, T. Shalaby, P. Rivera, K. Shin-ya and M.A. Grotzer, *Eur. J. Cancer*, 2005, **41**, 2873–2881.
46. R.J. Harrison, S.M. Gowan, L.R. Kelland and S. Neidle, *Bioorg. Med. Chem. Lett.*, 1999, **9**, 2463–2468.
47. M. Read, R.J. Harrison, B. Romagnoli, F.A. Tanious, S.M. Gowan, A.P. Reszka, W.D. Wilson, L.R. Kelland and S. Neidle, *Proc. Natl. Acad. Sci. USA*, 2001, **98**, 4844–4849.
48. (*a*) R.J. Harrison, J. Cuesta, G. Chessari, M.A. Read, S.K. Basra, A.P. Reszka, J. Morrell, S.M. Gowan, C.M. Incles, F.A. Tanious, W.D. Wilson, L.R. Kelland and S. Neidle, *J. Med. Chem.*, 2003, **46**, 4463–4476; (*b*) R.J. Harrison, A.P. Reszka, S.M. Haider, B. Romagnoli, J. Morrell, M.A. Read, S.M. Gowan, C.M. Incles, L.R. Kelland and S. Neidle, *Bioorg. Med. Chem. Lett.*, 2004, **14**, 5845–5489.
49. (*a*) C.M. Schultes, B. Guyen, J. Cuesta and S. Neidle, *Bioorg. Med. Chem. Lett.*, 2004, **14**, 4347–4351; (*b*) M.J.B. Moore, C.M. Schultes, J. Cuesta, F. Cuenca, M. Gunaratnam, F.A. Tanious, W.D. Wilson and S. Neidle, *J. Med. Chem.*, 2006, **49**, 582–599.
50. S. Haider, G.N. Parkinson and S. Neidle, *J. Mol. Biol.*, 2003, **326**, 117–125.
51. E. Gavathiotis, R.A. Heald, M.F. Stevens and M.S. Searle, *J. Mol. Biol.*, 2003, **334**, 25–36.
52. G.N. Parkinson, M.P.H. Lee and S. Neidle, *Nature*, 2002, **417**, 876–880.
53. J. Li, J.J. Correia, L. Wang, J.O. Trent and J.B. Chaires, *Nucleic Acids Res.*, 2005, **33**, 4649–4659.
54. (*a*) I.N. Rujan, J.C. Meleney and P.H. Bolton, *Nucleic Acids Res.*, 2005, **33**, 2022–2031; (*b*) L. Ying, J.J. Green, H. Li, D. Klenerman and S. Balasubramanian, *Proc. Natl. Acad. Sci. USA*, 2003, **100**, 14629–14634; (*c*) A.T. Phan and D.J. Patel, *J. Am. Chem. Soc.*, 2003, **125**, 15021–15027.
55. J.A. Stuart, C.M. Schultes and S. Neidle (unpublished observations).
56. A.M. Burger, F. Dai, C.M. Schultes, A.P. Reszka, M.J. Moore and S. Neidle, *Cancer Res.*, 2005, **65**, 1489–1496.
57. M.A. Shammas, H. Koley, D.G. Beer, C. Li, Goyal, R.K. Goyal and N.C. Munshi, *Gastroenterology*, 2004, **126**, 1337–1346.
58. D. Gomez, N. Aouali, A. Renaud, C. Douarre, K. Shin-Ya, S. Martinez, C. Trentesaux, H. Morjani and J.-F. Riou, *Cancer Res.*, 2003, **63**, 6149–6153.
59. J.M. Zhou, X.F. Zhu, Y.J. Lu, R. Deng, Z.S. Huang, Y.P. Mei, Y. Wang, W.L. Huang, Z.C. Liu, L.O. Gu and Y.X. Zeng, *Oncogene*, 2006, **25**, 503–511.
60. C.M. Incles, C.M. Schultes, H. Kempski, H. Koehler, L.R. Kelland and S. Neidle, *Mol. Cancer Ther.*, 2004, **3**, 1201–1206.
61. (*a*) M.T. Hemann, M.A. Strong, L.Y. Hao and C.W. Greider, *Cell*, 2001, **107**, 67–77; (*b*) Y. Zou, A. Sfeir, S.M. Gryaznov, J.W. Shay and W.E. Wright, *Mol. Biol. Cell*, 2004, **15**, 3709–3718.

62. J. Karlseder, A. Smogorzewska and T. de Lange, *Science*, 2002, **295**, 2446–2449.
63. W.E. Wright, V.M. Tesmer, K.E. Huffman, S.D. Levene and J.W. Shay, *Genes Dev.*, 1997, **11**, 2801–2809.
64. (*a*) G.Z. Li, M.S. Eller, R. Firoozabadi and B.A. Gilchrest, *Proc. Natl. Acad. Sci. USA*, 2003, **100**, 527–531; (*b*) S.A. Stewart, I. Ben-Porath, V.J. Carey, B.F. O'Connor, W.C. Hahn and R.A. Weinberg, *Nature Genet.*, 2003, **33**, 492–496.
65. (*a*) D. Gomez, R. Paterski, T. Lemarteleur, K. Shin-Ya, J.-L. Mergny and J.-F. Riou, *J. Biol. Chem.*, 2004, **279**, 41487–41494; (*b*) C. Granotier, G. Pennarun, L. Riou, F. Hoffschir, L.R. Gauthier, A. De Cian, D. Gomez, E. Mandine, J.-F. Riou, J.-l. Mergny, P. Maillet, B. Dutrillaux and F.D. Boussin, *Nucleic Acids Res.*, 2005, **33**, 4182–4190.
66. A.J. Zaug, E.R. Podell and T.R. Cech, *Proc. Natl. Acad. Sci. USA*, 2005, **102**, 10864–10869.
67. S.M. Gowan, J.R. Harrison, L. Patterson, M. Valenti, M.A. Read, S. Neidle and L.R. Kelland, *Mol. Pharmacol.*, 2002, **61**, 1154–1162.
68. (*a*) S. Neidle and G.N. Parkinson, *Nat. Rev. Drug Discov.*, 2002, **1**, 383–393; (*b*) E.M. Rezler, D.J. Bearss and L.H. Hurley, *Curr. Opin. Pharmacol.*, 2002, **2**, 415–423; (*c*) J.-L. Mergny, J.-F. Riou, P. Mailliet, M.-P. Teulade-Fichou and E. Gilson, *Nucleic Acids Res.*, 2002, **30**, 839–865.

CHAPTER 10

The Mechanism of Action of Telomestatin, a G-Quadruplex-Interactive Compound

DAEKYU SUN[a] AND LAURENCE H. HURLEY[a,b,c]

[a] University of Arizona,, College of Pharmacy, Tucson Arizona 85721, USA
[b] Arizona Cancer Center, 1515 N. Campbell Ave., Tucson Arizona 85724, USA
[c] BIO5 Institute for Collaborative Bioresearch, University of Arizona, 1657 E. Helen Street, Tucson Arizona 85721, USA

10.1 Introduction

10.1.1 Telomere Structure in Mammals and Telomerase

Telomeres, found at the ends of linear eukaryotic chromosomes, are unique nucleoprotein structures that consist of telomeric DNA and proteins that bind specifically to these sequences.[1] They play an essential role in preserving genomic integrity as protective caps at the ends of chromosomes, protecting essential coding sequences from cellular exonucleases and non-homologous end-joining.[2] Telomeric DNA in all vertebrates consists of a duplex (TTAGGG) repeat tract (about 4–14 kb long in humans) terminating in a 150–200 bp 3'-G-rich overhang.[3–5] Through every cell cycle, most somatic cells progressively lose telomeric DNA during routine DNA replication due to the end-replication problem.[6,7] The cumulative loss of telomeric DNA after many cell divisions is proposed to function as a biological clock, eventually limiting the proliferative life-span of somatic cells and leading to cellular, or replicative, senescence.[8–10] However, some cell types, such as germ line or most cancer cells, can extend their life-span through the reactivation of telomerase.[11,12] Telomerase is a unique ribonucleoenzyme that consists of RNA and protein components, with a short RNA motif in the RNA component serving as a template for the synthesis of telomeric DNA.[13] Telomerase replenishes telomeric DNA at the ends of chromosomal DNA, allowing the cells to obtain an unlimited replicative capacity.[11,12] To date, more than 90% of neoplastic cells are known to express telomerase activity, while normal somatic tissues, with the exception of germ line and hematopoietic stem cells, do not.[14-16] Thus, telomerase has been regarded as an ideal target for anticancer drug development because it is specifically expressed in human cancers and the inactivation of telomerase results in shortening of telomeres.[14,17] However, in the presence of telomerase inhibitors, cancer cells are known to still divide normally until the telomeres of the cancer cells shorten sufficiently to have detrimental effects on cellular proliferation.[18,19] This expected lag phase varies depending on the initial telomere length in cell lines treated with telomerase inhibitors. There is also the chance that cancer cells might become resistant to telomerase inhibitors or develop alternative mechanisms of telomere maintenance independent of telomerase, a phenomenon already observed in experimentally immortalized human cell lines.[20,21]

Traditionally, the human telomere was viewed as a linear structure of telomeric DNA complexed with a variety of telomere-binding proteins, including those shown in Figure 1(A).[22-24] Among these proteins, TRF1 and TRF2 are known to play major roles in maintaining telomere length and structure by directly binding to double-stranded telomere DNA and interacting with a number of other telomeric proteins.[25] Currently, TRF1 is regarded as a negative regulator, in cis, of telomere length through multiple mechanisms.[26] This protein is known to control telomerase access through its interaction with TIN2, PTOP/PIP1, PINX1, and the single-stranded telomere DNA-binding protein POT1, inhibiting a telomerase-dependent elongation of telomeric DNA.[25] Thus, overexpression of TRF1 is known to cause shortening of telomeres in human cells, whereas dominant negative TRF1 leads to elongated telomeres.[26] In comparison, the TRF2 protein has been implicated in the inhibition of telomere–telomere fusions through telomerase-independent mechanisms.[27] The overexpression of mutant TRF2 leads to the loss of TRF2 bound at the telomere, the loss of overhang, induction of the p53-dependent damage pathway, end fusion, and growth arrest, as in replication senescence.[27]

Several recent studies have suggested that the structure of telomeres is more complicated than was previous thought. The 3'-G-rich strand loops back and subsequently anneals to the duplex telomeric repeats to form a loop structure (t-loop) at the ends of telomeres (see Figure 1(B)).[28-30] A displacement (D) loop also can be formed as the G-rich strand displaces one strand of the duplex telomeric repeats. Therefore, the formation of a t-loop/D-loop structure at the ends of linear chromosomes could mask the overhang structure and confer some protection from exonucleases, preventing telomere–telomere fusion.[28-30] These t-loop structures have been identified by electron microscopic examination of telomeric DNA isolated from different species, including human, mouse, plant, avian, and multiple protist species, suggesting that t-loops are a common mechanism for chromosome protection among eukaryotes.[29] Telomere-associated proteins (*e.g.*, TRF1 and TRF2) may facilitate the formation and maintenance of both t- and D-loops. In fact, the TRF2 protein is known to be directly involved in t-loop formation by stabilizing the 3'-G-rich overhang.[29]

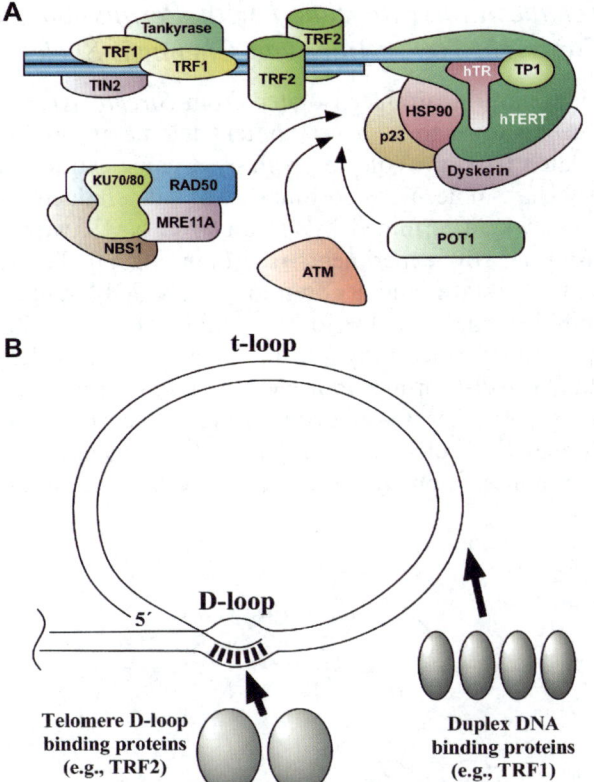

Figure 1 *(A) Schematic representation of the composition of telomeric complexes, telomerase, and proteins implicated in cellular signaling from the telomere (telomere repair). Telomerase comprises the RNA component hTR, which includes the template for telomere DNA synthesis, and the protein catalytic component hTERT. Additional molecules have been implicated in regulating the in vivo activity of hTR–hTERT and the maintenance of telomere structure. For example, the proteins TRF1, TRF2, tankyrase, TIN2, RAP1 (not shown), and POT1 are all involved in interacting with the telomere and might regulate the opening and closing of the free telomere end and access to the telomere by other protein complexes such as telomerase. A variety of proteins and ribonucleoproteins, including HSP90, as well as DKC1, L22, P23, and GAR1 (not shown), might also assist telomerase assembly and facilitate interactions between telomerase and the telomere. Other proteins, such as MRE11A, NBS1, KU70, KU80, DNAPK (not shown), and ATM might function in the detection of short telomeres and trigger DNA-damage response pathways or the repair of telomere sequences. Abbreviations and more information on each component can be found in ref. 89. (B) The classical view of telomere structure. A t-loop is proposed to form when the single-stranded 3'-G-rich overhang is looped back and annealed to the duplex telomere repeats*

(Figure 1(A) reprinted from ref. 89 with permission from Cambridge University Press. Figure 1(B) reprinted from ref. 28 with permission from Elsevier.)

10.1.1.1 Telomestatin (Reprinted with Permission from ref. 90. Copyright 2002, American Chemical Society)

Telomestatin is a natural product isolated from *Streptomyces anulatus* 3533-SV4 and has been shown to be a very potent telomerase inhibitor.[31] Telomestatin was isolated from the culture broth by organic extraction and purified by HPLC.[31] The molecular formula was determined by FAB-MS and the structure was determined by ^{1}H and ^{13}C NMR with direct ^{1}H–^{13}C correlations using HMBC experiments (see Figure 2(A)). The structural similarity between telomestatin and a G-tetrad (Figure 2(B)) suggested to us that telomerase inhibition might be due to its ability either to facilitate the formation of or trap out preformed G-quadruplex structures, and thereby sequester single-stranded d[T_2AG_3]$_n$ primer molecules required for telomerase activity.[32] Indeed a number of other G-quadruplex-interactive compounds, including the anthraquinones,[33–36] cationic porphyrins,[37–40] perylenes,[41–43] ethidium derivatives,[44] quinolones,[45] piperazines,[46] pentacyclic acridinium salts,[47] and

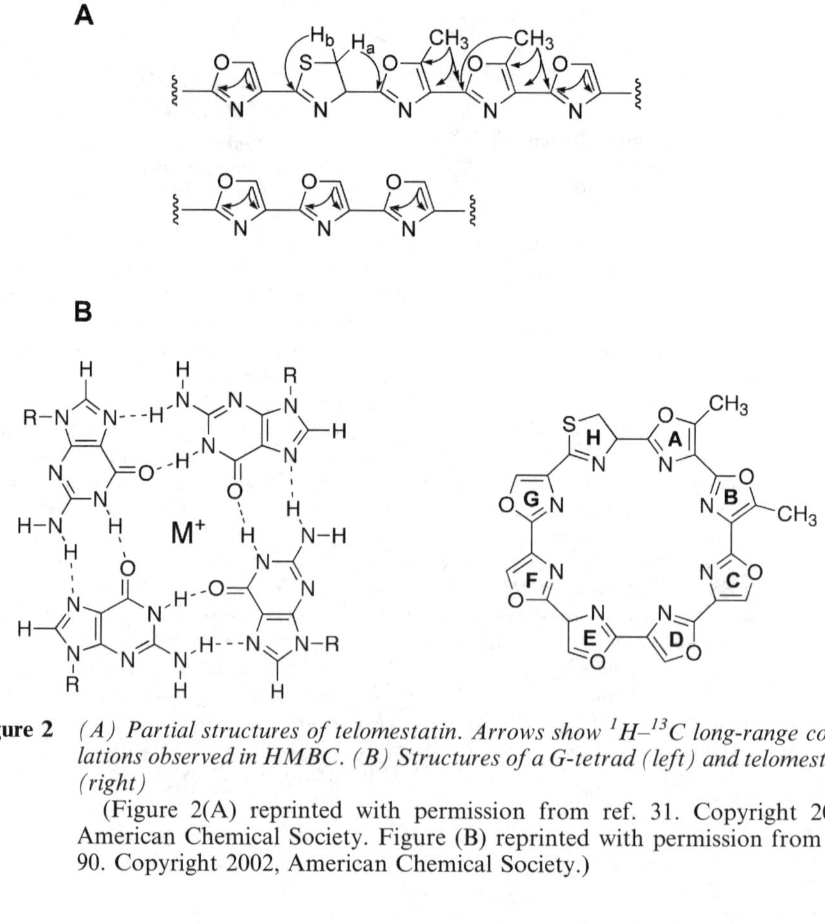

Figure 2 *(A) Partial structures of telomestatin. Arrows show [1]H–[13]C long-range correlations observed in HMBC. (B) Structures of a G-tetrad (left) and telomestatin (right)*

(Figure 2(A) reprinted with permission from ref. 31. Copyright 2001, American Chemical Society. Figure (B) reprinted with permission from ref. 90. Copyright 2002, American Chemical Society.)

fluoroquinophenoxazines,[48] have been shown to inhibit telomerase, most probably by this telomeric primer sequestration mechanism. Significantly, telomestatin appears to be a more potent inhibitor of telomerase (5 nM) than any of the described G-quadruplex-interactive molecules.[31]

10.1.2   Mechanism of Inhibition of Telomerase by Telomestatin

The initial experiments to determine whether telomestatin forms a complex with the human telomeric sequence were carried out using 12% native PAGE (Figures 3(A) and (B)). The results show that telomestatin stabilizes a band that migrates more slowly than the single-stranded linear DNA, most likely due to the formation of an intramolecular G-quadruplex.

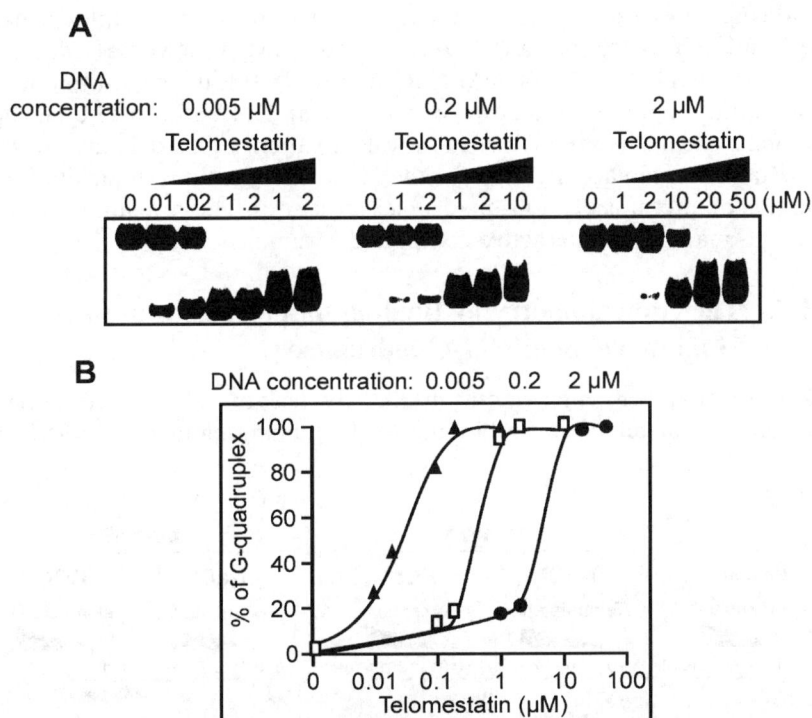

Figure 3   *Effects of telomestatin on the formation of intramolecular G-quadruplex. (A) End-labeled oligonucleotide (d[T$_2$AG$_3$]$_4$) at three different concentrations was incubated for 30 min with various concentrations of telomestatin in buffer, as previously described.[5] After incubation, samples were mixed with glycerol (final 5%) and run on a 12% native polyacrylamide gel with 1 × TBE. (B) G-quadruplex and linear DNA in (A) were quantified using ImageQuant 5.1 software from Molecular Dynamics*

(Reprinted with permission from ref. 90. Copyright 2002, American Chemical Society.)

10.1.2.1 *Telomestatin Can Replace the Need for Sodium or Potassium to Stabilize the Human Intramolecular G-Quadruplex Structure*

Monovalent cations, notably sodium and potassium, have been shown to stabilize human telomeric G-quadruplex structures, presumably by coordinating with the eight carbonyl oxygen atoms present between stacked tetrads.[49] To determine the importance of monovalent cations for the formation of G-quadruplex structures by telomestatin, the oligomer Hu4 (5'-[TTAGGG]₄-3') was incubated with increasing concentrations of telomestatin in the presence and absence of NaCl and KCl. Samples were run on a native polyacrylamide gel with 1 × TBE buffer without the addition of salt. Sodium and potassium ions were both found to act in synergy with telomestatin to stabilize the formation of intramolecular G-quadruplex structures, and the effect of sodium was slightly stronger than that of potassium, most notably at a concentration of 60 mM (Figure 4). EC_{50} values for the formation of the G-quadruplex structure were found to be 0.015 (no salt), 0.011, and 0.012 µM (10 mM NaCl and KCl), and <0.010 and 0.012 µM (60 mM NaCl and KCl). It is important to note that telomestatin is apparently able to convert linear DNA into a G-quadruplex structure, even in the absence of monovalent cations. This demonstrates that telomestatin can replace the need for the monovalent cations in facilitating the formation of intermolecular G-quadruplex structures. This is a unique property among G-quadruplex-interactive compounds examined to date.

10.1.3 The Stoichiometry of Binding of Telomestatin to the Human Telomeric G-Quadruplex

A CD titration was carried out in the absence of salt to determine the stoichiometry for telomestatin binding to the human telomeric G-quadruplex

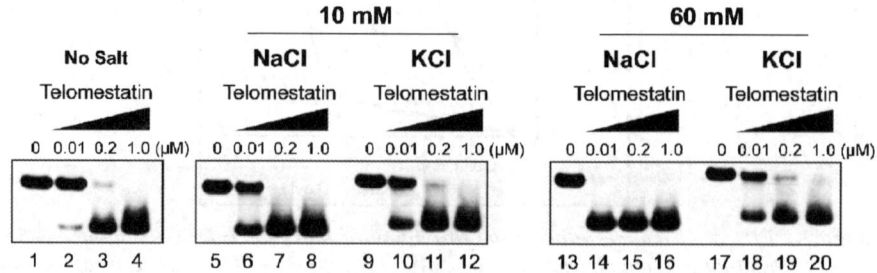

Figure 4 *Effects of telomestatin on the formation of intramolecular G-quadruplexes in NaCl and KCl. The end-labeled oligonucleotide Hu4 was incubated for 30 min with increasing concentrations of telomestatin in the absence and presence of NaCl and KCl. Lanes 1–4 contain no monovalent cations, lanes 5–8 contain 10 mM NaCl, lanes 9–12 contain 10 mM KCl, lanes 13–16 contain 60 mM NaCl, and lanes 17–20 contain 60 mM KCl.[57]*

structure. A sample of the $d[T_2AG_3]_4$ at a strand concentration of 25 μM was titrated with 0.5 mole equivalents of telomestatin (1 mM solution). After each addition of telomestatin, the reaction was allowed to equilibrate for at least 15 min (until no elliptic changes were observed) and CD spectra were collected. The resulting nine spectra were plotted and are shown inFigure 5(A) (where the arrow directions indicate the progressive evolution or disappearance of CD signal). The band at 257 nm decreased in ellipticity with each addition of

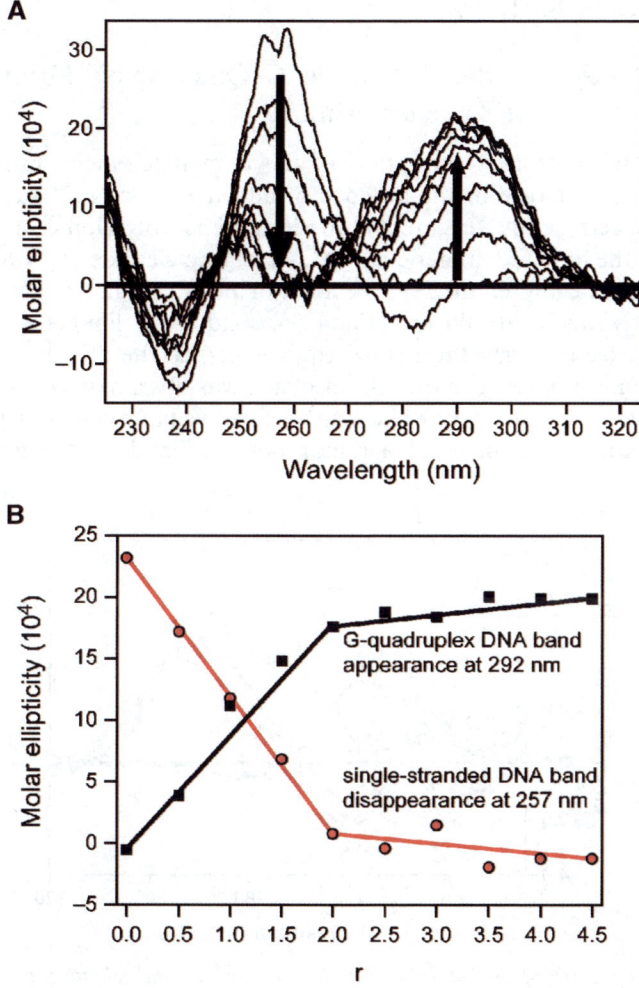

Figure 5 *(A) A stacked plot representing the CD titration of the $d[T_2AG_3]_4$ with half equivalents of telomestatin (spectra collected at 25 °C). (B) Two titration curves representing the changes in ellipticity with the addition of each telomestatin half equivalent as monitored at 257 and 292 nm. An inflexion point is observed for the two curves at r=2 (where r is the telomestatin/DNA fraction)*
(Reprinted with permission from ref. 91. Copyright 2005, American Chemical Society.)

telomestatin, while the band at 292 nm appeared and appreciably increased in ellipticity until a ratio of 2:1 for telomestatin to $d[T_2AG_3]_4$ was reached. The changes in ellipticity at these two wavelengths (257 and 292 nm) where the greatest changes in ellipticity (observed in non-maxima regions) as a function of telomestatin/DNA fraction (r) are plotted in Figure 5(B). An inflexion point observed in each graph when $r=2$ is consistent with the formation of a 2:1 telomestatin–oligonucleotide complex, and the CD spectrum of this complex is also consistent with an antiparallel G-quadruplex structure of the basket-type being adopted by the DNA.

10.1.4 Identity of the Telomeric G-Quadruplex Formed in the Presence of Telomestatin

The telomestatin-induced formation of the human telomeric intramolecular G-quadruplex structure in the absence of added K^+ or Na^+ was monitored using CD spectroscopy (Figure 6). In the absence of telomestatin, the CD spectrum of the human telomeric $d[T_2AG_3]_4$ oligonucleotide was found to have a negative band centered at 238 nm, a major positive band at 257 nm, and a minor negative band at 280 nm (Figure 6, solid black line). However, upon addition of excess telomestatin (5 M equivalents) to the $d[T_2AG_3]_4$ oligonucleotide, a dramatic change in the CD spectrum was observed. The bands at 238, 257, and 280 nm disappeared, while a major positive band centered at 292 nm, a negative band at 262 nm, and a minor positive band at 245 nm appeared

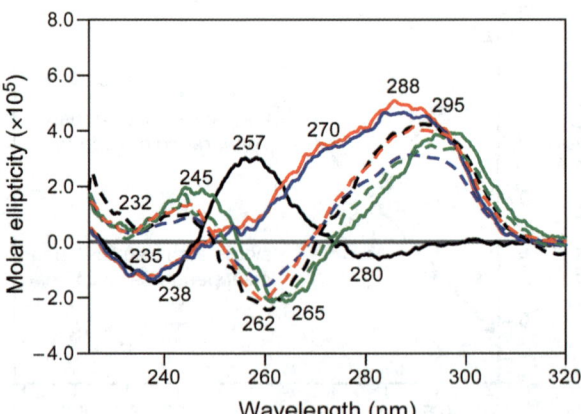

Figure 6   *CD spectra of $d[T_2AG_3]_4$ with Na^+ and/or K^+ and telomestatin. Line colors: solid black=$d[T_2AG_3]_4$ (no salt added); dashed black=$d[T_2AG_3]_4$ + telomestatin (no salt added); solid red=$d[T_2AG_3]_4$ + 100 mM KCl; dashed red=$d[T_2AG_3]_4$ + 100 mM KCl + telomestatin; solid green=$d[T_2AG_3]_4$ + 100 mM NaCl; dashed green=$d[T_2AG_3]_4$ + 100 mM NaCl + telomestatin; solid blue=$d[T_2AG_3]_4$ + 100 mM KCl + 100 mM NaCl; dashed blue=$d[T_2AG_3]_4$ + 100 mM KCl + 100 mM NaCl + telomestatin*

(Reprinted with permission from ref. 91. Copyright 2005, American Chemical Society.)

(Figure 6, dashed black line). Approximately 15 min after the addition of telomestatin, the new peaks and troughs ceased to increase in elliptic intensity, and the reaction probably reached, or was very close to, the equilibrium state. These dramatic changes in the CD spectrum of the human telomeric d[T$_2$AG$_3$]$_4$ oligonucleotide after the addition of telomestatin are consistent with this drug binding to the DNA and thus causing substantial change(s) in the conformation of the DNA. In fact, the CD spectrum of this new DNA conformation is virtually identical to the CD spectra of antiparallel G-quadruplexes described in previous studies, where the major positive band was usually observed around 290 nm with a negative band at 265 nm and a smaller positive band at 246 nm.[50–53] Importantly, in the previous studies the G-quadruplexes were characterized as intramolecular basket-type structures using CD as well as NMR, EMSA, and DMS footprinting and other chemical probing studies.[50–55] As anticipated, upon addition of 100 mM NaCl to the d[T$_2$AG$_3$]$_4$ telomeric oligomer in the absence of telomestatin, the CD spectrum exhibits a similar, but not identical, maxima-minima pattern to that described with addition of telomestatin; a major positive band around 290 nm, a negative band at 265 nm, and a minor positive band at 245 nm (Figure 6, solid green line).

10.1.5 Proposed Models for Telomestatin Binding to the Human Telomeric G-Quadruplex Structure

A simulated annealing docking combined with molecular dynamics was used to study the binding interactions of telomestatin with the intramolecular basket G-quadruplex structure.[56] A cartoon of this folding pattern is shown in Figure 7(A). Previous studies showed that stoichiometry for binding was 2:1. A 2:1 model for the telomestatin bound in the external stacking mode in an energy-minimized complex with the human telomeric basket-type G-quadruplex is shown inFigure 7(B) (unpublished results). During the dynamics simulations of the 1:1 diagonal loop complex, the telomestatin–G-quadruplex complex shows conformational changes resulting in favorable potential energy of the telomestatin and G-quadruplex complex.

In the final minimized complex model, one telomestatin molecule is oriented between the diagonal loop and the top G-tetrad exhibiting stacking interactions. This telomestatin molecule shows a binding energy of -117 kcal mol^{-1}. Analysis of dynamic trajectories reveals that the second telomestatin molecule interacts with the two lateral loops, containing the TTA bases and the G-tetrad, exhibiting an increased binding energy of -140.9 kcal mol^{-1}.

As an alternative, a 2:1 model for the telomestatin bound in the intercalation mode in an energy-minimized complex with the human telomeric basket-type G-quadruplex is shown in Figure 7(C) (unpublished results). During the dynamics simulations of the intercalation complex, the G-tetrads showed significant conformational changes to accommodate telomestatin molecules intercalating within the tetrads. In the final minimized complex model, one telomestatin molecule is oriented between the top and middle G-tetrads exhibiting stacking interactions. This telomestatin molecule shows a binding

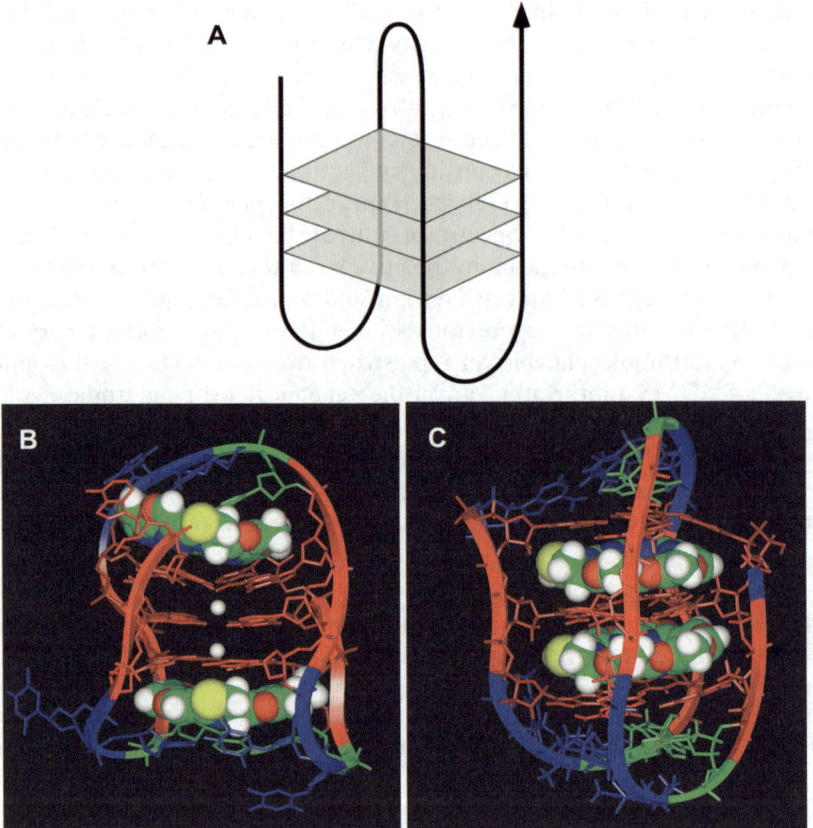

Figure 7   *Proposed model for telomestatin binding to the human telomeric G-quadruplex structure. (A) Schematic diagram showing a basket form of telomeric G-quadruplex. (B) A 2:1 model for the telomestatin bound in the external stacking mode. (C) A 2:1 model for the telomestatin bound in the internal stacking mode*

energy of -152.8 kcal mol^{-1}, while the second telomestatin molecule interacts with the middle and bottom G-tetrads, exhibiting a binding energy of -134.0 kcal mol^{-1}. At this point it is not possible to differentiate between these two models.

10.1.5.1   *Biological Effects of Telomestatin Associated with its Inhibition of Telomerase in Human Cancer Cells*

Telomestatin has been shown to be a very potent telomerase inhibitor, with an IC$_{50}$ of ~ 5 nM, and telomerase inhibition is believed to be attributable to its ability to interact directly with G-quadruplex structures, subsequently sequestering single-stranded d(TTAGGG)$_n$ primer molecules required for telomerase activity (see above). On the basis of this knowledge, several independent studies have been undertaken in preclinical settings to address whether telomerase inhibition by telomestatin is directly related to various biological activities shown by this compound against human cancers.[57-63] Telomestatin interacts preferentially with intramolecular over intermolecular G-quadruplex structures and has a 70-fold greater selectivity for intramolecular G-quadruplex structures over duplex DNA.[57] Telomestatin stabilizes G-quadruplex structures that can be directly formed from duplex human telomeric DNA as well as from single-stranded DNA.[57] We addressed whether the preference of telomestatin for the intramolecular *vs.* the intermolecular G-quadruplex structures and its selectivity for the G-quadruplex structure over a single-stranded or duplex DNA structure could be associated with various biological effects produced by telomestatin in human cancer cell lines.[57] TMPyP4 and TMPyP2 were used as control compounds (Figure 8(A)). TMPyP4 is known to preferentially facilitate the formation of intermolecular rather than intramolecular G-quadruplexes from human telomeric sequences and to induce anaphase bridges.[57] The long-term biological effects of telomestatin were evaluated using select human colon (SW26 and SW39) and pancreatic (MiaPaCa) (see Figure 9) cancer cell lines using nontoxic concentrations (Figure 8(B)). These concentrations of telomestatin were chosen because the biological effects were not anticipated until the telomere length became significantly shortened after multiple cell divisions. At a concentration of 0.15 μM, telomestatin suppresses the proliferation of telomerase-positive cells (SW39) within 3 weeks and completely stops their growth after 8 weeks (see Figure 8(C), left panel). Telomere shortening was observed after 39 days ($\sim$1 kb) with this concentration of telomestatin in the SW39 cell line (Figure 8(C), left panel), suggesting that the growth arrest in this cell line was associated with telomere shortening caused by telomestatin. In the study using the human pancreatic carcinoma cell line (MiaPaCa), we observed that telomestatin induced MiaPaCa cell growth arrest, senescence, apoptosis, and telomere length shortening within 12 weeks, but it failed to induce anaphase bridge formation (Figure 9).[58] TMPyP2 is a positional isomer of TMPyP4 that does not appreciably interact with G-quadruplexes.[38] The difference in lag phase between the cell lines (SW39 *vs.* MiaPaCa) could be due to the difference in the initial telomere length of the cell lines. In contrast, the presence of telomestatin at nontoxic concentrations did not affect the growth of telomerase-negative/alternative lengthening of telomeres (ALT)-positive cell line SW26 even after 9 weeks, while TMPyP4, a compound that preferentially facilitates the formation of intermolecular G-quadruplex structures, suppressed the proliferation of the same cell line within 2 weeks (Figure 8(C), right panel). We also observed that TMPyP4 induces anaphase bridges in MiaPaCa cells, whereas telomestatin does not (Figure 9(C)).[58] We speculate that TMPyP4, -but not telomestatin, might facilitate the formation of intermolecular G-quadruplex structures between the 3'-G-overhang from two different chromosomes, forming anaphase bridges and thereby preventing the strand invasion of the 3'-G-overhang into the telomeric duplex region during the ALT process. The results from our studies led us to conclude that the selectivity of telomestatin for intramolecular G-quadruplex structures and TMPyP4 for intermolecular G-quadruplex structures is important in mediating different

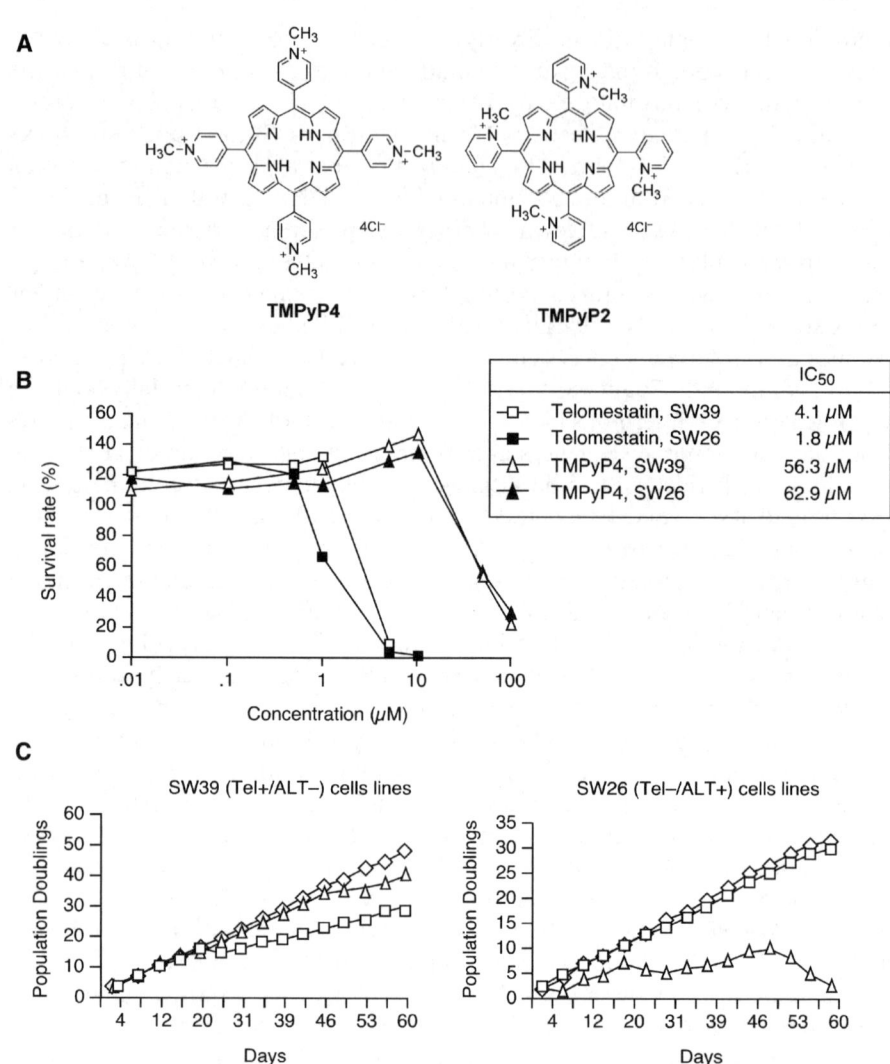

Figure 8 *(A) Structures of TMPyP4 and TMPyP2. Effects of telomestatin and TMPyP4 on the growth of SW39 (telomerase-positive/ALT-negative) and SW26 (telomerase-negative/ALT-positive) cell lines. (B) Short-term cytotoxicity. Cells were exposed to the indicated concentrations of compounds. Three days later the cytotoxicity was assessed and expressed as a percentage of the survivals of untreated cells (100%). Each experiment was performed four times at each point. (C) Long-term exposure with nontoxic concentrations. SW39 cells were exposed to 0.5 or 1 μM concentrations of telomestatin or TMPyP4, respectively. SW26 cells were exposed to 0.15 or 1 μM concentrations of telomestatin or TMPyP4, respectively. Each experiment was performed four times at each point[57]*

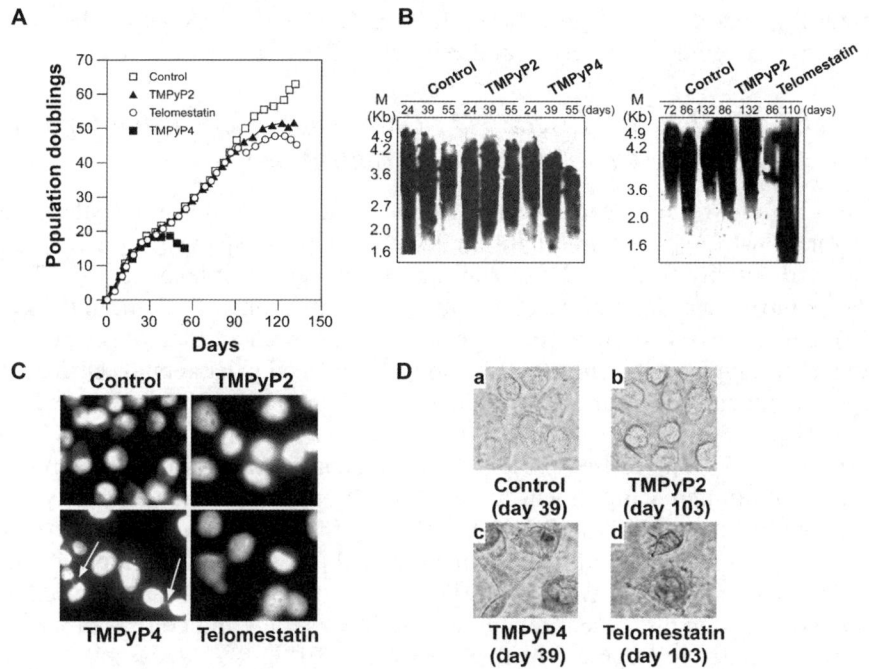

Figure 9 *(A) Growth arrest of MiaPaCa cells induced by long-term treatment with TMPyP4, TMPyP2, and telomestatin. Cells were treated with TMPyP4 (2.5 μM), TMPyP2 (2.5 μM), and telomestatin (0.05 μM), and the total number of cells was counted using a haemocytometer. (B) Effect of TMPyP4, TMPyP2, and telomestatin on telomere length. MiaPaCa cells were treated with TMPyP4 (2.5 μM), TMPyP2 (2.5 μM), and telomestatin (0.05 μM) and harvested at different times (days) of the culture. A Southern blot analysis was used to measure telomere length by using the TeloTAGGG telomeric length assay kit according to the manufacturer's protocol. (C) Anaphase bridges induced by TMPyP4, but not by TMPyP2 and telomestatin. Arrows indicate the anaphase bridges. After treatment with TMPyP4 (50 μM), TMPyP2 (50 μM), and telomestatin (0.5 μM) for 48 h, the PI-stained cells were observed under a fluorescence microscope (×400). (D) Senescent-like phenotype induced by long-term treatment with TMPyP4 and telomestatin. Expression of β-galactosidase activity as a biomarker of cell senescence was measured in MiaPaCa cells harvested at day 103 (a) and in the cells treated with 2.5 μM TMPyP2 for 103 days (b), 2.5 μM TMPyP4 for 39 days (c), and 0.05 μM telomestatin for 103 days (d). Treatment with TMPyP4 and telomestatin increased the number of senescent-like cells*[58]

biological effects: stabilization of intramolecular G-quadruplex structures produces telomerase inhibition and accelerated telomere shortening, whereas facilitation of the formation of intermolecular G-quadruplex structures induces the formation of anaphase bridges. In summary, these results further validate the current notion that the growth arrest of cells after prolonged exposure to telomerase inhibitors or to noncytotoxic concentrations of telomestatin is primarily caused by the progressive loss of telomere DNA and subsequent induction of cellular senescence and cell death.[58]

10.1.5.2 Biological Effects of Telomestatin on Hematopoietic Cancer Cells at High Concentrations

Recently, the results from several independent studies have revealed that treatment with telomestatin could limit the cellular life-span of human cancer cells, presumably through the disruption of telomere maintenance.[59-62] These studies have been done at relatively high concentrations of telomestatin (>2 µM), particularly with hematopoietic cancer cells, such as BCR-ABL-positive human leukemia cells, multiple myeloma cells, acute leukemia cells, acute myeloid leukemia cells, and freshly obtained primary acute leukemia cells from patients.

In one study among them, the ability of telomestatin to inhibit the proliferation of human cancer cells was evaluated using human acute myeloid leukemia cell lines (U937 and NB4), including freshly obtained acute myeloid leukemia cells from patients.[59] Treatment with 2 µM telomestatin reproducibly inhibited telomerase activity in U937 and NB4 cells and was followed by telomere shortening, when telomere length was determined by either the terminal restriction fragment method or by flow-FISH (see Figure 10). Telomestatin-treated U937 cells showed almost complete inhibition after 30 days (around PD20), while NB4 cells ceased to proliferate after treatment for 13 days (around PD8). Based on flow cytometric analysis, the frequency of apoptosis was about 80% in both cell lines that ceased to proliferate with telomestatin treatment. Moreover, telomere shortening associated with apoptosis by telomestatin was also observed in some freshly obtained leukemia cells from acute myeloid leukemia patients, regardless of sub-types of AML and post-myelodysplasia AML, when these cells were incubated in the presence of telomestatin for 10 days. Telomere shortening and induction of apoptosis in freshly isolated primary acute leukemia cells from leukemia patients were also observed after exposure to 5 µM of telomestatin for 10 days in another independent study,[60] suggesting that telomestatin could lead to more rapid onset of telomere shortening, chromosomal instability, cell crisis, and apoptosis at cytotoxic concentrations and that antitelomerase therapy might be useful in some acute leukemias.

Interestingly, enhanced apoptosis in response to chemotherapeutic agents such as daunorubicin and cytosine-arabinoside was observed in telomestatin-treated U937 cells before ultimate telomere shortening (see Figure 11). These results indicate the therapeutic possibility of telomestatin for acute myeloid leukemia in combination with other traditional chemotherapeutic agents, warranting further investigation of telomestatin as a therapeutic agent in the clinic.

In another recent study, the potential application of telomestatin in the treatment of multiple myeloma was explored using ARD, ARP, and MM1S as model myeloma cell lines.[61] Telomestatin treatment led to inhibition of

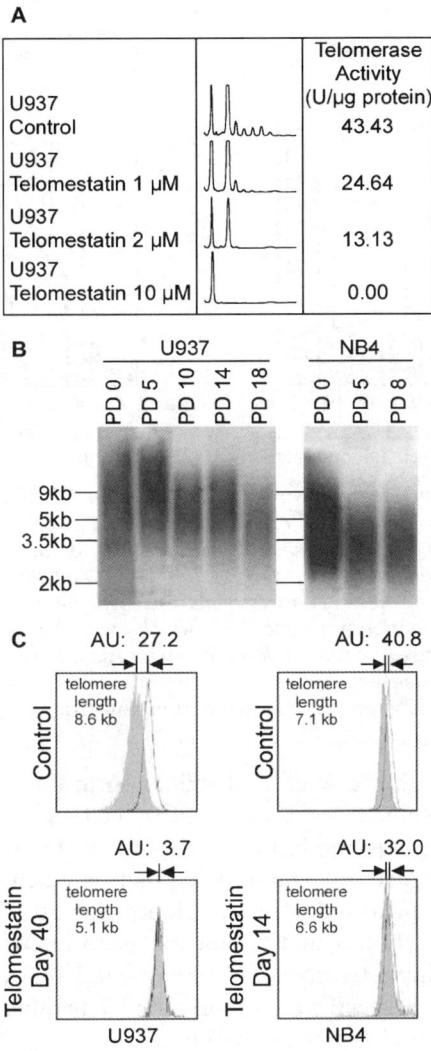

Figure 10 *Effects of telomestatin on telomerase activity and telomere length in leukemia cells. (A) Effect of telomestatin on telomerase activity in U937 cells. U937 cells were incubated with the indicated concentrations of telomestatin for 48 h, then telomerase activity was examined by a TRAP assay. Telomerase activity was suppressed by telomestatin in a dose-dependent manner. (B) Telomere length in U937 cells and NB4 cells was assessed for telomere restriction fragment size by Southern blot analysis with a telomeric probe. In both myeloid leukemia cell lines, telomere length was progressively shortened after telomestatin treatment at indicated population doubling (PD). (C) U937 cells and NB4 cells were cultured with or without 2 μM of telomestatin for the number of days indicated, and telomere length was determined by flow-FISH analysis. This indicates, in combination with (B), that the flow-FISH method is reproducible to determine telomere length in leukemia cells (AU=telomere fluorescence of sample with probe – telomere fluorescence of sample without probe)*

(Reprinted from ref. 59 with permission from the publisher.)

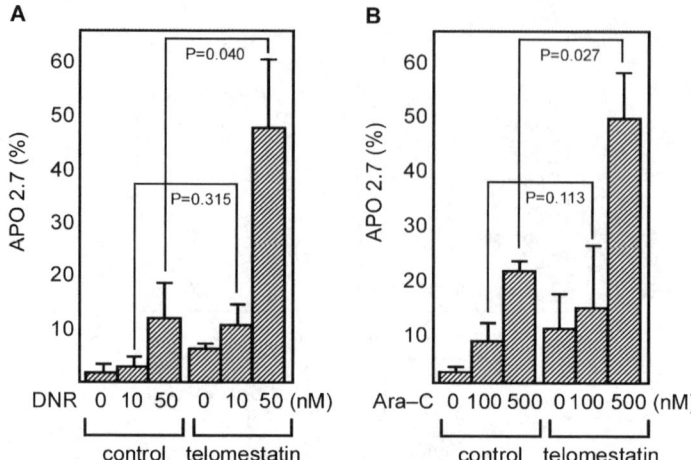

Figure 11 *Enhancement of apoptosis from DNR or Ara-C treatment in telomestatin-treated U937 cells. U937 cells cultured with 2 μM telomestatin at PD 10 were incubated with the indicated concentrations of daunorubicin DNR (A) or cytosine arabinoside Ara-C (B) for 48 h. The incidence of apoptosis was determined by flow cytometric analysis with an FITC-conjugated APO 2.7 monoclonal antibody (clone 2.7), which was raised against the 7A6 antigen and is expressed by cells undergoing apoptosis. Similar results were obtained in two independent experiments*
(Reprinted from ref. 59 with permission from the publisher.)

telomerase activity within a week and reduction in telomere length and apoptotic cell death within 3–5 weeks in these cell lines, demonstrating that telomestatin causes growth inhibition and apoptosis in myeloma cells predominantly through its effects on telomerase function, such as the inhibition of telomerase and subsequent reduction in telomere length. The results from this study support the conclusion that telomerase is an important potential therapeutic target for multiplying myeloma therapy and that other G-quadruplex-interactive agents with specificity for binding to telomeric sequences can be important agents for additional evaluation.

The effects of telomestatin on telomere dynamics and its potential clinical utility have also been evaluated by assessing its effects on BCR-ABL-positive human leukemia cells.[62] This study has revealed that treatment with telomestatin reproducibly inhibited telomerase activity in the BCR-ABL-positive leukemia cell lines OM9;22 and K562, disrupting telomere maintenance, and eventually resulting in telomere dysfunction. It is worthwhile to note that telomestatin effectively suppressed the plating efficiency of K562 cells at 1 μM (Figure 12(A)), whereas telomestatin had much less effect on colony formation from normal bone marrow cells (CD34+) (Figure 12(B)), suggesting that G-quadruplex-interactive agents, such as telomestatin, might have some degree of selectivity to cancer cells over normal hematopoietic stem cells. Furthermore, enhanced chemosensitivity toward imatinib (Figure 13) and chemotherapeutic agents[62] was observed in telomestatin-treated K562 cells. Also, telomestatin

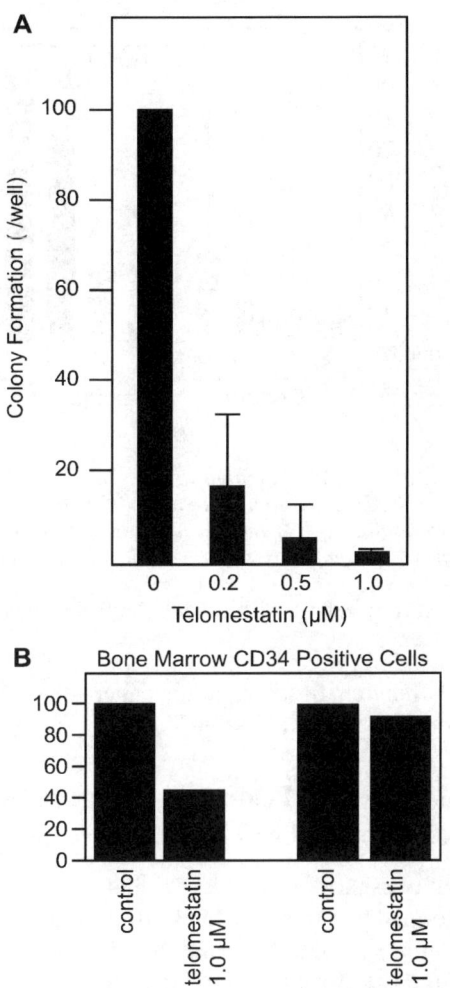

Figure 12 *The plating efficacy of K562 cells and the standard progenitor colony assay of normal CD34-positive bone marrow cells treated with telomestatin. In all, 1000 (K562) cells (A) or 1000 (bone marrow CD34-positive) cells (B) were treated with telomestatin and seeded in triplicate in conditional medium MethCult GF H4434 (Stem Cell Technologies, Vancouver, Canada). The leukemic colonies (>50 cells) of K562 were scored on day 14. Progenitor cell-derived colonies were scored on day 14 and classified as either BFU-Es (burst-forming units-erythroid) or CFU-GMs (colony-forming units granulocyte-macrophage)*
(Reprinted from ref. 62 with permission from Macmillan Publishers Ltd.)

more effectively inhibited hematopoietic colony formation by primary human chronic myelogenous leukemia cells when combined with imatinib. This study suggests a potential utilization of telomestatin or other type of G-quadruplex-interactive agent for the treatment of human leukemia in

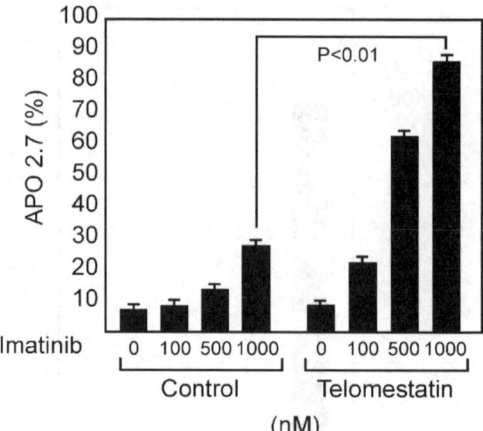

Figure 13   *Enhancement of apoptosis in telomestatin-treated K562 cells by imatinib. K562*
cells were cultured with 2 mm of telomestatin for 10 days, and subsequently
these cells were incubated with the indicated concentrations of imatinib for 72
h. The incidence of apoptosis was determined by flow cytometric analysis with
APO 2.7 mAb
        (Reprinted from ref. 62 with permission from Macmillan Publishers Ltd.)

combination with other chemotherapeutic agents, such as imatinib, daunorubicin, mitoxantrone, and vincristine.[62]

10.1.6   Potential Effect of Telomestatin on the Assembly of Telomeres into Higher-Order Structures

Since intact 3'-G-rich overhang plays a key role in remodeling the duplex form
of telomeres into t-loop structures, the facilitation and stabilization of secondary structures such as a G-quadruplex in the 3'-G-rich overhangs by telomestatin could disrupt the formation of a t-loop structure at telomeres.
Interestingly, several other studies using relatively high concentrations (>2
µM) of telomestatin have shown encouraging results beyond telomerase inhibition, including telomeric disruption and the induction of apoptosis
and cellular senescence after short-term treatment (Figure 14(A)).[59-62] These
biological effects of telomestatin could be attributed to its selectivity for
interaction with intramolecular G-quadruplex structures formed by telomeric
3'-G-rich overhangs rather than simple inhibition of the catalytic activity of
telomerase. A recent study by Riou and colleagues[63] has provided evidence that
telomestatin strongly binds to the telomeric overhang *in vitro* and *in vivo* and
impairs its single-stranded conformation (Figure 14(B)). The alteration of the
G-overhang conformation by telomestatin is proposed to alter chromosomal
capping mediated by telomeric proteins such as TRF2. Furthermore, the same
study used specific hybridization or telomeric oligonucleotide ligation assay
experiments to demonstrate that after long-term exposure of A549 human
lung cancer cells to telomestatin, the G-overhang size is greatly reduced

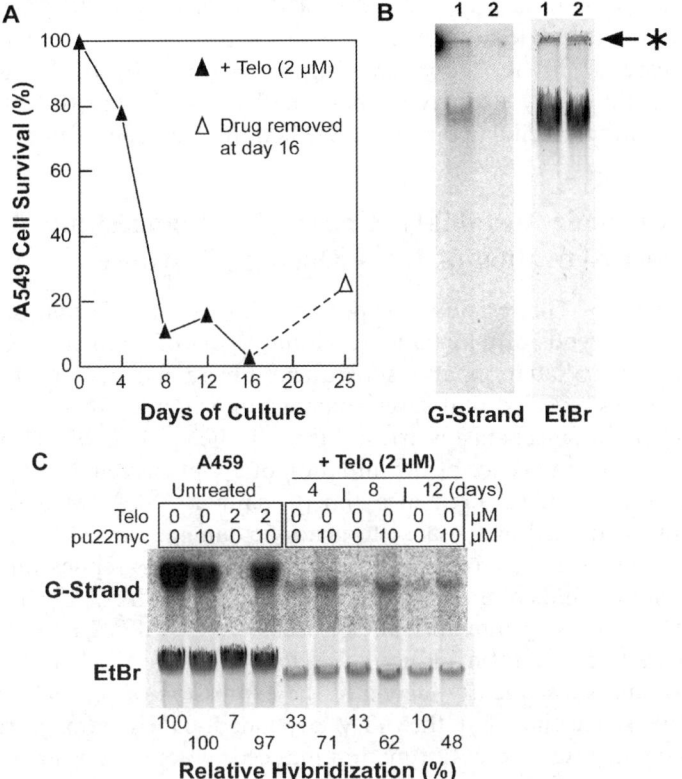

Figure 14 *Long-term treatment with telomestatin induced an effective G-overhang degradation. (A) Growth of A549 cells after telomestatin removal. A549 cells were induced to growth arrest by treatment with telomestatin (2 μM, solid triangle). Drug was removed at day 16 and cells were incubated for an additional 9 days (open triangle) and DNA was extracted. (B) Non-denaturing solution hybridization analysis of the 3'-telomeric overhang in A549 cells induced to growth arrest by telomestatin and maintained for 9 days in drug-free medium. Lane 1, control untreated A549; lane 2, A549 cells at day 25 after drug removal at day 16 (see A, open triangle). The asterisk indicates the position of the loading well at the top of the gel. (C) Non-denaturing solution hybridization analysis of the 3'-telomeric overhang in A549 cells untreated or treated by 2 μM telomestatin for 4, 8, and 12 days, as indicated. DNA from untreated cells (A549 untreated) was incubated with 2 μM telomestatin with or without 10 μM pu22myc quadruplex competitor, as indicated. DNAs from telomestatin-treated cells (Telo 2 μM) were incubated with or without pu22myc quadruplex competitor (10 μM), as indicated. Numbers indicate the relative hybridization signal (%) relative to untreated A549 DNA (upper numbers) or relative to untreated A549 DNA with 10 μM pu22myc (lower numbers).*

(Figure 14(B) reproduced with permission of the American Soceity for Biochemistry and Molecular Biology. Copyright 2004 by American Society for Biochemistry and Molecular Biology. In the format Other Book via Copyright Clearance Center.[63] Figure 14(C) reprinted by permission from Macmillan Publishers Ltd.)

(Figure 14(C)). Significantly, the onset of the delayed growth arrest in A549 cells with prolonged treatment with telomestatin is well correlated with a marked decrease in the G-overhang signal (Figure 14(A)). Altogether, the results from this study strongly support the hypothesis that the telomeric G-overhang is an intracellular target for the action of telomestatin.

10.1.7 Genomic Instability Caused by Telomestatin Treatment and Activation of DNA Damage Response

The loss of the 3'-G-overhang is expected to disrupt the formation of t-loop structures at the ends of telomeres, resulting in loss of telomeric sequences by exonucleolytic degradation and subsequent severe damage to the genome. Previous studies revealed that severe damage in genomic DNA could activate DNA checkpoint mechanisms irrespective of their mode of actions, which could lead to cell senescence or the initiation of apoptosis cell death pathways if DNA damage is not repaired in a timely manner.[64,65] ATM and Chk2 are known to be activated in response to genomic damage caused by direct DNA cleaving agents such as ionizing radiation,[65,66] while replication arresting agents, such as aphidicolin and hydroxyurea, activate Chk1.[65-67] Interestingly, a previous study using human leukemia cells (*e.g.*, OM9;22 and K562) demonstrated that telomestatin induced the activation of ATM and ChK2 (see Figure 15), and subsequently increased the expression of p21[Cip1] and p27[Kip1] (Figure 16), suggesting that the ATM-dependent DNA damage response is activated through telomere dysfunction induced by telomestatin.

10.1.8 Other Mechanisms of Telomestatin in Mediating its Biological Activity

10.1.8.1 *Potential Effect of Telomestatin on RNA Splicing*

There are some known genes that encode G-rich mRNA capable of forming G-quadruplex RNA.[68-72] For instance, the 5'-end of hTERT intron 6 contains the GGG repeat, possibly involved in alternative splicing of the hTERT gene.[71] Indeed, it has been observed that the G-quadruplex-interactive ligand 12459 causes an alteration of the hTERT splicing, resulting in the downregulation of telomerase activity, presumably through the facilitation of the folding of this region into a stable G-quadruplex by 12459.[72] Although telomestatin did not appear to alter splicing and transcript levels of hTERT,[72] there is still the possibility that telomestatin and other G-quadruplex-interactive compounds could facilitate the formation of the G-quadruplex structures in several unknown mRNAs containing G-repeats capable of forming G-quadruplex RNA, resulting in the alteration of the translation of those mRNAs. This could be an important therapeutic target for G-quadruplex-interactive agents such as telomestatin if it could be demonstrated that the biological consequence was specific to cancer cells.

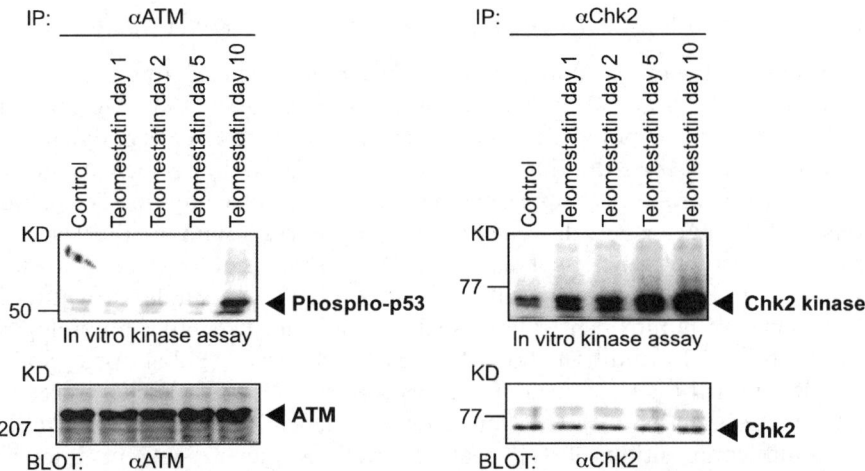

Figure 15 *Telomestatin induces the activation of ATM and Chk2. ATM and Chk2 were immunoprecipitated (IP) from untreated OM9;22 cells and telomestatin-treated OM9;22 cells for indicated days. Activation of ATM was examined by kinase assays in vitro with recombinant p53 proteins as a substrate (left panel). Autophosphorylation of Chk2 was examined by kinase assays in vitro without target substrate (right panel)*
(Reprinted from ref. 62 with permission from Macmillan Publishers Ltd.)

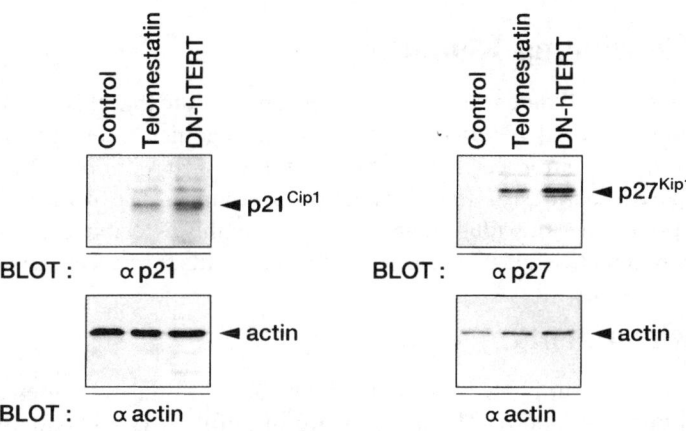

Figure 16 *Immunoblot analysis of p21^{Cip1} and p27^{Kip1} protein. Total cellular protein (10 mg per lane) from untreated OM9;22 cells, DN-hTERT-transfected OM9;22 cells (PD 10), and telomestatin-treated OM9;22 cells (10 days) was separated on 12.5% SDS–PAGE. p21^{Cip1} and p27^{Kip1} protein levels were detected by immunoblotting with antibodies directed against p21^{Cip1} and p27^{Kip1}*
(Reprinted from ref. 62 with permission from Macmillan Publishers Ltd.)

10.1.8.2   *Potential Effect of Telomestatin on Transcription*

The promoter region of many growth-related genes, such as Hmga-2, EGF-R, c-Myc, c-MycB, I-R, AR, c-Src, c-Ki-Ras, TGF-b, and PDGF A-chain consisting of polypurine/polypyrimidine tracts, are proposed to be very dynamic in conformation, easily adopting non-B-DNA conformations, such as melted DNA, hairpin structures, slipped helices, or others under physiological conditions.[73-81] In particular, the guanine-rich sequences found in the tracts of promoters of other genes, such as PDGF-A, c-myb, and Ki-ras, are known to form a wide range of stable parallel or antiparallel G-quadruplex structures in the presence of monovalent cations such as Na^+ and K^+ under physiological conditions.[82-84] In addition, both NMR and X-ray crystallographic studies have determined the structures of various forms of G-quadruplexes, revealing that folding patterns of these G-quadruplexes depend on the precise arrangement and composition of G-repeats as well as the size of the intervening sequences.[85-87] Our recent study provided further evidence that a specific G-quadruplex structure formed within the c-Myc promoter functions as a transcriptional repressor element, establishing the principle that c-Myc transcription can be controlled by ligand-mediated silencer element stabilization.[88] Even though the c-Myc expression does not seem to be affected by telomestatin, there is still a possibility that similar G-quadruplex structures might be formed in the promoter regions of other uncharacterized genes associated with growth and proliferation. Therefore, these G-quadruplexes formed in the promoter regions of several genes represent another potential molecular target for telomestatin beyond telomeric G-quadruplex.

10.2   Concluding Remarks

Many studies done to date using telomestatin demonstrate that this G-quadruplex-interactive agent could be effective in disrupting telomere maintenance mechanisms in human tumor cells by facilitating the formation of intramolecular G-quadruplex structures at the 3'-telomeric overhangs. The results from these studies support the previous suggestion that highly specific and potent G-quadruplex-interactive agents could be promising agents for cancer chemotherapy.

Acknowledgments

This research was supported by grants from the National Institutes of Health (CA94166 and CA95060). The authors are grateful to Dr. David Bishop for preparing, proof reading, and editing the final version of the text and figures.

References

1. R.K. Moyzis, J.M. Buckingham, L.S. Cram, M. Dani, L.L. Deaven, M.D. Jones, J. Meyne, R.L. Ratliff and J.R. Wu, *Proc. Natl. Acad. Sci. USA*, 1988, **85**, 6622.

2. F. Rodier, S.H. Kim, T. Nijjar, P. Yaswen and J. Campisi, *Int. J. Biochem. Cell Biol.*, 2005, **37**, 977.
3. T. de Lange, L. Shiue, R.M. Myers, D.R. Cox, S.L. Naylor, A.M. Killery and H.E. Varmus, *Mol. Cell Biol.*, 1990, **10**, 518.
4. V.L. Makarov, Y. Hirose and J.P. Langmore, *Cell*, 1997, **88**, 657.
5. W.E. Wright, V.M. Tesmer, K.E. Huffman, S.D. Levene and J.W. Shay, *Genes Dev.*, 1997, **11**, 2801.
6. A.M. Olovnikov, *J. Theor. Biol.*, 1973, **41**, 181.
7. J.D. Watson, *Nat. New Biol.*, 1972, **239**, 197.
8. C.B. Harley, A.B. Futcher and C.W. Greider, *Nature*, 1990, **345**, 458.
9. R.C. Allsopp, H. Vaziri, C. Patterson, S. Goldstein, E.V. Younglai, A.B. Futcher, C.W. Greider and C.B. Harley, *Proc. Natl. Acad. Sci. USA*, 1992, **89**, 10114.
10. M.Z. Levy, R.C. Allsopp, A.B. Futcher, C.W. Greider and C.B. Harley, *J. Mol. Biol.*, 1992, **225**, 951.
11. C.M. Counter, A.A. Avilion, C.E. LeFeuvre, N.G. Stewart, C.W. Greider, C.B. Harley and S. Bacchetti, *EMBO J.*, 1992, **11**, 1921.
12. C.B. Harley, N.W. Kim, K.R. Prowse, S.L. Weinrich, K.S. Hirsch, M.D. West, S. Bacchetti, H.W. Hirte, C.M. Counter, C.W. Greider, M.A. Piatyszek, W.E. Wright and J.W. Shay, *Cold Spring Harb. Symp. Quant. Biol.*, 1994, **59**, 307.
13. C.W. Greider, *Curr. Opin. Cell Biol.*, 1991, **3**, 444.
14. N.W. Kim, M.A. Piatyszek, K.R. Prowse, C.B. Harley, M.D. West, P.L. Ho, G.M. Coviello, W.E. Wright, S.L. Weinrich and J.W. Shay, *Science*, 1994, **266**, 2011.
15. J.W. Shay and S. Bacchetti, *Eur. J. Cancer*, 1997, **33**, 787.
16. A.A. Avilion, M.A. Piatyszek, J. Gupta, J.W. Shay, S. Bacchetti and C.W. Greider, *Cancer Res.*, 1996, **56**, 645.
17. B. Herbert, A.E. Pitts, S.I. Baker, S.E. Hamilton, W.E. Wright, J.W. Shay and D.R. Corey, *Proc. Natl. Acad. Sci. USA*, 1999, **96**, 14276.
18. J.W. Shay and W.E. Wright, *Cancer Cell*, 2002, **2**, 257.
19. L.R. Kelland, *Eur. J. Cancer*, 2005, **41**, 971.
20. T.M. Bryan, A. Englezou, J. Gupta, S. Bacchetti and R.R. Reddel, *EMBO J.*, 1995, **14**, 4240.
21. T.M. Bryan, A. Englezou, L. Dalla-Pozza, M.A. Dunham and R.R. Reddel, *Nat. Med.*, 1997, **3**, 1271.
22. L. Crabbe and J. Karlseder, *Curr. Mol. Med.*, 2005, **5**, 135.
23. D. Liu, M.S. O'Connor, J. Qin and Z. Songyang, *J. Biol. Chem.*, 2004, **279**, 51338.
24. K.A. Mattern, S.J. Swiggers, A.L. Nigg, B. Lowenberg, A.B. Houtsmuller and J.M. Zijlmans, *Mol. Cell Biol.*, 2004, **24**, 5587.
25. J.Z. Ye, J.R. Donigian, M. van Overbeek, D. Loayza, Y. Luo, A.N. Krutchinsky, B.T. Chait and T. de Lange, *J. Biol. Chem.*, 2004, **279**, 47264.
26. T. Iwano, M. Tachibana, M. Reth and Y. Shinkai, *J. Biol. Chem.*, 2004, **279**, 1442.
27. B. van Steensel, A. Smogorzewska and T. de Lange, *Cell*, 1998, **92**, 401.

28. C.W. Greider, *Cell*, 1999, **97**, 419.
29. R.M. Stansel, T. de Lange and J.D. Griffith, *EMBO J.*, 2001, **20**, 5532.
30. A.J. Cesare and J.D. Griffith, *Mol. Cell. Biol.*, 2004, **24**, 9948.
31. K. Shin-ya, K. Wierzba, K. Matsuo, T. Ohtani, Y. Yamada, K. Furihata, Y. Hayakawa and H. Seto, *J. Am. Chem. Soc.*, 2001, **123**, 1262.
32. A.M. Zahler, J.R. Williamson, T.R. Cech and D.M. Prescott, *Nature*, 1991, **350**, 718.
33. D. Sun, B. Thompson, B.E. Cathers, M. Salazar, S.M. Kerwin, J.O. Trent, T.C. Jenkins, S. Neidle and L.H. Hurley, *J. Med. Chem.*, 1997, **40**, 2113.
34. P.J. Perry, M.A. Read, R.T. Davies, S.M. Gowan, A.P. Reszka, A.A. Wood, L.R. Kelland and S. Neidle, *J. Med. Chem.*, 1999, **42**, 2679.
35. M.A. Read, A.A. Wood, R.J. Harrison, S.M. Gowan, L.R. Kelland, H.S. Dosanjh and S. Neidle, *J. Med. Chem.*, 1999, **42**, 4538.
36. M.A. Read, R.J. Harrison, B. Romagnoli, F.A. Tanious, S.M. Gowan, A.P. Reszka, W.D. Wilson, L.R. Kelland and S. Neidle, *Proc. Natl. Acad. Sci. U.S.A.*, 2001, **98**, 4844.
37. R.T. Wheelhouse, D. Sun, H. Han, F.X. Han and L.H. Hurley, *J. Am. Chem. Soc.*, 1998, **120**, 3261.
38. F.X. Han, R.T. Wheelhouse and L.H. Hurley, *J. Am. Chem. Soc.*, 1999, **121**, 3561.
39. H. Han, A. Rangan, D.R. Langley and L.H. Hurley, *J. Am. Chem. Soc.*, 2001, **123**, 8902.
40. D.-F. Shi, R.T. Wheelhouse, D. Sun and L.H. Hurley, *J. Med. Chem.*, 2001, **44**, 4509.
41. H. Han, C.L. Cliff and L.H. Hurley, *Biochemistry*, 1999, **38**, 6981.
42. O. Yu. Fedoroff, M. Salazar, H. Han, V.V. Chemeris, S.M. Kerwin and L.H. Hurley, *Biochemistry*, 1998, **37**, 12367.
43. A. Rangan, O. Yu Fedoroff and L.H. Hurley, *J. Biol. Chem.*, 2001, **276**, 4640.
44. F. Koeppel, J.-F. Riou, A. Laoui, P. Malliet, P.B. Arimondo, D. Labit, O. Petigenet, C. Hélène and J.-L. Mergny, *Nucleic Acids Res.*, 2001, **29**, 1087.
45. R.J. Harrison, S.M. Gowan, L.R. Kelland and S. Neidle, *Bioorg. Med. Chem. Lett.*, 1999, **9**, 2463.
46. J.-F. Riou, P. Mailliet, A. Laoui, E. Renou, O. Petigenet, L. Guittat and J.-L. Mergny, *Proc. Am. Assoc. Cancer Res.*, 2001, **42**, 837.
47. S.M. Gowan, L. Brunton, M. Valenti, R. Heald, M.A. Read, J.R. Harrison, M.F.G. Stevens, S. Neidle and L.R. Kelland, *Proc. Am. Assoc. Cancer Res.*, 2001, **42**, 86.
48. W. Duan, A. Rangan, H. Vankayalapati, M.-Y. Kim, Q. Zeng, D. Sun, O. Yu. Fedoroff, D. Nishioka, S.Y. Rha, E. Izbicka, D.D. Von Hoff and L.H. Hurley, *Mol. Cancer Ther.*, 2001, **1**, 103.
49. J.R. Williamson, M.K. Raghuraman and T.R. Cech, *Cell*, 1989, **59**, 871.
50. P. Balagurumoorthy and S.K. Brahmachari, *J. Biol. Chem.*, 1994, **269**, 21858.
51. R. Giraldo, M. Suzuki, L. Chapman and D. Rhodes, *Proc. Natl. Acad. Sci. USA*, 1994, **91**, 7658.

52. J.L. Mergny and J.C. Maurizot, *ChemBioChem.*, 2001, **2**, 124.
53. W. Li, P. Wu, T. Ohmichi and N. Sugimoto, *FEBS Lett.*, 2002, **526**, 77.
54. Y. Wang and D.J. Patel, *Structure*, 1993, **1**, 263.
55. F.W. Smith and J. Feigon, *Nature*, 1992, **356**, 164.
56. P.B. Arimondo, J.-F. Riou, J.-L. Mergny, J. Tazi, J.-S. Sun, T. Garestier and C. Hélène, *Nucleic Acids Res.*, 2000, **24**, 4832.
57. M.-Y. Kim, M. Gleason-Guzman, E. Izbicka, D. Nishioka and L.H. Hurley, *Cancer Res.*, 2003, **63**, 3247.
58. W. Liu, L.H. Hurley and D. Sun, *Nucleosides Nucleotides Nucleic Acids*, 2005, **24**, 1801.
59. M. Sumi, T. Tauchi, G. Sashida, A. Nakajima, A. Gotoh, K. Shin-ya, J.H. Ohyashiki and K. Ohyashiki, *Int. J. Oncol.*, 2004, **24**, 1481.
60. A. Nakajima, T. Tauchi, G. Sashida, M. Sumi, K. Abe, K. Yamamoto, J.H. Ohyashiki and K. Ohyashiki, *Leukemia*, 2003, **17**, 560.
61. M.A. Shammas, R.J. Shmookler Reis, C. Li, H. Koley, L.H. Hurley, K.C. Anderson and N.C. Munshi, *Clin. Cancer Res.*, 2004, **10**, 770.
62. T. Tauchi, K. Shin-ya, G. Sashida, M. Sumi, A. Nakajima, T. Shimamoto, J.H. Ohyashiki and K. Ohyashiki, *Oncogene*, 2003, **22**, 5338.
63. D. Gomez, R. Paterski, T. Lemarteleur, K. Shin-ya, J.L. Mergny and J.F. Riou, *J. Biol. Chem.*, 2004, **279**, 41487.
64. B.B. Zhou and S.J. Elledge, *Nature*, 2000, **408**, 433.
65. R.T. Abraham, *Genes Dev.*, 2001, **15**, 2177.
66. S. Matsuoka, G. Rotman, A. Ogawa, Y. Shiloh, K. Tamai and S.J. Elledge, *Proc. Natl. Acad. Sci. USA*, 2000, **97**, 10389.
67. C. Feijoo, C. Hall-Jackson, R. Wu, D. Jenkins, J. Leitch, D.M. Gilbert and C. Smythe, *J. Cell. Biol.*, 2001, **154**, 913.
68. J.C. Darnell, K.B. Jensen, P. Jin, V. Brown, S.T. Warren and R.B. Darnell, *Cell*, 2001, **107**, 489.
69. S. Bonnal, C. Schaeffer, L. Créancier, S. Clamens, H. Moine, A.C. Prats and S. Vagner, *J. Biol. Chem.*, 2003, **278**, 39330.
70. P. Sirand-Pugnet, P. Durosay, E. Brody and J. Marie, *Nucleic Acids Res.*, 2005, **23**, 3501.
71. D. Gomez, N. Aouali, A. Renaud, C. Douarre, K. Shin-ya, J. Tazi, S. Martinez, C. Trentesaux, H. Morjani and J.F. Riou, *Cancer Res.*, 2003, **63**, 6149.
72. D. Gomez, T. Lemarteleur, L. Lacroix, P. Mailliet, J.L. Mergny and J.F. Riou, *Nucleic Acids Res.*, 2004, **32**, 371.
73. A. Rustighi, M.A. Tessari, F. Vascotto, R. Sgarra, V. Giancotti and G. Manfioletti, *Biochemistry*, 2002, **41**, 1229.
74. A.C. Johnson, Y. Jinno and G.T. Merlino, *Mol. Cell. Biol.*, 1988, **8**, 4174.
75. G.A. Michelotti, E.F. Michelotti, A. Pullner, R.C. Duncan, D. Eick and D. Levens, *Mol. Cell. Biol.*, 1996, **16**, 2656.
76. D.S. Tewari, D.M. Cook and R. Taub, *J. Biol. Chem.*, 1989, **264**, 16238.
77. S. Chen, P.C. Supakar, R.L. Vellanoweth, C.S. Song, B. Chatterjee and A.K. Roy, *Mol. Endocrin.*, 1977, **11**, 3.

78. S. Ritchie, F.M. Boyd, J. Wong and K. Bonham, *J. Biol. Chem.*, 2000, **275**, 847.
79. D.G. Pestov, A. Dayn, E.U. Siyanova, D.L. George and S.M. Mirkin, *Nucleic Acids Res.*, 1991, **19**, 6527.
80. R. Lafyatis, F. Denhez, T. Williams, M. Sporn and A. Roberts, *Nucleic Acids Res.*, 1991, **19**, 6419.
81. Z. Wang, X.H. Lin, Q.Q. Qiu and T.F. Deuel, *J. Biol. Chem.*, 1992, **267**, 17022.
82. M.A. Keniry, *Biopolymers*, 2001, **56**, 123.
83. T. Simonsson, *Biol. Chem.*, 2001, **382**, 621.
84. S. Chowdhury and M. Bansal, *J. Biomol. Struct. Dyn.*, 2000, **18**, 11.
85. G.N. Parkinson, M.P. Lee and S. Neidle, *Nature*, 2002, **417**, 876.
86. A.T. Phan, Y.S. Modi and D.J. Patel, *J. Am. Chem. Soc.*, 2004, **126**, 8710.
87. A. Ambrus, D. Chen, J. Dai, R.A. Jones and D. Yang, *Biochemistry*, 2005, **44**, 2048.
88. A. Siddiqui-Jain, C.L. Grand, D.J. Bearss and L.H. Hurley, *Proc. Natl. Acad. Sci. USA.*, 2002, **99**, 11593.
89. W.N. Keith, A. Bilsland, T.R.J. Evans and R.M. Glasspool, *Exp. Rev. Mol. Med.*, 2002, 22 April, http://www.expertreviews.org/02004507h.htm.
90. M.-Y. Kim, H. Vankayalapati, K. Shin-ya, K. Wierzba and L.H. Hurley, *J. Am. Chem. Soc.*, 2002, **124**, 2098.
91. E.M. Rezler, V. Gokhale, J. Seenisamy, S. Bashyam, M.-Y. Kim and L.H. Hurley, *J. Am. Chem. Soc.*, 2005, **127**, 9339.

CHAPTER 11

Structural Features of the Specific Interactions between Nucleic Acids and Small Organic Molecules

ALEXANDER SERGANOV AND DINSHAW J. PATEL

Structural Biology Program, Memorial Sloan-Kettering Cancer Center, New York NY 10021, USA

11.1 Introduction

Nucleic acids play crucial roles in diverse biological processes, including the storage, transport, processing, and expression of genetic information. Most nucleic acid functions are carried out synergistically with proteins and other ligands, including small organic molecules; therefore, intervention in these processes can significantly affect cellular activity. Although much current drug research focuses on agents that inhibit the action of proteins, nucleic acids also represent attractive therapeutic targets. Small drug molecules can be directed towards known intermolecular interfaces or simply distributed randomly along the nucleic acid helices. Potentially, a disease can be controlled at the source, as demonstrated by widely used anticancer DNA-binding agents, or at later stages of gene expression as exemplified by numerous ribosome-binding antibiotics. The driving forces behind research in this field include severe cytotoxic effects of non-specific drugs and increasing prevalence of antibiotic resistance. Interaction of small organic molecules with nucleic acids can be achieved by different modes, including covalent linkage, intercalation between bases, positioning in the helical grooves and interaction with the sugar-phosphate backbone. These aspects of the recognition process have been extensively reviewed by Lown,[1] Neidle,[2] Pindur *et al.*[3] and many others. Non-specific binding has attractive alternatives, such as sequence-specific interaction and precise recognition of the nucleic acid fold. Specific modes of binding to DNA along with structural characterization of the complexes have been described by Shafer,[4] Han and Gao,[5] Neidle and Parkinson,[6] and Olecsi *et al.*[7] Many excellent

233

reviewers have also summarized the current progress on targeting folded RNA molecules, typically the ribosome and viral control elements, with small molecules (Walter et al.,[8] Sucheck and Wong,[9] Gallego and Varani,[10] Cheng et al.,[11] Tor,[12] Hermann,[13] Sutcliffe[14]). Here, we seek to complement this literature by focusing on the detailed characterization of several recent structures of complex-folded nucleic acids bound to small organic ligands as well as comparison of these structures with well-known RNA and DNA folds. Although some of these structures are a sideline to the major stream of therapeutic research, all the structures presented contain novel and interesting features that broaden our knowledge of the principles of intermolecular recognition between nucleic acids and small organic ligands.

11.2   Diels–Alder Ribozymes

The currently known natural ribozymes catalyse a narrow range of chemical reactions, namely the transesterification and hydrolysis of internucleotide bonds (Cech,[15] Doudna and Cech[16]) and peptide bond formation (Steitz and Moore[17]). *In vitro* selection and evolution, on the other hand, have demonstrated that ribozymes can accelerate a much broader spectrum of reactions (Wilson and Szostak[18]). Whereas high-resolution structures and biochemical investigations of several natural ribozymes provide a basic understanding of how RNA performs phosphodiester chemistry and peptide bond formation (Doudna and Cech[16]; Moore and Steitz[19]), little is known about how RNA can catalyse other reactions.

A first insight into the principles of substrate recognition and the molecular mechanism of the chemical reaction, which joins two small organic molecules was obtained by Serganov et al.[20] who determined three-dimensional structures for a Diels–Alder ribozyme (Seelig and Jäschke[21]) in the free form and bound to its reaction product. Two *in vitro* selections yielded ribozymes that accelerate the formation of carbon–carbon bonds by Diels–Alder reaction, a [4+2] cycloaddition reaction between a tethered diene and a biotinylated dienophile (Seelig and Jäschke[21]; Tarasow et al.[22]) (Figure 1A). This type of reaction creates two carbon–carbon bonds with up to four chiral carbons in one step, and therefore the reaction is broadly applied in organic chemistry for the stereospecific synthesis of complex molecules (Nicolaou et al.[23]). The Diels-Alderase selected by Tarasow et al.[22] contains chemically modified uridines and enhances the rate of reaction about 800-fold, while the Seelig-Jäschke ribozyme (Seelig and Jäschke[21]) consists of natural nucleotides and achieves an impressive reaction rate enhancement of approximately 20,000-fold when an anthracene substrate is tethered to the 5′ end of the RNA, and about 1100-fold in a true catalytic reaction.

The minimal bipartite Seelig–Jäschke ribozyme used for structural studies was composed of 38 and 11 nt RNA strands and a hexaethyleneglycol (HEG)-linked cycloaddition product covalently attached to the 5′ end of the 11-mer (Figures 1A and B). The structure (Figure 1C) showed that the ribozyme consists of three helical domains brought together by a central segment. The connecting segment contains opposing invariant fragments of the 38-mer

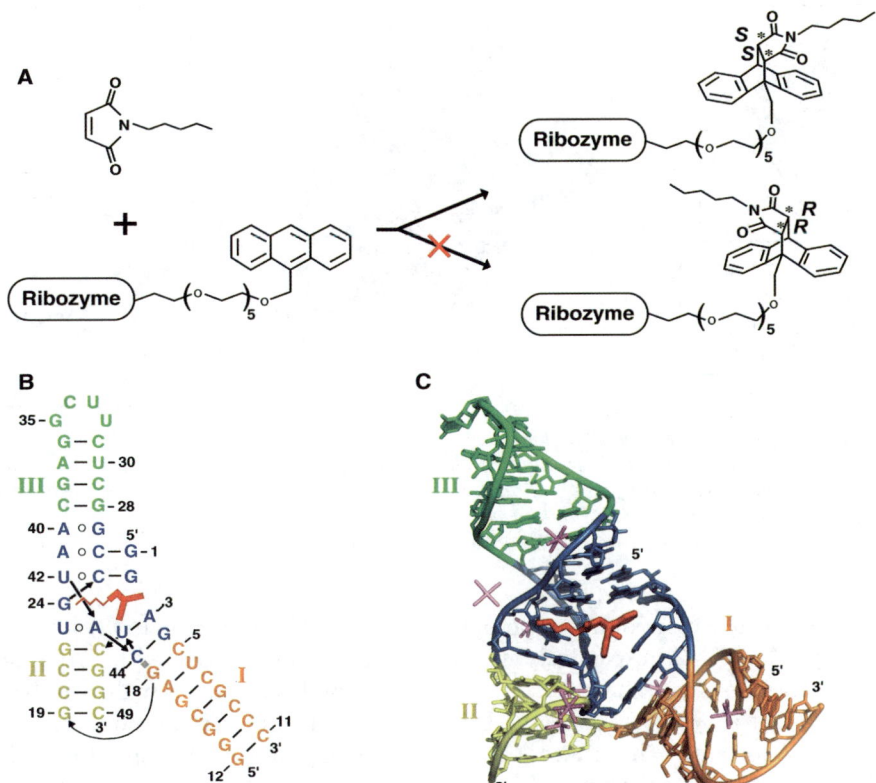

Figure 1 *Secondary and tertiary folds of the Diels–Alder ribozyme. (A) The ribozyme-catalysed Diels–Alder reaction between tethered anthracene and N-pentyl maleimide resulting in formation of a single enantiomer of the complex. (B) The ribozyme tertiary fold with stems I, II and III coloured orange, gold and green respectively, and residues found invariant in the selection procedure coloured in blue. The same colours are used throughout this figure. The ribozyme consists of an 11-mer (G1 to C11) RNA HEG-linked to anthracene and another 38-mer (G12 to C49) RNA. The bound product (coloured in red) is in stick representation. (C) Three-dimensional topology in the crystal structure of the ribozyme-product complex. The backbone is depicted with a ribbon, while the hydrated Mg^{2+} cations (coloured in violet) are in stick representation* (Adapted from ref. 20 with permission)

(nt 23–27 and 40–45), previously assigned as an internal asymmetric loop (Keiper *et al.*[24]), and an invariant 5′-terminal GGAG sequence of the 11-mer (Figures 1B and C). Stems II and III, bridged by a zipped-up central segment, are stacked co-linearly and stem I lies inclined at about 50° to the composite I-II helix, thus forming a λ-shaped topology for the whole molecule.

The catalytic pocket is built entirely by the central segment. The 5′-G1-G2-A3-G4 region of the 11-mer, found critical for catalysis (Keiper *et al.*[24]), forms Watson–Crick G1-C26, G2-C25, A3-U45, and G4-C44 base pairs with residues from the opposite strands of the asymmetric loop region, thereby generating a

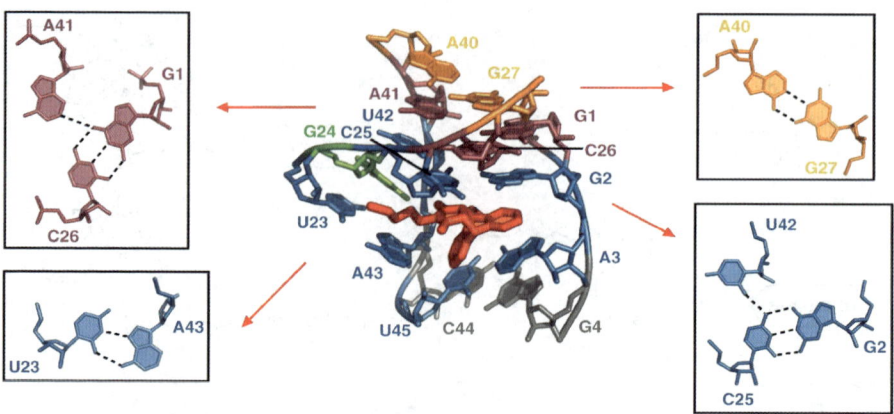

Figure 2   *Three-dimensional structure of the central part of the Diels–Alder ribozyme showing residues surrounding the bound reaction product (red sticks). Noncanonical base pairs and base triples are shown viewed from above. Hydrogen bonds are indicated by dotted lines*
(Adapted from ref. 20 with permission)

complex nested pseudoknot topology. Two top base pairs (Figure 2) interact with A41 and U42, respectively, forming three-base platforms A41·(G1-C26) and U42·(G2-C25). The U42·(G2-C25) triple builds a roof over the catalytic pocket, and a Watson–Crick A3-U45 base pair makes a foundation for the pocket. A reversed-Hoogsteen U23·A43 pair together with non-paired G24 bracket a canyon protruding from the central opening between G2-C25 and A3-U45 thereby forming a wedge-shaped catalytic pocket.

The cycloaddition product is positioned within the catalytic pocket with one of its aromatic rings sandwiched between purines G2 and A3, and the other wedged between the base and sugar components of A43 and U45 (Figures 2 and 3). The HEG-linked bridgehead position is directed inwards into the RNA fold, while the unmodified bridgehead position points outwards towards the solvent. The maleimide ring is stacked over C25, positioned through a pair of hydrogen bonds made by one of its two carbonyl oxygens with the NH_2 group of G24 and the 2'-OH of U42 (Figure 3). The maleimide side chain runs inside a hydrophobic canyon created by bases C25 (top), A43 (bottom), and G24 (back) (Figure 3). The unpaired G24 residue is sandwiched between U23 and U42 (Figure 2), and in addition to the hydrogen bond with maleimide, it forms two hydrogen bonds with the phosphate oxygens of A43 and its 2'-OH is hydrogen bonded with U42 (Figure 3).

A Mg^{2+} cation locks G24 in place. Remarkably, the catalytic pocket is primarily formed by base edges with minimal contribution from the sugar-phosphate backbone, and there are no Mg^{2+} cations or phosphates located in the immediate vicinity of the pocket. Such hydrophobic composition of the pocket is clearly required for the rather tight (10 μM range) binding of the predominantly hydrophobic reaction product (Stuhlmann and Jäschke[25]), which can only use 1 or 2 atoms for hydrogen bonding. The pocket virtually

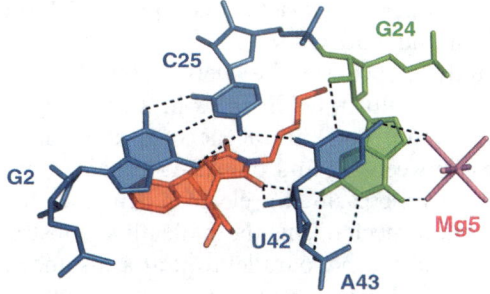

Figure 3 *Specific interactions of the reaction product with the RNA segment aligned above the product. The view is rotated 180° along the vertical axis relative to the schematic in Figure 1. The colour code is the same as in Figure 1. Coordination bonds to a magnesium atom are represented by violet sticks. Oxygen and nitrogen atoms in the product are shown, respectively, in pink and blue*
(Adapted from ref. 20 with permission)

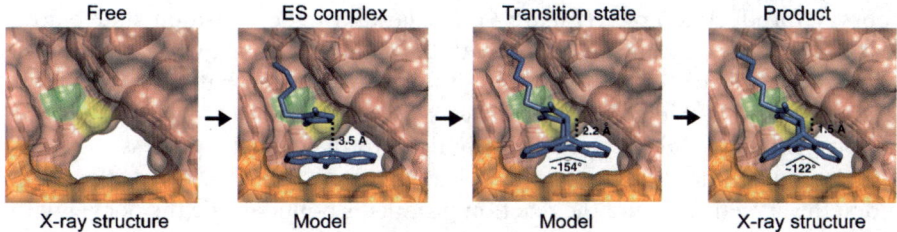

Figure 4 *Proposed model for the catalytic mechanism of the Diels–Alder ribozyme. The surfaces of the A3-U45 base pair, amino group of G24 and 2'-OH of U42 are orange, green and yellow, respectively. Angles showing the buckling of the anthracene ring system and key distances during formation of the carbon-carbon bonds are indicated. The reaction substrates and product are shown in blue*
(Adapted from ref. 20 with permission)

repeats the shape of the reaction product, providing maximal stacking, van der Waals and hydrophobic interactions, and allowing binding of only one product enantiomer. A striking feature of the catalytic pocket is a large opening towards the front, which can be used for substrate binding and product release without significant changes in the pocket shape. Indeed, the RNA structure of the free ribozyme (Serganov *et al.*[20]) was found to be almost identical to the RNA structure in the ribozyme-product complex, strongly suggesting pre-formation of the catalytic pocket within the RNA scaffold (Figure 4). Independent chemical probing experiments (Keiper *et al.*[24]) as well as subtle changes in NMR spectra (Serganov *et al.*[20]) support this conclusion, indicating that the overall RNA-folding is not changed upon addition of the anthracene substrate or the cycloaddition product. The observation of a preformed catalytic pocket for the Diels–Alder ribozyme is reminiscent of the hepatitis delta virus ribozyme (Ferré-D'Amaré *et al.*[26]), which also folds as a nested pseudoknot and has a preformed active site, and contrasts with many examples of adaptive recognition found in other ligand-RNA complexes (Hermann and Patel,[27] Williamson,[28] Leulliot and Varani[29]).

The preformed catalytic pocket of the Diels–Alder ribozyme suggests that the ribozyme should bind its substrates in *trans* in a precisely defined orientation and in an ultimate position. The anthracene substrate is likely to be bound by stacking interactions between G2 and the A3-U45 pair (Figure 4), which could accelerate the reaction by increasing the electron density of the diene. According to the proposed reaction mechanism, the maleimide substrate should then be stacked on top of the anthracene, parallel to it at a distance of about 3.5 Å. In this position, the maleimide side chain can be fitted within the hydrophobic canyon thus approaching the anthracene only from one direction and dictating the stereoselectivity of the reaction. In addition, the maleimide can form hydrogen bonds with G24 and U42, which may change the electron density distribution of the dienophile, facilitating the reaction. A snug fit along the short edges of the anthracene may also contribute towards bending of this planar molecule in the transition state of the reaction, further reducing the minimal unoccupied space underneath the bridgehead carbons seen in the structure of the ribozyme-product complex (Figure 4). Remarkably, the transition state enjoys near-perfect shape complementarity with the catalytic pocket, which might be a major driving force for the reaction, although it appears that ribozyme-based catalysis of Diels–Alder reactions reflects a combination of proximity, shape complementarity, and energetic contributions to the catalytic process.

Interestingly, the catalytic pocket has a narrower opening towards the back, and in the crystal structure the reaction product is bound inside the pocket with the sterically more demanding long edge directed inward into the RNA fold, contrary to what had been expected (Stuhlmann and Jäschke[25]). Such positioning indicates that the anthracene substrate could enter the catalytic pocket through the back opening, and, since it is not symmetrical, this could theoretically lead to the synthesis of both stereoisomers depending on the entrance side. Follow-up biochemical experiments (Wombacher *et al.*[30]) have demonstrated that, indeed, the Diels-Alderase can synthesize both stereoisomers. However, the synthesis does depend on the format of the reaction and only one stereoisomer is predominant in most cases. In the true catalytic reaction (lacking covalent attachment of RNA to the anthracene), the anthracene edge carrying the HEG linker is oriented outward towards the solvent and the reaction produces the *R,R* product. In the tethered reaction format, the short flexible linker must be threaded through the narrow disfavouring 'back door', yielding the *S,S* stereoisomer and leading to the orientation of the reaction product observed in the crystal. An extension of the tether such that it can reach the 'front door' folding around the RNA molecule leads to increased production of the *R,R* product (Wombacher *et al.*[30]).

11.3   Theophylline and Flavin Mononucleotide Binding

Specific binding of aromatic ligands to RNA has previously been structurally characterized for a few complexes. The best-known examples of such

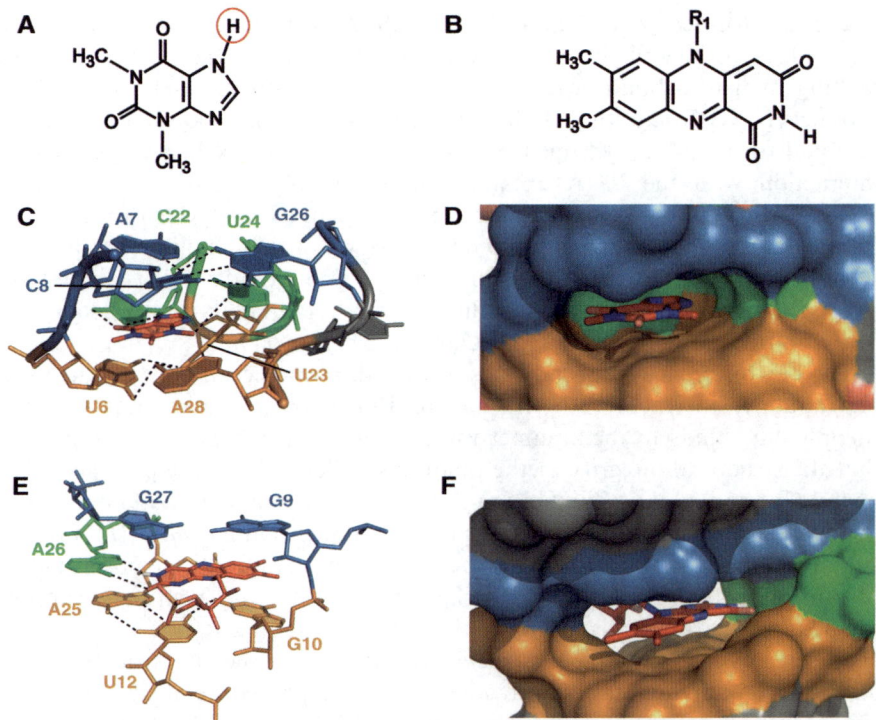

Figure 5 *Molecular recognition of planar aromatic ligands by nucleic acids. (A) The-ophylline. (In caffeine the circled hydrogen is replaced by a methyl group). (B) Flavin mononucleotide (FMN), an isoalloxazine derivative where R1 represents the ribose-phosphate moiety. The ligand-binding pockets and surface views are shown in C and D for theophylline (Zimmermann et al.[33]) and for FMN in E and F (Fan et al.[31])*

interactions come from structural studies of RNA aptamers that bind the-ophylline (Zimmermann *et al.*[33]) (Figure 5A) and flavin mononucleotide (FMN; Fan *et al.*[31]) (Figure 5B).

The theophylline-binding RNA aptamer displays an affinity for its cognate ligand 10,000-fold higher than for caffeine, which differs from theophylline by a single methyl group (Figure 5A) (Jenison *et al.*[32]). The structure (Zimmermann *et al.*[33]) demonstrates that the RNA folds as a helical domain where two internal loops interact in an unusual way creating a theophylline binding pocket. Theophylline is nearly completely enclosed within the RNA scaffold and forms multiple contacts that direct the specificity of binding. As in the Diels-Alderase structure, the upper part of the pocket is built from a base triple [A7·(C8-G26), Figure 5C], while the bottom part is made of another triple (U6·U23·A28), not two base pairs. Formation of the triples and accommodation of the ligand require the RNA to fold irregularly involving a few particular structural motifs, such as an S-turn, 1-3-2 stacking, an A-platform and others. Since theophylline is flat, the pocket is narrow and the bound ligand is perfectly sandwiched between these two layers of triples. The sides of the pocket are formed by C22 and U24, which make hydrogen bonds with both long edges of theophylline, somewhat reminiscent of the non-paired G27 making hydrogen bonds with the maleimide component of the Diels–Alder product. Nevertheless, theophylline is richer in hydrogen bond donors/acceptors and forms more hydrogen bonds, which provides exclusive specificity of interaction with the RNA aptamer and, probably, affords better binding affinity. Discrimination against caffeine is explained by the hydrogen bonds formed between the aptamer and the N7 position of theophylline. The presence of a methyl group at the N7 position of caffeine would directly disrupt two hydrogen bonds and probably indirectly disrupt other bonds, affecting the formation of the core of the molecular complex.

The key difference from the Diels–Alder ribozyme lies in how the theophylline aptamer interacts with the ligand. Biochemical studies indicated that theophylline binds to the aptamer much better than its analogues, suggesting that the structural integrity of the binding site depends on the presence of the ligand (Jenison *et al.*,[32] Zimmermann *et al.*[33]). The finding that almost 90% of the ligand is buried within the RNA pocket (Zimmermann *et al.*[33]) (Figure 5D), supported by NMR titration experiments that reveal dramatic changes in imino proton spectra upon addition of the ligand (Jenison *et al.*[32]), strongly supports the adaptive nature of the aptamer-theophylline binding.

Another typical example of specific recognition between an aromatic ligand and an RNA aptamer has come from the selection procedure of Burgstaller and Famulok[34] who identified a 35-mer sequence binding to FMN with an affinity of about 0.5 μM. The NMR study of Fan *et al.*[31] revealed that the FMN-binding pocket is located within the RNA helical segment and is built by several residues of an internal asymmetric loop (Figure 5E). Similarly to the theophylline-binding aptamer, the planar ring system of FMN is positioned between two layers formed by a non-canonical G27 · G9 base pair and a G10 · U12 · A25 base triple thus providing maximal stacking interactions. Specific recognition of FMN is achieved through hydrogen bond alignment of a non-paired A26 with the uracil-like edge of the isoalloxazine ring. Although the surface representation of the FMN aptamer shows two wide openings (Figure 5F), FMN is deeply penetrated into the binding pocket, and, in contrast to Diels-Alderase, the pocket is unlikely to be preformed for FMN recognition. This suggestion was strongly supported by dramatic changes in the NMR spectrum of the free RNA upon addition of FMN, indicative of the folding of the central RNA region around the ligand (Fan *et al.*[31]).

11.4 Purine Riboswitches

The known structures of complexes between small organic molecules and nucleic acids have recently been complemented by studies on celluar metabolites bound to their natural mRNA targets termed riboswitches (Batey *et al.*,[35] Serganov *et al.*[36]).

Riboswitches have recently been identified as non-coding mRNA elements that utilize metabolite binding to regulate gene expression profiles for numerous bacterial biosynthetic pathways, thereby controlling up to 2% of bacterial genomes (Mandal and Breaker,[37] Nudler and Mironov,[38] Soukup and Soukup[39]). Riboswitches exhibit a modular architecture composed of metabolite-sensing domains linked to expression platforms. The metabolite-sensing domain of each riboswitch undergoes conformational stabilization upon docking of its corresponding ligand. This conformational switch in the sensing domain typically alters the base-pairing arrangements in the adjoining expression platform, thus modulating transcription, translation or RNA processing events according to the metabolic state of the cell (Mandal and Breaker,[37] Nudler and Mironov[38]). Riboswitches usually exhibit remarkable affinity for their cognate ligands and can discriminate against even closely related analogues (Winkler *et al.*,[40,41] Mandal *et al.*[42]), as is typical of metabolite-sensing protein factors. However, the known riboswitches do not use proteins for molecular sensing, and therefore must rely exclusively on ligand-binding pockets formed by nucleotides in collaboration with metal ions and water. To date, riboswitches have been identified that respond to 10 fundamental metabolites including coenzymes, nucleobases, amino acids, and a sugar-phosphate compound (reviewed by Breaker[43]).

Purine riboswitches represent one of the most intriguing classes of riboswitches (Mandal *et al.*,[42] Mandal and Breaker[44]). They contain a metabolite-binding domain initially termed a G-box (Mandal *et al.*[42]). The G-box is composed of three stems connected by a junction (Figure 6A) and contains many evolutionarily conserved nucleotides within the junction, hairpin loops (L2, L3), and stem P1 residues. The G-box typically precedes the genes coding for proteins involved in metabolism of guanine and its analogues and can interact with nanomolar affinity with guanine, hypoxanthine, and xanthine if their concentration exceeds a threshold value. Upon binding of guanine, the recognition module alters the conformation of the downstream expression platform and represses gene expression by transcriptional or translation control (Mandal *et al.*[42]). Remarkably, the guanine-specific riboswitch (G-riboswitch) demonstrated ten-fold reduced affinity for hypoxanthine and xanthine compared with that of guanine, while adenine binding was reduced by six orders of magnitude.

G-box-like sequences were also identified next to the genes involved in metabolism or transport of adenine, and surprisingly, despite significant similarity to the G-riboswitches, those mRNA regions responded to adenine but not to guanine or its analogues (Mandal and Breaker[44]). Analysis of the multiple A- and G-riboswitch sequences revealed that the major difference occurs at position 74, which is C in the guanine- and U in the adenine-sensing mRNAs. Replacement of this C by U in the G-riboswitch RNA alters its specificity from guanine to adenine, whereas replacing the corresponding U by C in the A-riboswitch converts it to a G-riboswitch (Mandal and Breaker[44]). Thus, remarkably, single nucleotide substitutions can be used to interchange the guanine/adenine specificity within the natural aptamer domains.

Recently, the first crystal structures have been reported for a guanine-sensing riboswitch bound to hypoxanthine (Batey *et al.*[35]) and guanine (Serganov

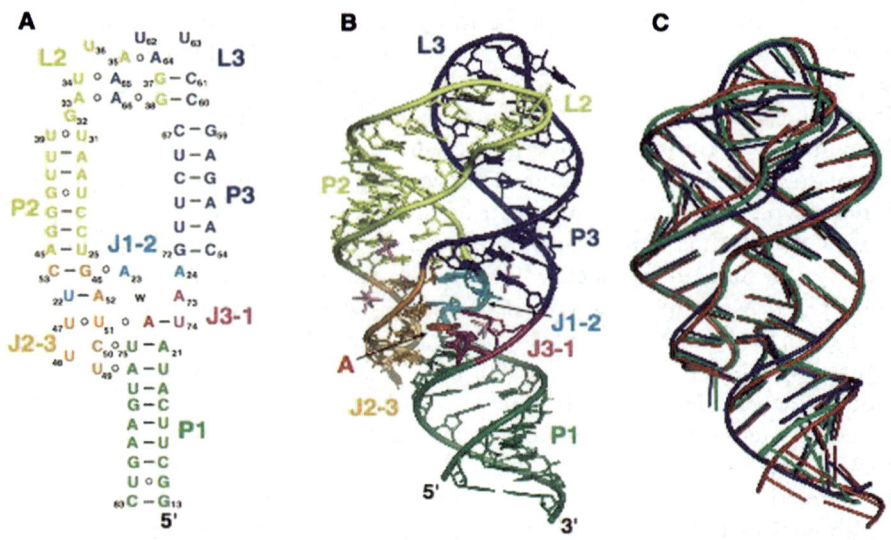

Figure 6   *Schematic illustration and structure of riboswitch-purine complexes. (A)*
Schematic highlighting tertiary interactions in the folded structure of the
A-riboswitch:adenine complex. RNA segments are colour-coded. (B) Three-
dimensional structure of the A-riboswitch: Adenine complex shown in the same
colour-coding scheme. The bound adenine is shown in red and discriminatory
nucleotide 74 is shown in a thick stick representation. (C) Superposition of the
complexes A-riboswitch:adenine (red), G-riboswitch:guanine (green) and
G-riboswitch:hypoxanthine (blue)
      (Adapted from ref. 36 with permission)

et al.[36]), and for an adenine-sensing riboswitch bound to adenine (Serganov *et
al.*[36]). The adenine-bound 71-mer A-riboswitch adopts a tuning fork-like
compact fold (schematic in Figure 6A; structure in Figure 6B) where stem P1
forms the handle of the tuning fork, and stems P2 and P3, which form the
prongs, are aligned parallel to each other and anchored at their tips through
extensive interaction between their hairpin loops L2 and L3. Since the majority
of the differences between G- and A-riboswitches occur in non-essential helical
regions, it is not surprising that the G-riboswitches bound to hypoxanthine and
guanine display a remarkable similarity to the overall folding of the A-
riboswitch (Figure 6C).

   The central RNA region zips up through stacked base triple alignments
between the three junction-connecting segments J1-2, J2-3, and J3-1, and two
junctional base pairs of stem P1, thereby generating five layers of stacked triples
(Figure 7) that constitute an adenine-binding pocket. Two triples, A23 · (G46-
C53) and water-mediated A73 · (A52-U22), located above the adenine-binding
site, involve adenines A23 and A73, which are positioned in the minor groove
of their respective Watson–Crick base pairs. The same minor groove positioning is also observed for C50 and U49 in the two additional triples, C50 · (U75-
A21) and U49 · (A76-U20), located below the adenine-binding site. Finally,
adenine recognition occurs through formation of a U51 · (U47 · adenine-U74)

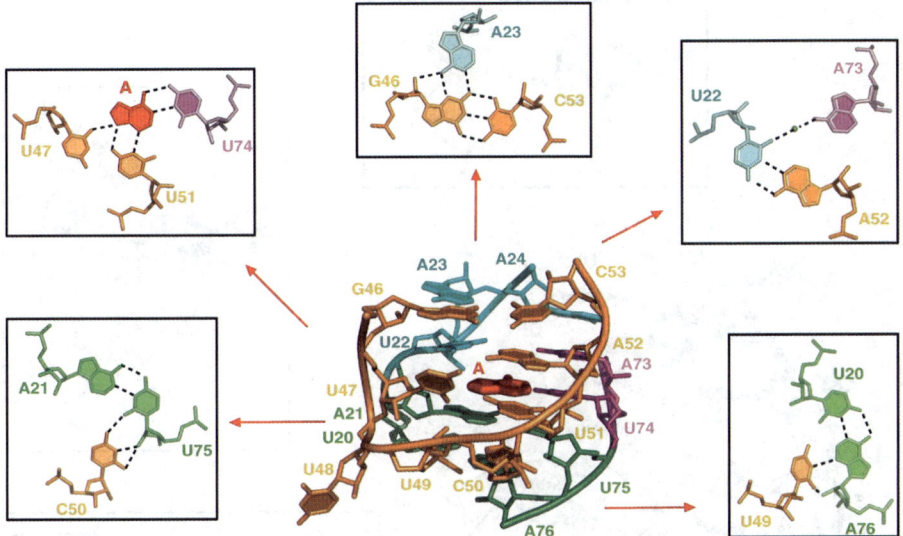

Figure 7 *Details of the tertiary interactions within the adenine-binding pocket of the A-riboswitch:adenine complex*
(Adapted from ref. 36 with permission)

platform, where U51 lies positioned in the minor groove of a Watson–Crick adenine-U74 base pair.

A striking feature of the purine riboswitch structures is stabilization of the overall fold by kissing loop interactions between conserved hairpin loops L2 and L3 anticipated from complementarity between the loop sequences (Mandal *et al.*[42]; Mandal and Breaker[44]) (Figure 8). The loop–loop interactions are formed by multiple hydrogen bonds which position stems P2 and P3 parallel to each other thereby locking up the purine-binding pocket. The kissing loops are held by five base pairs, which involve two Watson–Crick pairs G38-C60 and G37-C61, and three non-canonical pairs, the *trans* U34 · A65, *trans* A33 · A66 and *trans* A35 · A64 pairs. Further alignment amongst the pairs results in the formation of (A33 · A66) · (C60-G38) and (U34 · A65) · (C61-G37) tetrad platforms, additionally anchoring the kissing interaction between the hairpin loops. A mixed tetrad has been reported previously in the structure of a viral pseudoknot (Su *et al.*[45]) where it participated in the docking of the loop into a helical segment.

The bound purines are held in position through formation of direct hydrogen bonds with three base edges and a sugar 2'-OH group (Figure 9). In both riboswitches, the bound purines are surrounded by uridine residues U22, U47, and U51, which form common hydrogen bonds to N^7, N^9H, and N^3 positions of the ligands. The specificity of purine recognition is defined by Watson–Crick hydrogen bonding with a single discriminatory nucleotide: U74 in the A-riboswitch and C74 in the G-riboswitch. These nucleotides form two hydrogen bonds in the A-riboswitch:adenine and G-riboswitch:hypoxanthine

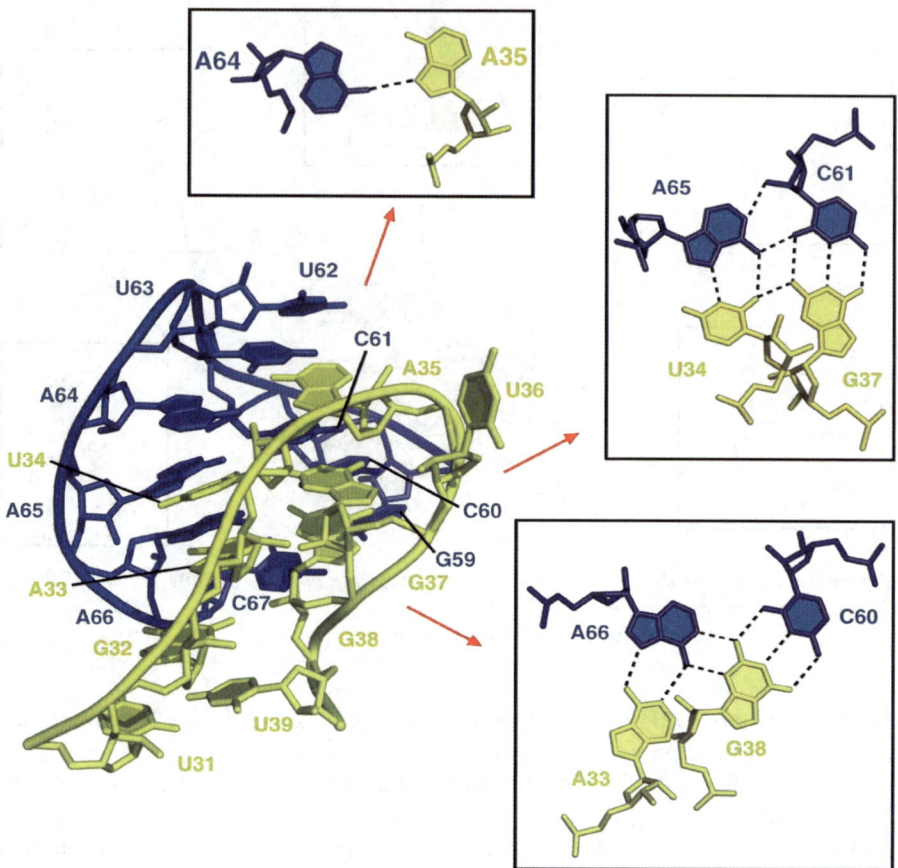

Figure 8   *Details of the tertiary interactions between the kissing hairpins in the structure*
of the A-riboswitch: Adenine complex
     (Adapted from ref. 36 with permission)

complexes, and three hydrogen bonds in the G-riboswitch:guanine complex.
U51 also forms a hydrogen bond with N^2H_2 of guanine in the G-riboswitch-
guanine complex. The two additional hydrogen bonds likely account for the
higher binding affinity of the G-riboswitch:guanine complex compared to the
other purine riboswitch-ligand interactions. The ligand-binding nucleotides are
stabilized and their positions are well-defined by multiple stacking interactions
and hydrogen bonding, thus preventing non-canonical pairing of the bound
purines with discriminatory nucleotides. This explains the dramatic difference
in binding energy between cognate and non-cognate purine ligands and riboswitches. On the other hand, the structures predict that a single mutation of the
nucleotide at position 74 is sufficient to change the specificity of the riboswitch
in excellent agreement with the biochemical data (Mandal and Breaker[44]).

   The bound purines are almost completely enveloped by the riboswitch RNA
strongly suggesting at least local conformational re-arrangements in the RNA

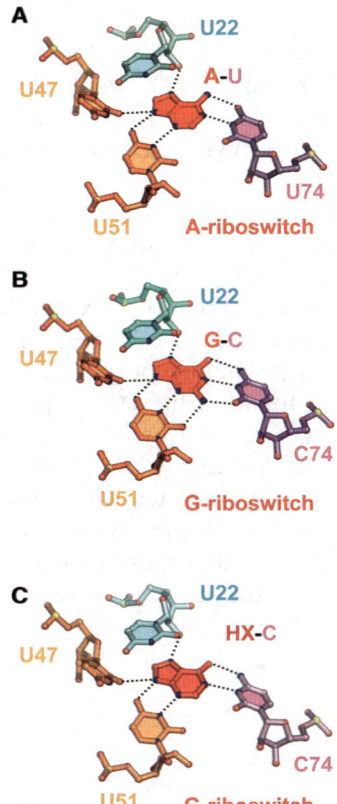

Figure 9 *Recognition of bound purines by riboswitches.(A) A-riboswitch bound to adenine. (B) G-riboswitch bound to guanine. (C) G-riboswitch bound to hypoxanthine. Oxygen, nitrogen, and phosphorus atoms are shown, respectively, as pink, blue and yellow balls*
(Adapted from ref. 36 with permission)

and induced-fit binding of the ligand. Significant changes in the NMR spectrum of the free RNA upon addition of purines support this conclusion (Serganov *et al.*[36]).

11.5   Adenosine Monophosphate Binding

Remarkably, the logical choice of uridine as an adenine-binding partner observed in the A-riboswitch is *not* realized in the recognition schemes for adenosine monophosphate (AMP) used by DNA (Huizenga and Szostak[46]) and RNA aptamers (Sassanfar and Szostak[47]). Two NMR structures (Jiang *et al.*,[48] Dieckmann *et al.*[49]) of RNA aptamers and a structure of a DNA aptamer (Lin and Patel[50]) bound to AMP demonstrate that, despite their distinct sequences, secondary structures and overall tertiary folds, the molecular details of ligand recognition by AMP-RNA and AMP-DNA aptamers attest to convergent binding strategies in both types of complexes.

In the DNA aptamer (Lin and Patel[50]), two AMP molecules are intercalated at adjacent sites within a widened minor groove. As is evident from Figure 10A and B, showing interaction of a single AMP molecule with DNA, the AMP ring is sandwiched between G8·G19 (which is a part of a three-base platform G8·G19·A20) and adenosine A10 (which forms a non-canonical base pair A10·G18). In addition, the AMP pairs through its Watson–Crick edge with the minor groove edge of guanosine G9.

A similar AMP recognition pattern is observed in the structures of the AMP-binding RNA aptamer (Jiang *et al.*,[48] Dieckmann *et al.*[49]). Although the overall GRNA loop-like fold of the binding pocket in the RNA aptamer differs substantially from the fold found in the DNA aptamer (Figure 10C and D), AMP makes the same hydrogen bonds with non-paired guanosine G8, and is intercalated between adenine A10 and a non-canonical G7·G11 base pair.

In contrast to the adenine encapsulated within the A-riboswitch, the AMP molecules in both AMP-specific aptamers are not completely inserted into the binding pockets (Figure 10B and D), exposing their sugar-phosphate moiety to the solvent. Nevertheless, as in the purine riboswitches, recognition of AMP by aptamers requires an adjustment in the nucleic acid conformation (Lin and Patel,[50] Jiang *et al.*[48]).

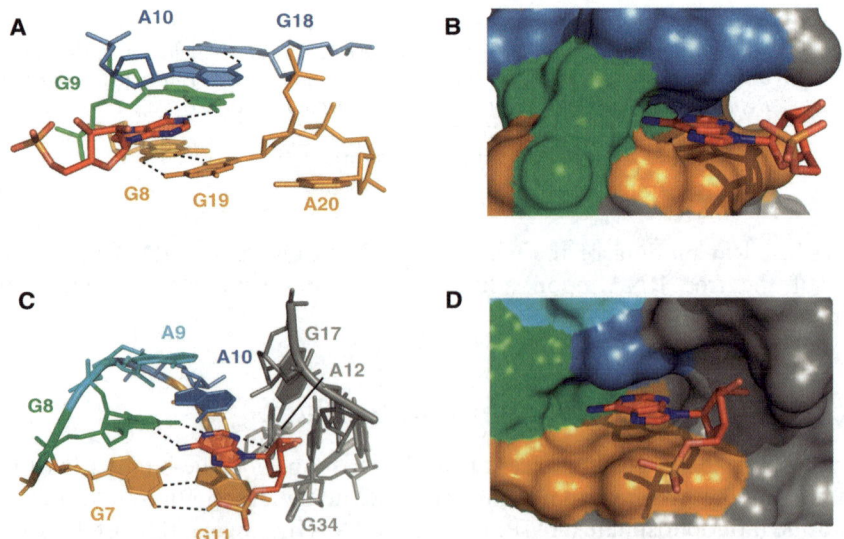

Figure 10   *Recognition of adenosine monophosphate by DNA and RNA aptamers. Binding pockets are shown in stick and surface representation for the DNA aptamer in A and B (Lin and Patel[50]), and for the RNA aptamer in C and D (Jiang et al.[48])*

11.6 Conclusions

Three-dimensional structural analyses have provided valuable insights into key problems of molecular recognition of small organic ligands by nucleic acids. It is now apparent how nucleic acids can discriminate between related ligands and what are the differences between ligand-binding pockets formed by *in vitro* selected and natural nucleic acids. In most cases, the enclosure of a large part of the ligand by the RNA provides the basis for specific recognition of the cognate molecule. The specificity is achieved through multiple hydrogen bonding together with precise positioning of the ligand and bases for maximal stacking interactions. This typically occurs *via* an induced-fit mechanism whereby nucleic acids change their conformations, at least locally, to adapt to their cognate ligands. The principle of adaptive recognition, previously assigned as a characteristic of *in vitro*-selected aptamers (Hermann and Patel[27]), is extended towards natural RNAs with the discovery of riboswitches. In rare instances of preformed-binding pockets, such as in the Diels–Alder ribozyme, the ligand-binding affinity correlates directly with the goodness-of-fit into the pocket. A remarkable feature shared by the majority of nucleic acids interacting with small ligands is an extensive use of base triples to create a binding pocket for wholly planar ligands or their flat parts (Zimmermann *et al.*,[33] Fan *et al.*,[31] Serganov *et al.*,[20] Baugh *et al.*[51]). Given the abundance of triples in the natural RNA structures (Nissen *et al.*[52]), this is not totally surprising. Base triples can easily be formed by approximating a single nucleotide to a base pair, allowing many triples to mediate loop–loop or loop–helix interactions (Nissen *et al.*,[52] Doherty *et al.*[53]). This property of base triples is fully exploited in the ligand-binding pockets, which are often formed or stabilized by the docking of an additional strand into the groove of the helix. In addition, triples can simply provide more room in the pocket, and, using their nucleotide backbones, maximally protect the ligand from exposure to the solvent.

Discrimination between related ligands is probably the most important and difficult task of the nucleic acid that interacts with small molecules. As illustrated by purine riboswitches, a single mutation within the 80-mer metabolite-sensing domain changes the specificity of the riboswitch and can cause repression or activation of the wrong set of genes. A specific hydrogen bonding scheme is a key element for discrimination between many ligands, such as purines (Batey *et al.*[35]; Serganov *et al.*[36]) and AMP (Lin and Patel[50]; Jiang *et al.*[48]; Dieckmann *et al.*[49]). Steric hindrance due to a methyl group plays a major role in discriminating theophylline from caffeine (Jenison *et al.*,[32] Zimmermann *et al.*[33]). Other discriminatory factors, including stacking, shape complementarity, electrostatic interactions, and their interplay with hydrogen bonding are chiefly involved in recognition of more complex ligands.

The planarity of the nucleotide bases favours stacking interactions in complexes between nucleic acids and small organic ligands. Stacking plays a pivotal role in all complexes described here and in most other complexes. As suggested by close examination of the Diels–Alder ribozyme structure (Serganov *et al.*[20]), stacking interactions are not only necessary for holding the ligand; they may also participate directly in the catalytic process, redistributing electron density around substrates in the course of the chemical reaction.

A comparison of complex ligand-binding sites in artificial aptamers and natural riboswitches reveals unique structural features attesting the different characteristics of evolutionary pressure acting on these two families of macro-molecules. The only function of those aptamers for which three-dimensional structures are available is the binding to a given ligand. Metabolite-sensing domains of riboswitches, however, in addition to highly specific ligand binding, must undergo conformational rearrangement or stabilization, which implies particular folding in the adjacent expression platforms. The expression plat-form should typically fold into a hairpin that terminates transcription or prevents translation initiation, or, alternatively, form a hairpin that disrupts other regulatory elements. Therefore, the ligand-binding sites of riboswitches are composed of nucleotides belonging to a few subdomains and must contain additional elements not directly involved in the ligand binding. Aptamers often comprise unpaired loop regions, which are disordered in the free nucleic acid and acquire a defined conformation by adaptive folding around the ligand. Purine riboswitches seem to fit this profile, although the purine binding site is formed within the three-way junction loop, not in the middle of a helical region. Further studies are required to address this issue.

11.7   Perspectives

Structural data on small molecule-nucleic acid complexes will be especially helpful for the rational exploration of nucleic acids as a drug target. The key roles that DNA and RNA play in all steps of cellular metabolism mark them out as prime targets for therapeutic intervention. Since a large number of natural antibiotics target the ribosome and the ribosome structure is now available at high resolution, without doubt ribosomal RNAs are uniquely promising candidates for drug design approaches. Recent studies (Sudarsan et al.[54]) have shown that some riboswitches can also be externally controlled by drug-like metabolite analogues. Given the important functional role of ribo-switches in numerous microorganisms and the fact that riboswitches have not been detected to date in the human genome, the high-resolution structures should enable researchers to employ rational drug discovery strategies to create novel classes of antibacterial and antifungal compounds by targeting ribo-switches.

In addition to therapeutic implications, detailed understanding of the rec-ognition of small molecules by nucleic acids could play an important role in medical diagnostics (Famulok and Mayer,[55] Famulok[56]), as biosensors (Bruno and Kiel,[57] Lim et al.[58]), and molecular switches (Steele et al.[59]). Although the structures of many interesting small molecule binders have not yet been determined, the available structural information combined with molecular modelling techniques is crucial for successful exploitation of nucleic acid molecules as targets for small ligand binding.

References

1. J.W. Lown, DNA recognition by lexitropsins, minor groove binding agents, *J. Mo. Recognit.*, 1994, **7**, 79–88.
2. S. Neidle, DNA minor groove recognition by small molecules, *Nat. Prod. Rep.*, 2001, **18**, 291–309.
3. U. Pindur, M. Jansen and T. Lemster, Adavances in DNA-ligands with groove binding, intercalating and/or alkylating activity: chemistry, DNA-binding and biology, *Curr. Med. Chem.*, 2005, **12**, 2805–2847.
4. R.H. Shafer, Stability and structure of model DNA triplexes and quadruplexes and their interactions with small ligands, *Prog. Nucleic Acid Re. Mol. Biol.*, 1998, **59**, 55–94.
5. X. Han and X. Gao, Sequence specific recognition of ligand-DNA complexes studied by NMR, *Curr. Med. Chem.*, 2001, **8**, 551–581.
6. S. Neidle and G. Parkinson, Telomere maintenance as a target for anticancer drug discovery, *Nat. Rev. Drug. Discov.*, 2002, **1**, 383–393.
7. A. Oleksi, A.G. Blanco, R. Boer, I. Uson, J. Aymami, A. Rodger, M.J. Hannon and M. Coll, Molecular recognition of a three-way DNA junction by a metallosupramolecular helicate, *Angew. Chem. Int. Ed. Engl.*, 2006, **45**, 1227–1231.
8. F. Walter, Q. Vicens and E. Westhof, Aminoglycoside-RNA interactions, *Curr Opin. Chem. Biol.*, 1999, **3**, 694–704.
9. S.J. Sucheck and C.H. Wong, RNA as a target for small molecules, *Curr. Opin. Chem. Biol.*, 2000, **4**, 678–686.
10. J. Gallego and G. Varani, Targeting RNA with small-molecule drugs: Therapeutic promise and chemical challenges, *Acc. Chem. Res.*, 2001, **34**, 836–843.
11. A.C. Cheng, V. Calabro and A.D. Frankel, Design of RNA-binding proteins and ligands, *Curr. Opin. Struct. Biol.*, 2001, **11**, 478–484.
12. Y. Tor, Targeting RNA with small molecules, *Chembiochem*, 2003, **4**, 998–1007.
13. T. Hermann, Drugs targeting the ribosome, *Curr. Opin. Struct. Biol.*, 2005, **15**, 355–366.
14. J.A. Sutcliffe, Improving on nature: Antibiotics that target the ribosome, *Curr. Opin. Microbiol.*, 2005, **8**, 534–542.
15. T.R. Cech, Ribozymes, the first 20 years, *Biochem. Soc. Trans.*, 2002, **30**, 1162–1166.
16. J.A. Doudna and T.R. Cech, The chemical repertoire of natural ribozymes, *Nature*, 2002, **418**, 222–228.
17. T.A. Steitz and P.B. Moore, RNA, the first macromolecular catalyst: The ribosome is a ribozyme, *Trends Biochem. Sci.*, 2003, **28**, 411–418.
18. D.S. Wilson and J.W. Szostak, *In vitro* selection of functional nucleic acids, *Annu. Rev. Biochem.*, 1999, **68**, 611–647.
19. P.B. Moore and T.A. Steitz, The ribosome revealed, *Trends Biochem. Sci.*, 2005, **30**, 281–283.

20. A. Serganov et al., Structural basis for Diels–Alder ribozyme-catalyzed carbon-carbon bond formation, *Nat. Struct. Biol.*, 2005, **12**, 218–224.

21. B. Seelig and A. Jäschke, A small catalytic RNA motif with Diels-Alderase activity, *Chem. Biol.*, 1999, **6**, 167–176.

22. T.M. Tarasow, S.L. Tarasow and B.E. Eaton, RNA-catalyzed carbon–carbon bond formation, *Nature*, 1997, **389**, 54–57.

23. K.C. Nicolaou, S.A. Snyder, T. Montagnon and G.E. Vassilikogiannakis, The Diels–Alder reaction in total synthesis, *Angew. Chem. Int. Ed.*, 2002, **41**, 1668–1698.

24. S. Keiper, D. Bebenroth, B. Seelig, E. Westhof and A. Jäschke, An architecture of a Diels-Alderase ribozyme with a preferred catalytic pocket, *Chem. Biol.*, 2004, **11**, 1217–1227.

25. F. Stuhlmann and A. Jäschke, Characterization of an RNA active site: Interactions between a Diels-Alderase ribozyme and its substrates and products, *J. Am. Chem. Soc.*, 2002, **124**, 328–344.

26. A.R. Ferré-D'Amaré, K. Zhou and J.A. Doudna, Crystal structure of a hepatitis delta virus ribozyme, *Nature*, 1998, **395**, 567–574.

27. T. Hermann and D.J. Patel, Adaptive recognition by nucleic acid aptamers, *Science*, 2000, **287**, 820–825.

28. J.R. Williamson, Induced fit in RNA-protein recognition, *Nat. Struct. Biol.*, 2000, **7**, 834–837.

29. N. Leulliot and G. Varani, Current topics in RNA-protein recognition: Control of specificity and biological function through induced fit and conformational capture, *Biochemistry*, 2001, **40**, 7947–7956.

30. R. Wombacher et al., Control of stereoselectivity in an enzymatic reaction by backdoor access, *Angew. Chem. Int. Ed. Engl.*, 2006, **45**, 2469–2472.

31. P. Fan, A.K. Suri, R. Fiala, D. Live and D.J. Patel, Molecular recognition in the FMN-RNA aptamer complex, *J. Mol. Biol.*, 1996, **258**, 480–500.

32. R.D. Jenison, S.C. Gill, A. Pardi and B. Polisky, High-resolution molecular discrimination by RNA, *Science*, 1994, **263**, 1425–1429.

33. G.R. Zimmermann, R.D. Jenison, C.L. Wick, J.P. Simorre and A. Pardi, Interlocking structural motifs mediate molecular discrimination by a theophylline-binding, *Nat. Struct. Biol.*, 1997, **4**, 644–649.

34. P. Burgstaller and M. Famulok, Isolation of RNA aptamers for biological cofactors by *in vitro* selection, *Angew. Chem. Int. Ed. Engl.*, 1994, **33**, 1084–1087.

35. R.B. Batey, S.D. Gilbert and R.K. Montagne, Structure of a natural guanine-response riboswitch complexed with the metabolite hypoxanthine, *Nature*, 2004, **432**, 411–415.

36. A. Serganov et al., Structural basis for discriminative regulation of gene expression by adenine- and guanine-sensing mRNAs, *Chem. Biol.*, 2004, **11**, 1729–1741.

37. M. Mandal and R.R. Breaker, Gene regulation by riboswitches, *Nature Reviews Mol. Cell Biol.*, 2004a, **5**, 451–463.

38. E. Nudler and A.S. Mironov, The riboswitch control of bacterial metabolism, *TRENDS in Biochem. Scis.*, 2004, **29**, 11–17.

39. G.A. Soukup and J.K. Soukup, Riboswitches exert genetic control through metabolite-induced conformational change, *Curr. Opin. Struct. Biol.*, 2004, **14**, 344–349.

40. W. Winkler, A. Nahvi and R.R. Breaker, Thiamine derivatives bind messenger RNAs directly to regulate bacterial gene expression, *Nature*, 2002, **419**, 952–956.

41. W.C. Winkler, A. Nahvi, A. Roth, J.A. Collins and R.R. Breaker, Control of gene expression by a natural metabolite-responsive ribozyme, *Nature*, 2004, **428**, 281–286.

42. M. Mandal, B. Boese, J.E. Barrick, W.C. Winkler and R.R. Breaker, Riboswitches control fundamental biochemical pathways in *Bacillus subtilis* and other bacteria, *Cell*, 2003, **113**, 577–586.

43. R.R. Breaker, Riboswitches and the RNA World, in *The RNA World*, 3rd edn, R.F. Gesteland, T.R. Cech and J.F. Atkins (eds), Cold Spring Harbor Laboratory Press, Cold Spring Harbor, NY, 2006.

44. M. Mandal and R.R. Breaker, (2004b). Adenine riboswitches and gene activation by disruption of a transcription terminator, *Nat. Struct. & Mol. Biol.*, 2004, **11**, 29–35.

45. L. Su, L. Chen, M. Egli, J.M. Berger and A. Rich, Minor groove RNA triplex in the crystal structure of a ribosomal frameshifting viral pseudo-knot, *Nat. Struct. Biol.*, 1999, **6**, 285–292.

46. D.E. Huizenga and J.W. Szostak, A DNA aptamer that binds adenosine and ATP, *Biochemistry*, 1995, **34**, 656–665.

47. M. Sassanfar and J.W. Szostak, An RNA motif that binds ATP, *Nature*, 1993, **364**, 550–553.

48. F. Jiang, R.A. Kumar, R.A. Jones and D.J. Patel, Structural basis of RNA folding and recognition in an AMP-RNA aptamer complex, *Nature*, 1996, **382**, 183–186.

49. T. Dieckmann, E. Suzuki, G.K. Nakamura and J. Feigon, Solution structure of an ATP-binding RNA aptamer reveals a novel fold, *RNA*, 1996, **2**, 628–640.

50. C.H. Lin and D.J. Patel, Structural basis of DNA folding and recognition in an AMP-DNA aptamer complex: Distinct architectures but common recognition motifs for DNA and RNA aptamers complexed to AMP, *Chem. Biol.*, 1997, **4**, 817–832.

51. C. Baugh, D. Grate and C. Wilson, 2.8 A crystal structure of the malachite green aptamer, *J. Mol. Biol.*, 2000, **301**, 117–128.

52. P. Nissen, J.A. Ippolito, N. Ban, P.B. Moore and T.A. Steitz, RNA tertiary interactions in the large ribosomal subunit:The A-minor motif, *Proc. Natl. Acad. Sci. USA*, 2001, **98**, 4899–4903.

53. E.A. Doherty, R.T. Batey, B. Masquida and J.A. Doudna, A universal mode of helix packing in RNA, *Nat. Struct. Biol.*, **8**, 339–343.

54. N. Sudarsan, S. Cohen-Chalamish, S. Nakamura, G.M. Emilsson and R.R. Breaker, Thiamine pyrophosphate riboswitches are targets for the antimicrobial compound pyrithiamine, *Chem. Biol.*, **12**, 1325–1335.

55. M. Famulok and G. Mayer, Aptamers as tools in molecular biology and immunology, *Curr. Top. Microbiol. Immunol.*, 1999, **243**, 123–136.
56. M. Famulok, Allosteric aptamers and aptazymes as probes for screening approaches, *Curr. Opin. Mol. Ther.*, 2005, **7**, 137–143.
57. J.G. Bruno and J.L. Kiel, *In vitro* selection of DNA aptamers to anthrax spores with electrochemiluminescence detection, *Biosens. Bioelectron.*, 1999, **14**, 457–464.
58. D.V. Lim, J.M. Simpson, E.A. Kearns and M.F. Kramer, Current and developing technologies for monitoring agents of bioterrorism and bio-warfare, *Clin. Microbiol. Rev.*, 2005, **18**, 583–607.
59. D. Steele, A. Kertsburg and G.A. Soukup, Engineered catalytic RNA and DNA: New biochemical tools for drug discovery and design, *Am. J. Pharmacogenomics*, 2003, **3**, 131–144.

Subject Index